INTERNATIONAL GCSE (9–1)

ERICA LARKCOM
ROGER DELPECH

Biology
for Edexcel International GCSE

SECOND EDITION

HODDER
EDUCATION
AN HACHETTE UK COMPANY

Although every effort has been made to ensure that website addresses are correct at time of going to press, Hodder Education cannot be held responsible for the content of any website mentioned. It is sometimes possible to find a relocated web page by typing in the address of the home page for a website in the URL window of your browser.

Hachette UK's policy is to use papers that are natural, renewable and recyclable products and made from wood grown in well-managed forests and other controlled sources. The logging and manufacturing processes are expected to conform to the environmental regulations of the country of origin.

Orders: please contact Hachette UK Distribution, Hely Hutchinson Centre, Milton Road, Didcot, Oxfordshire, OX11 7HH. Telephone: +44 (0)1235 827827. Email education@hachette.co.uk. Lines are open from 9 a.m. to 5 p.m., Monday to Friday. You can also order through our website: www.hoddereducation.co.uk

© Erica Larkcom and Roger Delpech 2017

First published in 2017 by

Hodder Education,

An Hachette UK Company

Carmelite House

50 Victoria Embankment

London EC4Y 0DZ

Impression number   13

Year   2024

Cover photo © imageBROKER / Alamy Stock Photo

Illustrations by Aptara Inc. and Elektra Media Ltd.

Typeset in ITC Legacy Serif by Elektra Media Ltd.

Printed and bound in Great Britain by Bell & Bain Ltd, Glasgow

Project managed by Elektra Media Ltd.

A catalogue record for this title is available from the British Library.

ISBN 978 1 5104 0519 6

MIX
Paper | Supporting responsible forestry
FSC™ C104740

Contents

Contents

Acknowledgements

Every effort has been made to trace the copyright holders of material reproduced here. The following material is reproduced with kind permission:

- Earth System Research Laboratories, Graph of monthly mean CO_2 at Mauna Loa from *http://www.esrl.noaa.gov/gmd/ccgg/trends* US Department of Commerce, National Oceanic and Atmospheric Administration.
- 'Bleeding Canker of Horse Chestnut', map of geographical locations reported to Forest Research Disease Diagnosis Advisory Service from *http://www.forestry.gov.uk/fr/INFD-6KYBGV*. © Crown copyright 2012.
- Anne Bebbington, Table 6.5, data extracted from SAPS-FSC Plants for primary pupils 6 (Plants in their natural environment) from *http://www.saps.org.uk/attachments/article/1378/SAPS_book_6_-_Plants_in_their_natural_environment_-_2016.pdf*
- Fred Pearce, Graph (Mann, Bradley & Hughes, Nature 1998) from 'Variations of the Earth's surface temperature', from http://www.guardian.co.uk/environment/2010/feb/02/hockey-stick-graph-climate-change,*The Guardian* (2 February, 2010).
- Climate Choices & Children's Voices, Graph of 'Northern hemisphere. Departures in temperature (celcius) from the 1961 to 1990 average' (adapted) from *http://climatechoices.co.uk/pages/cchange3.htm*

Permission for re-use of all © Crown copyright information is granted under the terms of the Open Government Licence (OGL).

Getting the most from this book

Welcome to the Edexcel International GCSE (9–1) Biology Student Book. This book has been divided into seven Sections, following the structure and order of the Edexcel specification, which you can find on the Edexcel website for reference.

Each Section has been divided into a number of smaller Chapters to help you manage your learning.

The following features have been included to help you get the most from this book.

TO THINK ABOUT . . .

Try the activity before you start, and then have a look at it again once you have completed the Section, to see if your responses are different before and after learning more about the topics.

PRACTICAL

Practical boxes provide hints on key things to remember, or alternative practical work that you can do to help you learn more about that topic.

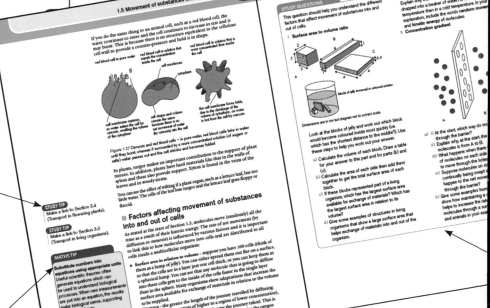

STUDY TIP

Study tips throughout the book will guide you in your learning process.

MATHS TIP

Maths tips and skills give you additional help with the maths in the book so you can avoid losing valuable marks in the exam.

At the end of each Section, you will find a summary checklist, highlighting the key facts that you need to know and understand, and key skills that you learnt in the Section.

Summary

I am confident that:

✓ I know that all living organisms require nutrition, that they respire and excrete their waste, that they respond to their surroundings and move, that they control their internal conditions, and are able to reproduce, grow and develop.

✓ I understand the meaning of the terms eukaryotic and prokaryotic.

✓ I understand that plants, animals, fungi and protoctists are eukaryotic organisms and that bacteria are prokaryotic organisms.

✓ I understand how the characteristics listed above are carried out by all living organisms, but sometimes in different ways by plants, animals and other living organisms.

✓ I know that plants and animals are multicellular organisms but that they differ in their method of obtaining food and I can list differences in the structure of their cells.

✓ I can describe features of fungi, bacteria and protoctists, and know why they are not included with plants and animals.

✓ I can name examples of all the main groups and can describe their structure.

✓ I understand why, in some classifications, viruses are not included as living organisms.

✓ I can describe some examples of pathogens from the different groups of fungi, bacteria, protoctists and viruses.

✓ I can draw and label a diagram of a plant cell and of an animal cell and use this to show the features they contain.

✓ I can describe the functions of the structures shown in the cell diagrams.

✓ I know the ways that a plant cell differs from an animal cell, in both structures and functions.

✓ I know that carbohydrates, proteins and lipids (fats and oils) contain the elements carbon, hydrogen and oxygen and that proteins also contain nitrogen and sometimes sulfur.

✓ I can describe how smaller basic units are built up into larger molecules and that starch and glycogen are made from simple sugar units, that proteins are made from amino acids and that lipids are made from fatty acids and glycerol.

✓ I understand that enzymes are proteins and that they act as catalysts in metabolic reactions.

✓ I understand that enzymes are affected by temperature and pH, and can describe how to do experiments to illustrate the effect of temperature on the activity of an enzyme.

✓ I can describe how to test a food sample for the presence of glucose, starch, proteins and of fats (oils).

✓ I know definitions for the terms diffusion, osmosis and active transport and understand how each of these contributes to the movement of substances into and out of cells.

✓ I can describe simple experiments that illustrate diffusion and osmosis in living and non-living systems.

✓ I understand that different factors can affect the rate of movement of substances into and out of cells.

✓ I can describe how surface area to volume ratio distance, temperature and concentration gradient can affect the rate of movement and describe examples that illustrate these effects.

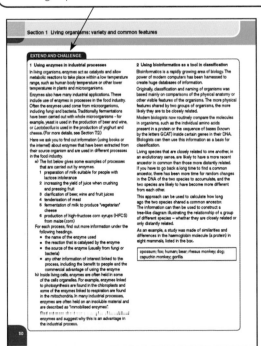

You will find *Exam-style questions* at the end of each Section covering the content of that Section and the different types of question you will find in an examination.

Before you try the *Exam-style questions*, look at the sample answers and expert's comments to see how marks are awarded and common mistakes to avoid.

Exam-style questions

1 The table gives the content of glucose, urea and calcium ions in the blood entering the kidney, in the glomerular filtrate and in the urine of a person. Values are given in mg per 100 cm³.

Component	Component mg / 100 cm³		
	Blood	Glomerular filtrate	Urine
glucose	100	100	0
urea	26	26	1 820
calcium ions	4	4	6

a) i) Which of the components provides energy for the body? [1]

ii) Which of the components is a metabolic waste product? [1]

b) Explain why the figures for each component are the same for the blood concentration and glomerular filtrate. [3]

c) i) If the person drank a large volume of water, far more than was needed by the body, predict what would happen to the figures in the urine column. [2]

ii) Describe the processes taking place in the body to support your answer to c)i). [3]

[Total = 10]

2 a) Give the meaning of the term **homeostasis** and use examples to explain why it is important in the human body. [4]

b) i) Explain how sweating helps to cool the body. [2]

ii) Suggest why, on a hot day in a dry atmosphere, you feel more comfortable than in a humid atmosphere when the temperature is the same. [3]

c) Mountaineers may experience extreme conditions of cold and wind. Often they wear clothing that uses several light layers rather than thick heavy garments, with a waterproof outer layer.

i) Suggest the advantage of several thin layers rather than thick heavy garments. [2]

ii) Suggest the importance of the waterproof outer layer. [2]

[Total = 13]

3 Auxin is a plant growth regulator and it can affect the growth of wheat coleoptiles (the first shoot of young seedlings of wheat plants).

The diagram shows an experiment into the effects of auxin on the growth of a wheat coleoptile. Lanolin is a grease that sticks to plant surfaces. In the experiment, auxin was mixed with some lanolin and the auxin could then diffuse into the cells of the coleoptile. The seedlings were held at the top of tubes containing water, but only the seedling is shown in the diagram.

A 'blob' of the mixture of lanolin and auxin was stuck to one side of a coleoptile, as shown in the diagram on the left. The seedlings were examined after 3 hours.

Possible results of the experiment are shown in diagrams A, B, C and D.

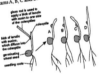

a) i) Which of the results, A, B, C or D, would you expect after 3 hours? [1]

ii) Give reasons to support your answer and explain what is happening with respect to growth in the coleoptile to produce this response. [3]

b) If the lanolin with auxin mixture had been placed in a complete circle around the coleoptile, which result would you expect? Give a reason for your answer. [2]

c) During the experiment, what light conditions should have been used? Explain your answer. [2]

d) Describe a control that should be included in the experiment. [2]

[Total = 10]

173

EXTEND AND CHALLENGE

When you have completed all the Exam-style questions for the Section, try the Extend and Challenge questions.

Section 1 Living organisms: variety and common features

EXTEND AND CHALLENGE

1 Using enzymes in industrial processes

In living organisms, enzymes act as catalysts and allow metabolic reactions to take place within a low temperature range, such as human body temperature or other lower temperatures in plants and microorganisms.

Enzymes also have many industrial applications. These include use of enzymes in processes in the food industry. Often the enzymes used come from microorganisms, including fungi and bacteria. Traditionally fermentations have been carried out with whole microorganisms – for example, yeast is used in the production of beer and wine, or *Lactobacillus* is used in the production of yoghurt and cheese. (For more details, see Section 7.2.)

Here we ask you to find out information (using books or the internet) about enzymes that have been extracted from their source organism and are used in different processes in the food industry.

a) The list below gives some examples of processes that are carried out by enzymes.

1 preparation of milk suitable for people with lactose intolerance
2 increasing the yield of juice when crushing and pressing fruit
3 clarification of beer, wine and fruit juices
4 tenderisation of meat
5 fermentation of milk to produce 'vegetarian' cheese
6 production of high-fructose corn syrups (HFCS) from maize (corn)

For each process, find out more information under the following headings.
- the name of the enzyme used
- the reaction that is catalysed by the enzyme
- the source of the enzyme (usually from fungi or bacteria)
- any other information of interest linked to the process, including the benefit to people and the commercial advantage of using the enzyme

b) Inside living cells, enzymes are often held in some of the cells organelles. For example, enzymes linked to photosynthesis are found in the chloroplasts and some of the enzymes linked to respiration are found in the mitochondria. In many industrial processes, enzymes are often held on an insoluble material and are described as 'immobilised enzymes'.

Find out more about the use of immobilised enzymes and suggest why this is an advantage in the industrial process.

2 Using bioinformatics as a tool in classification

Bioinformatics is a rapidly growing area of biology. The power of modern computers has been harnessed to create huge databases of information.

Originally, classification and naming of organisms was based mainly on comparisons of the physical anatomy or other visible features of the organisms. The more physical features shared by two groups of organisms, the more likely they are to be closely related.

Modern biologists now routinely compare the molecules in organisms, such as the individual amino acids present in a protein or the sequence of bases (known by the letters GCAT) inside certain genes in their DNA. Biologists can then use this information as a basis for classification.

Living species that are closely related to one another, in an evolutionary sense, are likely to have a more recent ancestor in common than those more distantly related. If you have to go back a long time to find a common ancestor, there has been more time for random changes in the DNA of the two species to accumulate, and the two species are likely to have become more different from each other.

This approach can be used to calculate how long ago the two species shared a common ancestor. The information can then be used to construct a tree-like diagram illustrating the relationship of a group of different species – whether they are closely related or only distantly related.

As an example, a study was made of similarities and differences in the haemoglobin molecule (a protein) in eight mammals, listed in the box.

opossum; fox; human; bear; rhesus monkey; dog; capuchin monkey; gorilla

50

ANSWERS

Answers for all questions and activities in this book can be found online at www.hoddereducation.co.uk/igcsebiology

STUDY QUESTIONS

At the end of each Chapter you will find Study Questions. Work through these in class or on your own for homework.

Picture credits

The authors and publishers would like to thank the following for permission to reproduce copyright illustrations:

p.1 *l* © Kirill Kurashov – Fotolia.com; *tc* © Herve Conge, ISM / Science Photo Library; *bc* © Monty Rakusen / Getty Images; *r* © Dr Kari Lounatmaa / Science Photo Library; **p.2** *tr* © Modis / NASA; *br* © Kenneth Libbrecht / Science Photo Library; *bl* © Pascal Goetgheluck / Science Photo Library; *tl* 01_02d © Bjanka Kadic / Science Photo Library; **p.8** *tl* © Perry – Fotolia.com; *tr* © Erica Larkcom; *ctl* © Nigel Pavitt / JAI / Getty Images; *ctr* © Sinclair Stammers / Science Photo Library; *cb* © Oxford Scientific / Getty Images; *b* © Roger Delpech; **p.9** *cbl* © Centre For Bioimaging, Rothamsted Research / Science Photo Library; *b* © CDC / Science Photo Library; *cbr* © James Cavallini / Science Photo Library; *tl* © Dr Kari Lounatmaa / Science Photo Library; *tr* © BSIP / Science Photo Library; *ct* M. I. Walker / Science Photo Library; *ccr* © LSHTM / Science Photo Library; *ccl* © Sinclair Stammers / Science Photo Library; **p.11** *b* © Natural History Museum, London / Science Photo Library; *t* © Andre – Fotolia.com; **p.13** © Ed Reschke / Getty Images; **p.14** © Roger Delpech; **p.16** © Roger Delpech; **p. 17** © Roger Delpech; **p.19** *t* © Image Source Plus / Alamy; *b* © Thomas Deerinck, NCMIR / Science Photo Library; **p.22** *t* © Dr Keith Wheeler / Science Photo Library; *b* © Biophoto Associates / Science Photo Library; **p.32** *t* © David Scharf / Science Photo Library; *b* © Dorling Kindersley / Getty Images; **p.50** *l* © John Deveries / Science Photo Library; *r* © Serghei Velusceac – Fotolia.com; **p.51** *t* © NASA; *b* © molekuul.be – Fotolia.com; **p.56** *t* © Colin Varndell / Science Photo Library; *b* © John Adds; **p.60** *t* © BlueOrange Studio – Fotolia.com; *b* © D. Roberts / Science Photo Library; **p.62** © Elena Schweitzer – Fotolia.com; **p.63** *t* © Simone van den Berg – Fotolia.com; *b* © www.schurr-fotografie.de – Fotolia.com; **p.70** *tl* © Ted Kinsman / Science Photo Library; *tr* © Ted Kinsman / Science Photo Library; *b* © Dr Jeorg Szarzynski; **p.71** © Dr Gopal Murti / Science Photo Library; **p.85** *tl* © SCIEPRO / Getty Images; *tr* © Ed Reschke / Getty Images; *b* © Kallista Images / Getty Images; **p.87** © John Adds; **p.88** © John Adds; **p.92** *l* © Aaron Amat – Fotolia.com; *c* © Robert Kneschke – Fotolia.com; *r* © InfinityPhoto – Fotolia.com; **p.98** © Biophoto Associates / Science Photo Library; **p.99** © Martinan – Fotolia.com; **p.100** *t* © Gerd Guenther / Science Photo Library; *c* © Charles Krebs / Getty Images; *b* © 2436digitalavenue – Fotolia.com; **p.101** *l* © Science Vu, Visuals Unlimited / Science Photo Library; *r* © Sinclair Stammers / Science Photo Library; **p.102** *t* © Dr Keith Wheeler / Science Photo Library; *b* 03_23 © Davidpstephens – Fotolia.com; **p.104** *t* © urosr – Fotolia.com; *b* © Mark Moffett / Getty Images; **p.105** © Ed Reschke / Getty Images; **p.106** *t* © CJ Runions; *c* © Dr Keith Wheeler / Science Photo Library; *b* 03_31b © Dr David Furness, Keele University / Getty Images; **p.107** © John Adds; **p.109** *t* © John Adds; *b* © Erica Larkcom; **p.112** *tl* © Alliance Images / Alamy; *cl* © Alexander Raths – Fotolia.com; *r* © Michelle Del Guercio / Science Photo Library; *bl* © Ed Reschke / Getty Images; **p.118** © Deep Light Productions / Science Photo Library; **p.121** *l* © Andres Rodriguez – Fotolia.com; *r* © Image Source IS2 – Fotolia.com; **p.135** *l* © Georgette Douwma / Science Photo Library; *r* © Georgette Douwma / Science Photo Library; **p.137** *tl* © Fotokon – Fotolia.com; *tr* © soniccc – Fotolia.com; *b* © Erica Larkcom; **p.138** *t* © Erica Larkcom; **p.140** © Visuals Unlimited, Inc. / Alex Wild / Getty Images; **p.146** *l* © bono – Fotolia.com; *c* © Netzer Johannes – Fotolia.com; *r* © Power And Syred / Science Photo Library; **p.147** *t* © Christine Glade / iStockphoto; *b* © Mr. Markin – Fotolia.com; **p.150** *t* © Andrey Smirnov – Fotolia.com; *c* 04_10b © Martin Shields / Alamy; *b* © Erica Larkcom; **p.155** *t* © kyslynskyy – Fotolia.com; *c* © Cathy Keifer – Fotolia.com; *b* © tomatito26 – Fotolia.com; **p.173** © Nigel Cattlin / Alamy; **p.175** © OMIKRON / Science Photo Library; **p.176** *t* © amenic181 – Fotolia.com; *c* © tsuppyinny – Fotolia.com; *bl* © Ian Howard – Fotolia.com; *br* © Melba / Getty Images; **p.177** *t* © Ingo Arndt / Minden Pictures / Getty Images; *b* © CrazyD / Wikipedia Commons; **p.180** *all* © Erica Larkcom; **p.181** *t* Erica Larkcom; *b* Erica Larkcom; **p.183** *t* Erica Larkcom; *b* © Biology Media / Science Photo Library; **p.184** *tl, tc* and *tr* © Erica Larkcom; *b* © John Bebbington; **p.190** *t* © Ernest F / Wikipedia Commons; *b* © Kotomiti – Fotolia.com; **p.194** © Steve Gschmeissner / Science Photo Library; **p.198** *l* © Evgeny Terentev / Getty Images; *c* © G. Wanner / Getty Images; *r* © Dr Torsten Wittmann / Science Photo Library; **p.199** © SMC Images / Getty Images; **p.200** © Herve Conge, ISM / Science Photo Library; **p.206** *l* © Bluestone / Science Photo Library; *r* © EDELMANN / Science Photo Library; **p.217** *t* © Alison Wright / National Geographic Society / Getty Images; *c* © Hbarrison / Wikipedia Commons (http://commons.wikimedia.org/wiki/File:Espanola_2010_09_29_0949.jpg); *b* © FPG / Getty Images; **p.219** *t* © Monkey Business – Fotolia.comp.220; *b* © Tyler Olson – Fotolia.com; **p.222** *t* © Andrew Darrington / Alamy; *b* © Rob & Ann Simpson / Getty Images; **p.236** © David Scharf / Getty Images; **p.238** *tl* © Kovalenko Inna – Fotolia.com; *tr* © Wong Sze Fei – Fotolia.com; *b* © kikkerdirk – Fotolia.com; **p.239** *all* © Erica Larkcom; **p.240** Erica Larkcom; **p.214** © Martyn F. Chillmaid / Science Photo Library; **p.243** *l* © Roger Delpech; *r* © Roger Delpech; **p.247** *tl*© Erica Larkcom; *tr* © Erica Larkcom; *c* © Thomas Lehne / lotuseaters / Alamy; *b* © Jacek Chabraszewski – Fotolia.com; **p.255** *both* © Erica Larkcom; **p.259** © Nigel Cattlin / Alamy; **p.261** *t* © Cimmerian / Shutterstock; *t* © Cozyta – Fotolia.com; *b* © Erica Larkcom; **p.262** © Aleksandar Nikolov – Fotolia.com; **p.267** © Luca Tettoni / Getty Images; **p.269** *l* © Stéphane Bidouze – Fotolia.com; *r* © Paul Edmondson / Getty Images; **p.281** © Ashley Cooper, Visuals Unlimited / Science Photo Library; **p.282** *tl* © lapas77 – Fotolia.com; *tr* © sergbob – Fotolia.com; *b* © JRstock – Fotolia.com; **p.283** *both* © Erica Larkcom; **p.284** Erica Larkcom; **p.285** © Wadsworth Controls; **p.287** *all* © Erica Larkcom; **p.290** *t* © Ivan Nesterov / Alamy; *c* © Meutia Chaerani / Wikipedia Commons; *b* © Phototake Inc. / Alamy; **p.292** © Roger Delpech; **p.295** © ewwwgenich1 – Fotolia.com; **p.297** *l* © Erica Larkcom; *r* © Vangelis Thomaidis – Fotolia.com; **p.298** *tl* © Erica Larkcom; *tr* © defun – Fotolia.com; *b* © Erica Larkcom; **p.299** © Cultura Creative / Alamy; **p.300** © Huon Aquaculture (www.hounaqua.com.au); **p.301** *tl* © Hemis / Alamy; *tc* © Erica Larkcom; *tr* © Simon Greig – Fotolia.com; *c* © rekemp – Fotolia.com; *b* © Bert Hoferichter / Alamy; **p.302** © Valery Shanin – Fotolia.com; **p.303** *l* © Gary Adams – Fotolia.com; *c* © janifest – Fotolia.com; *r* © Hagen Graebner / Wikipedia Creative Commons; **p.305** *t* © janifest – Fotolia.com; *c* © Martin Shields / Science Photo Library; *b* © Makoto Iwafuji / Eurelios / Science Photo Library; **p.311** *l* © Sara Blancett / Alamy; *r* © A3386 Uli Deck / dpa / Getty Images; **p.313** *t* © Phase4Photography – Fotolia.com; *c* © Big Cheese Photo LLC / Alamy; *b* © blickwinkel / Alamy; **p.314** *l* © maksymowicz – Fotolia.com; *r* © Derrick Neill – Fotolia.com; **p.316** *t* © National Pictures / TopFoto; *ct* © Nigel Cattlin / Holt Studios / Science Photo Library; *cb* © Erica Larkcom; *b* © Nigel Cattlin / Alamy

l = left, *r* = right, *c* = centre, *t* = top, *b* = bottom

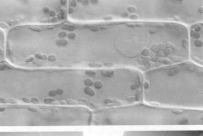

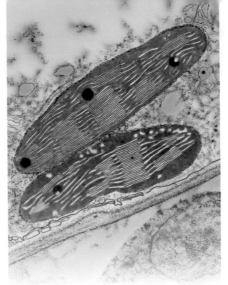

1

Living organisms: variety and common features

TO THINK ABOUT . . .

- Make a list of things you know occur inside cells.
- Name some organisms that are made up of one cell only.
- Make a list of some of the different types of cells that are found in plants and in animals.

TO THINK ABOUT . . .

Try to answer these questions at the start of this section, then come back to them when you have completed the section to see how far you have progressed in your understanding.

Seeing inside living organisms

The two images of cells and their internal structures were obtained using different microscopes. The first microscope is one that you might use in your biology course and it gives an image of a plant cell with chloroplasts. With this microscope, a beam of light is shone through the specimen, and then the light passes through a system of lenses that magnify the specimen and let you see inside the cell. The maximum useful magnification with this light microscope is around 1500. If the magnification is any higher, the image becomes too blurred to see any detail.

The second image of a cell was taken with an electron microscope, shown with two people using one. The electron microscope uses a beam of electrons rather than light and magnifications of 100 000 are quite common. That is how the detail inside a chloroplast can be revealed. A disadvantage of the electron microscope is that specimens have to be killed and sliced, so living material cannot be observed.

By using microscopes, we know that living organisms are made up of 'cells' and scientists have also learnt about the smaller structures inside the cells, such as the chloroplasts you see in these images. But are all cells the same? Or are there different kinds of cells in the diverse range of living organisms we can observe: plants, animals, fungi protoctists and bacteria? How do cells work together within an organism and how do cells connect with each other, allowing materials to move between them?

1.1 Characteristics of living organisms

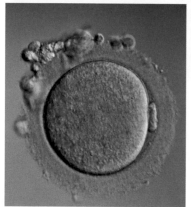

Is it alive?

Biology is the study of life. But what exactly do we mean when we say that something is living? Imagine for a moment, that you are an intelligent alien species visiting planet Earth – but with no humans to talk to. How would you decide whether an object was alive (like a mouse) or not (like a pebble)? You might decide that movement is a characteristic of life – but does that mean that a falling snowflake is alive? Or you might decide that growth is a characteristic of life – but does that mean that a growing crystal of salt is alive? In these images the snowflake is not living but the human egg cell and flower are both parts of living organisms.

This problem faces space scientists who wonder whether life has ever existed on nearby planets like Mars. Just because a sample of Martian soil absorbs oxygen, this does not by itself prove that there are living organisms using up the oxygen in the soil – the oxygen might just be reacting with minerals in the rocks. How can we be sure that something is living?

■ Introducing living organisms

One solution to the problem of what we mean when we say that something is 'living' is to produce a list of **characteristics** which is true for all forms of life on planet Earth. Unless an object does all of the things on the list, it cannot be considered alive. It is either **dead** or it was not a living organism in the first place.

Section 1.1 acts as an introduction to the rest of the book. It provides a list of what makes living organisms different from non-living things and establishes the special features (characteristics) of living organisms. It emphasises the common features possessed by all living organisms (plants, animals and microorganisms) that make them distinct from non-living things, such as a rock. These common characteristics link living organisms together, through a common origin and shared history since life began on Earth 3.5 billion years ago.

This introduction aims to give a brief view into the extended and more detailed discussions in the later sections of the book, in which you explore how living things work – first within individual cells, then (for multicellular organisms) inside larger whole organisms. Later you have a chance to look at ecological relationships, which show how living organisms interact with each other in the places where they live (their habitats). In the final sections of the book, you take a closer look at human beings – their influence on the natural environment and how humans use biological resources. We see how humans now (and in the past) use and apply biological principles – for example, in the production of food.

In Section 1.1, the focus is on green (flowering) plants and humans, though the characteristics apply to the whole range of living organisms (outlined in Section 1.2). As you work through Section 1.1, you can use the Study tips to guide you forwards to the relevant sections where the particular topics are discussed in more detail. Or, when you have studied other sections in the book, you can refer back to this introduction and see how far you have improved your understanding of what living organisms are and how they work.

Nutrition

Nutrition is the means of obtaining food, which provides the energy that living organisms need for the processes they carry out. It also provides the molecules that are built into other molecules that make up the substances of living material.

Green plants obtain their food by taking in simple molecules. They take in carbon dioxide from the air and water and mineral ions from the soil. Green plants trap energy from the Sun and use this to build complex molecules from these simple molecules. This process is called **photosynthesis** and is of fundamental importance as the route by which energy is captured by living organisms then distributed through the living world to allow life as we know it. Plants are described as **autotrophic** because they build up their own food ('auto' = 'self' and 'trophic' = 'feeding').

Animals eat other living organisms (plants or animals) and derive their energy from this food. The food contains complex molecules so the animal has to break these down into simpler units before building them up again into the complex molecules that form part of the animal. Animals are described as **heterotrophic** because they obtain food from a range of different sources ('hetero' = 'different'). Originally, the energy contained in the food comes from the Sun through plants, as indicated above, though often there is a food chain (or food web) that connects the plant and the animal (or human).

Respiration

Respiration is the process that releases the energy locked within substances such as glucose. Chemical energy has been stored in the glucose and is released when the molecular bonds are broken. Respiration takes place in the cells of all living organisms. It is a series of chemical reactions and usually oxygen is required and carbon dioxide and water are produced as waste products. The energy released is transferred to a molecule (such as ATP) that acts like an energy battery, which stores the energy until it is needed to power some process in the living cell.

STUDY TIP

In Biology, you have to understand the links between sections.

STUDY TIP

Note that the **Study tips** in Section 1.1 all direct you to other parts of the book where the topics are developed in greater detail.

STUDY TIP

Refer to Section 2.1 for nutrition in flowering plants, Section 2.2 for human nutrition and Section 6.2 for feeding relationships between organisms in an ecosystem.

STUDY TIP
Refer to Section 2.3 for the topic of respiration in plants, animals and some microorganisms.

Respiration must not be confused with gas exchange, which describes the processes involved in getting oxygen from the air to the cells and getting waste carbon dioxide out of the organism. In humans, breathing is part of the mechanism for gas exchange and is the way of getting air in and out of the lungs. You must be clear that breathing is not part of the process of respiration.

Excretion of waste

Many chemical reactions take place inside the cells of living organisms. These are described as **metabolic reactions**. These reactions form pathways in which each step involves changing substances into other materials.

Many of these metabolic reactions that take place in cells produce waste products that are not required by the cells and may even be toxic if they accumulate. Excretion describes the processes by which these substances are eliminated from the plant or animal. Examples of waste products include carbon dioxide from respiration and oxygen from photosynthesis. Water is also produced as a waste product in many reactions. In humans (and other animals), when proteins break down they produce a waste product that contains the element nitrogen – usually as urea. In humans, urea is eliminated from the body from the kidneys in urine.

STUDY TIP
Refer to Section 4.1 for excretion in flowering plants and Section 4.2 for excretion in humans.

In humans, you must not confuse egestion and excretion. Egestion describes the way that faeces are eliminated from the body. Most of the materials in faeces have never been absorbed from the intestines into the cells of the body.

STUDY TIP
Refer to Section 4.3 for general information about coordination and response, then to Section 4.4 for response in flowering plants and Sections 4.3 and 4.5 for details relating to response in humans.

Response to surroundings

Living organisms share the characteristic of sensitivity to different stimuli in their surroundings and an ability to respond. This means that living organisms can avoid problems (such as being eaten or becoming dried out) and can gain access to resources (such as light or food).

Plants and animals differ in the way that they respond to a stimulus. As examples, humans have special organs (known as receptors) that detect the stimulus (such as the eye for light or the ear for sound). They have a nervous system that coordinates a quick response, using muscles (that enable the person to run away from danger). Plants often respond much more slowly, by growing in a particular direction in response to the direction of light or gravity.

Movement

All living organisms move in some way, though we more often associate movement from one place to another with animals.

STUDY TIP
Refer to Sections 3.3 to 3.6 for details about movements of substances within organisms under the heading of 'transport' and to Section 4.4 for ways that parts of flowering plants move. There is no further discussion of locomotion in animals.

In plants, movement usually involves a part of the plant, rather than the whole organism. Some plants (such as climbing beans or peas) produce tendrils. These may twist around until they make contact with an object, like a stick, to which they can anchor.

Most animals are able to move from place to place and this is known as **locomotion**. Some do not move very far, but for many we are familiar with characteristic movements such as an athlete running, a bird flying, a worm wriggling or a fish swimming.

Control of internal conditions

The conditions inside an organism, such as temperature and water content, must be kept within strict limits. If these limits are exceeded, the processes that occur in the organism may suffer, having an effect on its health. The term **homeostasis** is used to describe the maintenance of the internal environment of an organism within the required limits.

For example, in the human body, if the internal temperature gets too high, there are mechanisms for losing heat and allowing the body to cool. Similarly, if the body loses too much water without taking in enough to compensate or if the level of carbon dioxide in the blood is too high, there are systems in the body to control these situations, bringing the internal environment back to normal.

In plants, various mechanisms exist that help control the internal environment. On a very hot day, a plant might lose a lot water from its leaves and then begin to wilt. The stomatal pores on the leaves may close as a result and this prevents any further loss of water vapour. On the other hand, in a plant with a good water supply, the stomata remain open encouraging water loss from the leaf. This process cools the leaves in the same way as sweating cools the skin of a mammal.

> **STUDY TIP**
>
> Look in Section 4.3 for details about control of internal conditions, though this refers mainly to humans. You can find more details (for humans) in Sections 4.2 and 4.5.

Reproduction

Every living thing ultimately dies. But if they can reproduce first, they ensure that there are some offspring to take their place in the world.

Living organisms all share the ability to reproduce and make more like themselves and so continue the species from one generation to the next. The information that controls an offspring's characteristics is contained in the DNA. This essential information divides when a cell divides, so that some is passed on to each of the cells that build up into the next generation.

For some, particularly for microscopic organisms, reproduction occurs simply by dividing into two. In other organisms, a piece breaks off and is able to grow into a new individual. This can occur in many plants and some animals. These are methods of **asexual reproduction**.

> **STUDY TIP**
>
> Refer to Section 5.1 then to Section 5.2 for flowering plants and Section 5.3 for humans. Section 5.4 gives details about DNA.

Most plants and animals reproduce by **sexual reproduction**. This occurs as a result of two special cells joining together. These special cells are the sex cells – one from the male and one from the female. The offspring receive some DNA from the male parent and some from the female parent and this DNA determines the characteristics of the offspring and ensures there is continuity to the next generation.

Growth and development

In sexual reproduction, the new individual starts as a single fertilised egg cell (**zygote**). The zygote divides to form the many different types of cells that make up the tissues and organs of an adult individual.

During development from zygote to adult, the organism gets larger and usually changes in shape and proportion. Animals reach a maximum size and then generally do not increase further, nor do they continue to develop new structures in the way that plants do.

STUDY TIP

Section 5.2 includes some information for flowering plants and Section 5.3 for humans, though you might also make links with Section 2.2 (human nutrition).

Plants continue to grow throughout their lives. For example, every year a tree puts out new growth from its buds, including new shoots and new leaves. This means it grows in size and complexity each year. In flowering plants, flowers develop into fruits and seeds and these may be dispersed to start the next generation.

You could try to find some photographs of yourself at different ages, and use these to trace your own growth and development from the time you were quite small to see how have you changed throughout your life so far.

STUDY QUESTION

1 The table below lists some descriptions or statements about processes carried out in living organisms. Some are carried out only by plants and some only by animals. Some processes are carried out by both plants and animals.

Copy and complete the table. Tick in the boxes to show whether the processes are carried out by plants or by animals (or both). In the final column, name the process involved.

(It will help you to think of a bean plant and a human as examples of a plant and an animal.)

Description or statement	Plants	Animals	Process
synthesise sugars and give out waste oxygen			
release energy from sugars and use oxygen			
grow faster on one side when exposed to light from one side			
eat other living organisms as food			
get rid of waste nitrogenous material from the breakdown of protein molecules			
produce more organisms that are the same or similar to themselves			
use muscles to run and jump			
get larger and change during life			

1.2 Variety of living organisms

From simple origins to the diversity of today's living organisms

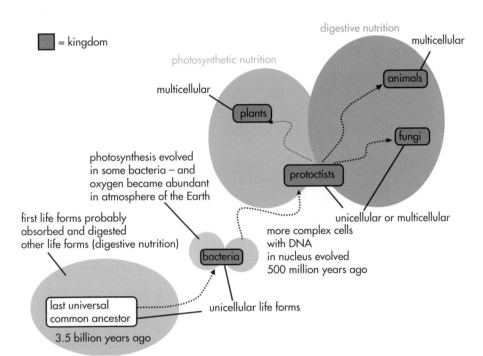

= kingdom

digestive nutrition

multicellular

photosynthetic nutrition

multicellular

plants

animals

fungi

protoctists

photosynthesis evolved
in some bacteria – and
oxygen became abundant
in atmosphere of the Earth

unicellular or multicellular

first life forms probably
absorbed and digested
other life forms (digestive nutrition)

more complex cells
with DNA
in nucleus evolved
500 million years ago

bacteria

last universal
common ancestor

unicellular life forms

3.5 billion years ago

The sweep through history and pre-history gives an overview of life from its possible origins, perhaps 3.5 billion years ago. It shows what scientists have pieced together from the probable beginnings of life to the diversity of living organisms we see on Earth today. The timescale is approximate but is backed by geological evidence in the rocks and later by fossil evidence showing the forms of some ancestors of plants and animals that exist today – and some that have become extinct.

Today's world has an enormous diversity of living organisms. To help us make sense of all this information, we find it useful to sort (or classify) them into groups. If you were in charge of a library, how would you organise the books? Would you arrange them according to the colour of their covers or when they were published or would you classify them on the basis of their authors or topics? Which would be most useful and what would you do if new books were added – where would you fit them in?

For living organisms, the diagram can be described as an outline of a sequence with branches, reflecting relationships between different groups and how they have changed or diverged over time. In present-day organisms, those with similar features are considered to be closely related, having at one time shared common ancestors.

A widely accepted classification is the 'Five kingdom' classification and the 'groups' described in this section are based on this, but what other systems of classification have been used in the past or are currently considered? Who did a lot to establish the binomial system for naming organisms (into genus and species)? Why do classification systems change and have to be altered as new species are found? And how do modern DNA studies confirm or raise questions about existing classification systems?

◼ Eukaryotic and prokaryotic organisms and their features

Cells that contain a nucleus with a distinct membrane are described as eukaryotic (see page 11). Eukaryotic organisms may be multicellular or single-celled. Plants, animals, fungi and protoctists are all eukaryotic organisms.

Prokaryotic cells do not have a nucleus and the nuclear material (a single circular chromosome) lies in the cytoplasm of the cell. Prokaryotic cells are much smaller (they have a thousand-fold smaller volume) than eukaryotic cells. In fact they are too small to contain chloroplasts or mitochondria (see page 12). Bacteria are prokaryotic organisms.

◼ Eukaryotic organisms

Plants and animals are two of the five main groups of living organisms. Both are multicellular. Some key features that distinguish plants from animals are listed.

Plants

- contain chloroplasts (and so carry out photosynthesis)
- have cellulose cell walls
- store carbohydrates as starch or sucrose.

Plants: Maize (a cereal) **Plants:** Pea (a herbaceous legume)

Animals

- do not contain chloroplasts (so cannot carry out photosynthesis)
- do not have cellulose cell walls
- often store carbohydrates as glycogen
- usually have nervous coordination
- are able to move from place to place.

Animals: A human **Animals:** A housefly (an insect)

Fungi

- cannot carry out photosynthesis
- have cell walls made of chitin
- some are single-celled (e.g. yeast)
- the body, for most, is made up of thread-like **hyphae**, containing many nuclei and organised into a **mycelium** (e.g. *Mucor*)
- feed by secreting extracellular digestive enzymes (outside the mycelium) onto the food and then absorbing the digested molecules
- this method of feeding is described as **saprotrophic** (feeding on decaying matter)
- some fungi are **parasitic** (feeding on living material)
- may store carbohydrate as **glycogen**.

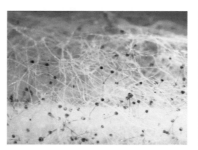 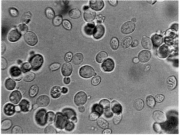

Fungi: *Mucor* (bread mould), showing mycelium, seen with a light microscope. The hyphae are between 10 and 25 μm in diameter.

Fungi: Yeast (a single-celled fungus), seen with a light microscope. The yeast cells are between 2 and 10 μm in diameter.

Protoctists

- are a rather diverse group of organisms, held together by the fact that they don't fit in any of the other groups
- are microscopic and single-celled, though some aggregate into larger forms (such as colonies or chains of cells that become filaments)
- are usually aquatic
- some have features like animal cells (e.g. *Amoeba* and *Plasmodium*, the parasite that causes malaria)
- some have features like plant cells, including chloroplasts (e.g. *Chlorella*). Algae are included in this group.

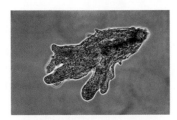

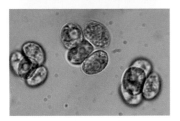

Protoctists: *Amoeba*, single-celled, size 0.01 to 0.1mm (may just be visible with the eye)

Prokaryotic organisms

Bacteria

- are single-celled, though have different shapes (e.g. rod-shaped, spherical)
- cell structure includes cell wall, cell membrane, cytoplasm, and plasmids
- have no nucleus but contain circular chromosome of DNA
- feed in different ways:
  - some can carry out photosynthesis
  - most feed off other organisms, living or dead. (If feeding off a living organism, the bacteria are described as **parasites**; if feeding off dead organisms, the bacteria are described as **saprobionts** or **decomposers**.)

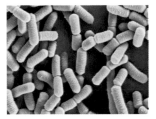

Protoctists: *Chlorella*, single-celled, containing chloroplast (size 8μm in diameter)

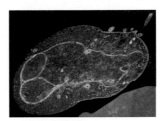

Protoctists: *Plasmodium*, a parasite (size 15μm in length)

Bacteria: *Lactobacillus* (a rod-shaped bacterium), size 2μm (2000nm) in length

Bacteria: *Streptococcus* (a spherical bacterium)

Pathogenic organisms, including viruses

The term **pathogen** is used to describe a microorganism that causes disease, for example in plants and animals. Fungi, protoctists and bacteria all include some pathogens.

All **viruses** are pathogenic and can exist only inside a living cell. Viruses can reproduce but do not carry out other characteristics of living things. For this reason, often viruses are not included in a classification of living organisms. Features of viruses are listed below.

Viruses

- very small particles (smaller than bacteria), variety of shapes and sizes
- lack a cellular structure
- consist of a protein coat that surrounds either DNA or RNA
- live and reproduce only inside living cells
- can reproduce (by instructing the host cell to make more of them), but do not carry out any other characteristics of living organisms
- all are parasitic and infect every type of living organism (including plants, animals and bacteria; viruses that infect bacteria are called bacteriophages).

Viruses: Tobacco mosaic virus (prevents formation of chloroplasts in plants), size 300nm in diameter

Viruses: HIV virus (causes AIDS), size 120nm in diameter

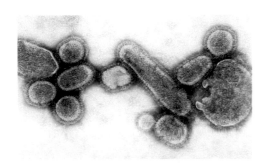

Viruses: Influenza virus (causes 'flu'), size 100nm in diameter

STUDY QUESTIONS

1   Section 1.2 gives a broad picture of the range of living organisms and the common features that have led to a classification system and the recognition of groups within this system. In Section 1.2, very few named examples have been included from the enormous number of living organisms that exist today (and have existed in the past).

For each group, list **three** more examples of organisms that would fit in that group. Check that your examples show the given features. Then include some information about each organism in your list (such as why they are important to humans or where they can be found). Give your information under the following headings:

- ■ Plants
- ■ Animals
- ■ Fungi
- ■ Protoctists
- ■ Bacteria
- ■ Viruses.

For the last four groups (fungi, bacteria, protoctists and viruses) include at least **one pathogen** in your list.

Explain why viruses are often not included in a biological classification of living organisms.

2   The **Study tips** are given in the box below as cross-references. These all make connections with topics in other parts of the book and which you study during your course.

Follow each of the links to help you make the connections and understand about the range of living organisms and the common features they show. You can check the links either when you study Section 1.2 or as you study different topics of your biology course. Devise your own chart to summarise the groups of organisms and the connections with other aspects of their biology.

Cross-references to other parts of the book

- For more information about plants look at Section 1.3 (cells and their organisation), Section 1.4 (biological molecules), and Section 2.1 (plant nutrition).
- For more information about animals look at Section 1.4 (biological molecules), Section 4.5 (coordination and response in humans), and Section 1.1 (characteristics of living organisms).
- For more information about fungi, read about glycogen in Section 1.4 (biological molecules), yeast in Section 7.2 (using microorganisms to produce food) for its role in fermentations, including making bread, and in Section 6.3 (cycles within ecosystems) for the role of fungi in extracellular digestion and in nutrient cycles and decomposition.
- For more information about bacteria, look at Section 1.3 to compare cell structure in bacteria with that of plant and animal cells, and read about DNA in Section 5.4 (genes and choromosomes) and use of plasmids in GM in Section 7.5 (genetic modification). You can also look at Section 7.2 for cultivation of bacteria in fermenters (for particular products), and make a link to pathogenic bacteria and immune response in Section 3.5 (transport in humans (1) - blood, structures and functions) and their role in decomposition in natural cycles in Section 6.3.
- For protoctists, make a link to eutrophication.
- For more information about viruses read about the use of bacteriophages as vectors for genetic modification in Section 7.5.
- Finally, you can look at Section 5.6 (variation, change and evolution) to help you understand how the variety of living organisms has arisen, and how species and groups have changed over time (and will continue to change).

1.3 Cells and their organisation

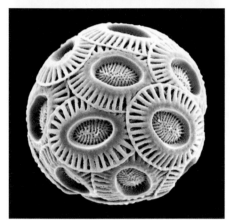

The importance of tiny cells

The white cliffs of Dover, a well-known landmark in England, are enormous and easily visible from 30 miles away across the sea. Yet they are mainly built from the shells of single-celled organisms called phytoplankton. The second photo shows an example of one phytoplankton cell, covered in plates of calcium carbonate (chalk), and its name is *Emiliania huxleyi*. These tiny cells swirl around the oceans of the world and, if they are not eaten, they die and settle on the bottom as a white layer of chalk. The white cliffs of Dover were once on the floor of the sea, and their chalk is made up of such phytoplankton shells, only visible individually with the help of an electron microscope.

The fundamental unit of all living matter is the cell, and life depends upon whole cells. A fragment of a cell cannot be regarded as living matter. So what are the key features shared by all cells? What are the differences between animal and plant cells? How do all the processes of living organisms go on inside cells? How are cells organised inside to make these living processes work and how do cells work together in larger multicellular organisms?

■ Cells and their internal structures

You need to use a microscope to see cells. With a simple light microscope you can see the outlines of cells and some of the structures inside. With more powerful microscopes (including electron microscopes), giving higher magnification, scientists are able to observe more details of the internal structures and find out a lot about these structures and how they function in a cell.

The internal structures are described as organelles and they provide organised 'compartments' inside the cell. Each organelle carries out a particular task and it is important that they all work together to carry out the functions of the whole cell.

■ Features of plant cells and animal cells

Figure 1.1 shows key features of a generalised plant cell and a generalised animal cell. The plant cell is similar to a palisade mesophyll cell in a leaf and the animal cell is similar to a 'cheek' cell from the lining of your mouth. Some features are common to all cells, whereas some features are found only in plant cells or only in animal cells. Many cells differ from this generalised view as they are specialised to carry out particular functions. We look at some examples of specialised cells later in Section 1.3.

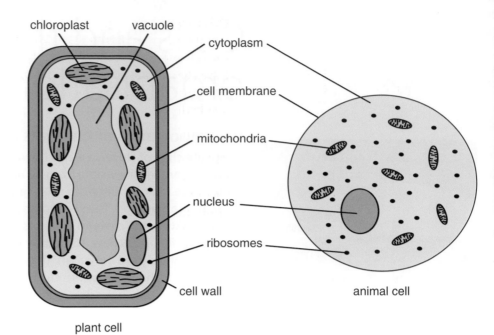

chloroplast   vacuole
cytoplasm
cell membrane
mitochondria
nucleus
ribosomes
cell wall
animal cell
plant cell

Figure 1.1 Features of a generalised plant cell (left) and generalised animal cell (right). The mitochondria are organelles where aerobic respiration takes place and are found in both plant and animal cells.

The list below links features labelled in the diagram (Figure 1.1) with their functions in the cell. Features found in both plant and animal cells are given first.

- **Cytoplasm** – This is like a thick watery liquid and many of the activities of cells take place in the cytoplasm. Important molecules here include **enzymes**, which catalyse many metabolic reactions, such as those of respiration and synthesis of proteins. A number of smaller structures, the **organelles**, are found floating in the cytoplasm – these look like particles when viewed with a light microscope. The organelles are not fixed in position and can move within the cytoplasm. Sometimes, for example, chloroplasts can be seen 'streaming' (i.e. moving round) in a cell.
- **Mitochondria** – These are sausage-shaped organelles, scattered throughout the cytoplasm, where key reactions of aerobic respiration take place. They are just visible as small specks under the light microscope.
- **Ribosomes** – These are very small complex particles (much too small to be seen with a light microscope), found in the cytoplasm. They are responsible for the synthesis of proteins, including enzymes, specified by genes in the nucleus of the cell (see Section 5.4). A single cell contains millions of ribosomes.
- **Nucleus** – This contains the genetic information of the cell. This information is located in the **DNA**, which forms the chromosomes. The chromosomes become visible as thread-like structures when the cell is dividing, but most of the time the nucleus just appears denser than the cytoplasm and the chromosomes cannot be seen. Information about each feature of the cell (and the whole organism) is held in the genes, along the length of the chromosome. The nucleus (through the genes) determines the substances the cell makes and controls the activities of the cell.

STUDY TIP
Look in Section 1.4 for more details about enzymes.

STUDY TIP
You can find information about DNA, genes and chromosomes and protein synthesis in Section 5.4.

- **Cell membrane** – This is the boundary between the cytoplasm of a cell and its surroundings. It keeps the cytoplasm and the organelles inside the cell and the cell membrane also controls which materials enter and leave the cytoplasm. Some materials may pass across the cell membrane by diffusion and others by active transport. Many materials are kept inside the cell, while others are prevented from entering.

These features are found only in plant cells.

- **Chloroplasts** – These are green disc-shaped organelles found in the photosynthetic cells of plants. (Chloroplasts are also found in some bacteria and in some protoctists.) This is where the chemical reactions of **photosynthesis** occur. The green colour is due to the presence of the pigment **chlorophyll**. The chloroplast is the second largest organelle in the plant cell, after the nucleus, and is easily seen under the light microscope. Many plant cells do not contain chloroplasts – such as those in roots.

- **Cell wall** – Plant cells are all surrounded by a tough cell wall. The plant cell wall is composed of a polymer of glucose called cellulose. The cell wall provides strength and protection to plant cells. It is strong and resists forces that could change the shape of the cell. Cell walls make the outline of plant cells easy to see under the light microscope. Fungi and bacteria also have cell walls, but made of different materials. In fungi, the cell walls are made of chitin, a substance also found in insects. In bacteria, the cell walls are made of a polymer of sugars and amino acids.

- **Vacuole** – This is a large region at the centre of a plant cell, separated by a membrane from the cytoplasm. It contains a watery solution known as cell sap. Often the cell sap pushes the cytoplasm outwards against the cell wall, giving the plant cell rigidity. Animal cells do not have a large vacuole though sometimes they contain small vacuoles.

STUDY TIP

Look in Section 1.5 and make the link to diffusion and active transport.

STUDY TIP

It is important to understand the role of chloroplasts in photosynthesis (see Section 2.1 for flowering plants and Section 1.2 for a wider variety of organisms).

STUDY TIP

Make a link to the cellulose molecule in Section 1.4.

STUDY TIP

Look in Section 1.5 and Section 3.1 to find out about the importance of turgid cells in plants.

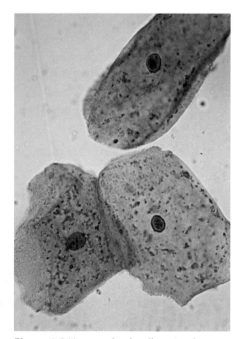

Figure 1.2 Human cheek cells stained with methylene blue, seen with a light microscope. Note the numerous mitochondria within the cytoplasm.

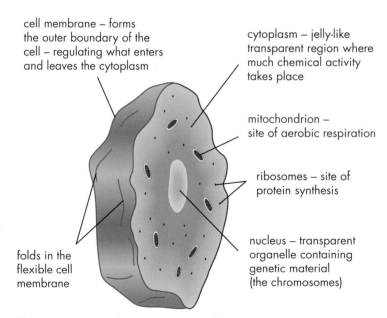

cell membrane – forms the outer boundary of the cell – regulating what enters and leaves the cytoplasm

cytoplasm – jelly-like transparent region where much chemical activity takes place

mitochondrion – site of aerobic respiration

ribosomes – site of protein synthesis

folds in the flexible cell membrane

nucleus – transparent organelle containing genetic material (the chromosomes)

Figure 1.3 Diagram showing structures of a human cheek cell. Compare this with the microscope view (Figure 1.2).

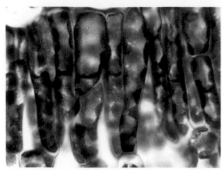

Figure 1.4 Some palisade cells in a leaf, seen with a light microscope. The staining helps to show the chloroplasts.

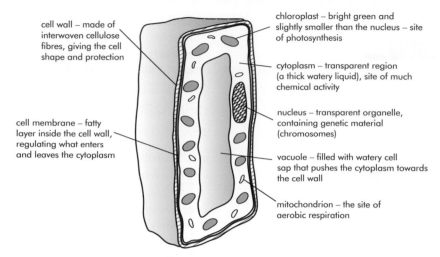

cell wall – made of interwoven cellulose fibres, giving the cell shape and protection

cell membrane – fatty layer inside the cell wall, regulating what enters and leaves the cytoplasm

chloroplast – bright green and slightly smaller than the nucleus – site of photosynthesis

cytoplasm – transparent region (a thick watery liquid), site of much chemical activity

nucleus – transparent organelle, containing genetic material (chromosomes)

vacuole – filled with watery cell sap that pushes the cytoplasm towards the cell wall

mitochondrion – the site of aerobic respiration

Figure 1.5 Diagram of a palisade mesophyll cell, cut in longitudinal section and showing the structures of the cell. Compare this with the microscope view (Figure 1.4).

Differences between plant and animal cells

Common features in all cells have been listed above. Differences between plant and animal cells are summarised in Table 1.1.

Table 1.1 Differences between plant and animal cells.

Plant cells	Animal cells
have cellulose cell wall	no cellulose cell wall
some cells contain chloroplasts	never contain chloroplasts
usually have large vacuoles	do not have large vacuoles though sometimes small vacuoles are present

■ Diversity and specialisation of cells

The generalised cells (Figure 1.1) show you the basic features of cells. But many cells, in both plants and animals, become specialised for the function they carry out. Here are just a few examples of specialised cells that you study in other parts of this book:

- root hair cell – an elongated 'root hair', increases the surface area for absorption of water
- guard cells – shape alters when turgid and flaccid, controls the opening and closing of stomata on the leaf surface
- red blood cell – has no nucleus, contains haemoglobin, carries oxygen in the human body
- nerve cell – an elongated 'nerve fibre', transmits nerve impulses.

STUDY TIP

For more details about these cells, look in Section 3.4 for root hair cells; in Section 3.1 for guard cells; in Section 3.5 for blood cells; in Section 4.5 for nerve cells.

Levels of organisation

For simple single-celled organisms, the cell is the whole organism. These organisms can only be seen using a microscope. For the many larger organisms that you can see and are familiar with, the body of each individual organism is made up of millions or billions of individual cells. These cells are organised in a way that allows them each to do a particular 'job' and contribute to the survival of the individual.

The human body is estimated to contain about 250 different types of cells, whereas a plant is made up of a more limited range of about 25 types of cells. These different types of cells are organised at different levels – first into different types of tissues then the tissues into organs. All of these are organised into a limited number of major organ systems.

The list summarises the levels of organisation and Table 1.2 gives some examples that illustrate these levels.

- **Organelle** – a component inside a cell that carries out a particular task.
- **Cell** – the basic functional unit of living organisms.
- **Tissue** – a collection of similar cells, carrying out a particular function within an organ.
- **Organ** – a collection of different tissues that work together to perform a common function.
- **System** – groups of organs working together to perform an overall function for the organism.

STUDY TIP
Make links with a range of examples showing the arrangement of tissues within a plant organ – see cross-section of leaf (Section 2.1), and cell and tissue structure in the small intestine (Section 2.2).

Table 1.2 Examples of organisation levels in a plant (e.g. bean) and an animal (e.g. human).

Level of organisation	Plant example	Animal example
organelle	chloroplast – where photosynthesis reactions occur in mesophyll cells of leaf	cilia – beating hairs projecting from cells lining the trachea (windpipe)
cell	guard cell – a cell that controls the size of the stomata in the lower epidermis of a leaf	neurone – nerve cell that transmits electrical signals (nerve impulses)
tissue	palisade mesophyll – photosynthetic cells near the top of a leaf	retina – nerve cells that detect and respond to light rays
organ	the flower – produces the seeds as part of reproduction	the eye – responsible for detecting light rays and sending nerve impulses to the brain
system	transport system moves materials around the organism, from organ to organ	circulatory system – moves materials around the organism, from organ to organ

Stem cells and cell differentiation

All multicellular organisms begin life as a single cell – the fertilised egg cell (zygote). This cell divides by mitosis to form two daughter cells, and these then divide to form more cells, and so on. Eventually an individual organism may be made up of several billion differentiated cells – grouped in the ways listed in Table 1.2.

As new cells are produced in the growing human embryo, specific genes are switched on (active) and off (inactive) inside each cell. This pattern of active and inactive genes cause a cell to commit itself to begin **differentiating** into one of the 250 possible types of cell in the adult human body. In the initial stage of this differentiation process, it is possible for the initial pathway of development to be switched to any other pathway – at this stage a cell is called a **totipotent stem cell** (as it has the potential to differentiate to become any kind of specialised cell). Eventually, however, cells become stuck on a fixed pathway of development, and can only become a specific kind of cell (e.g. red blood cell, neurone, etc).

Our bodies maintain a small population of stem cells in most of our organs. These stem cells are used to produce new cells as adult tissue cells are damaged or killed by accidents and disease. Our bone marrow, for instance, contains **stem cells** which divide to produce new red and white blood cells during our life.

Stem cells are the subject of current biomedical research, and they have been used to treat injured and diseased organs (for example heart muscle, brain, retina). In such **stem cell therapy**, the stem cells are injected into the damaged area. Here they divide to form fully differentiated replacement cells, restoring the function of the organ.

Current sources of therapeutic stem cells are human embryos (often discarded as part of IVF treatments) and umbilical cords of newborns. This creates an ethical dilemma: would it be ethical to create an embryo specifically to provide stem cells for a patient? One solution to this dilemma is to remove some adult (differentiated) cells from the patient and chemically trigger these cells (cultivated in a petri dish) to become stem cells ('induced' stem cells). These induced stem cells could then be reintroduced to the patient's body to help their problem.

There is another reason for caution with using stem cell therapy – once the cells have been injected into the patient they are expected to divide and grow and differentiate under the influence of the natural chemical signals of the body. If they do not respond to these chemical signals in the body, and continue to reproduce, then the stem cell therapy treatment might lead to cancer.

It is hoped that, in the near future, induced adult stem cells in tissue culture could be chemically triggered into forming whole replacement parts for a patient (e.g. retina, skin, etc.). If the induced stem cells are developed from adult cells removed from the patient, then a whole replacement organ, grown outside the body in tissue culture (*in vitro*), might be transplanted into the patient without any risk of rejection by the patient's immune system.

■ Practical activity – making a slide to observe moss leaf cells

1 Using fine forceps, carefully pull a single moss leaf (or two) from the shoot of a moss plant.
2 Place the whole leaf in a drop of water on a microscope slide, making sure that the leaf is not folded.
3 Add a second drop of water on top of the leaf.
4 Add a coverslip carefully to the slide, lowering it slowly so as to avoid trapping bubbles.
5 Examine the slide using a microscope and note the shape of the cells and their contents.

STUDY TIP

In vitro means 'in glass' - i.e. in a test tube or culture dish rather than in the living organism. You can find out more about tissue culture in Section 7.6.

PRACTICAL

This activity allows you to see some cells in the leaf of a simple plant. These cells are filled with green chloroplasts that give plant leaves their colour. The cells of a moss leaf form a single layer, which helps to make it easy to see the contents. Another source of a single sheet of plant cells (large cells, without chloroplasts) is onion epidermis. Onion cells can also be a useful way to observe the effects of osmosis (see Section 1.5).

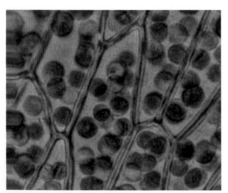

Figure 1.6 Unstained living moss leaf cells under the microscope.

This practical activity allows you to view some cells and if you stain with a drop of iodine solution, you can see starch grains (blue) and cellulose cell walls (colourless) side by side – two large molecules that are polymers of glucose (see Section 1.4).
Remember to wear eye protection.

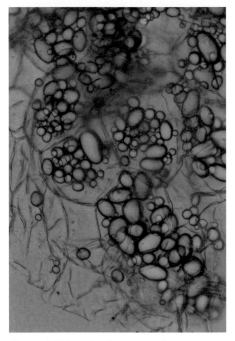

Figure 1.7 Unstained potato cells showing unstained starch grains clustered inside the cytoplasm of individual cells, under the microscope.

■ Practical activity – making a slide to observe starch in potato cells

Starch is stored in potatoes and can be observed with a microscope. Iodine solution goes blue–black when starch is present.

1 Take a whole potato and use a knife to cut a small block of potato from it. Then cut a tapering wedge from the block of potato tissue.
2 Cut a tiny slice from the thin edge of the wedge. (You could use a potato peeler.) Mount this in a drop of water on a microscope slide and add a coverslip.
3 Examine the slide using a microscope. Starch grains should be visible in the cells as clear grains.
4 Add a few drops of iodine solution to the edge of the coverslip. Use a paper towel to draw the iodine solution across the specimen, as shown in Figure 1.8. The starch grains should now be visible inside the cells and are a blue–black colour. This shows that they are starch.

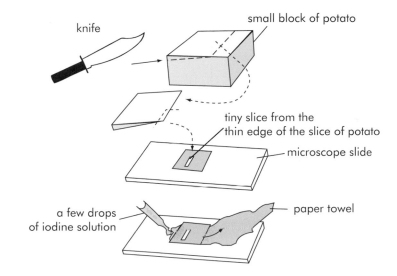

Figure 1.8 Making a slide to observe starch in potato cells.

■ Practical activity – making a slide to observe human cheek cells

Make sure you follow correct Health and Safety regulations.

1 Using the tip of a cotton bud (or clean wooden spatula), rub the inside lining of your cheek. Thousands of cheek cells rub off onto the cotton bud.
2 Rub the cotton bud over the central part of a clean microscope slide. This transfers hundreds of the cheek cells onto the glass.
3 Add a drop of stain, such as methylene blue. Lower the coverslip and avoid trapping bubbles of air in the stain.
4 Examine the slide using a microscope. The cheek cells should be visible (the colour depends on the stain used). After the experiment, place your cotton bud and slide into disinfectant; do not touch anyone else's.

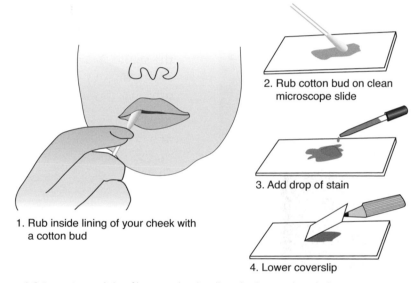

1. Rub inside lining of your cheek with a cotton bud

2. Rub cotton bud on clean microscope slide

3. Add drop of stain

4. Lower coverslip

Figure 1.9 Preparing a slide of human cheek cells to look at under a light microscope.

STUDY QUESTION

1 a) Draw and label a typical plant and a typical animal cell.
   b) Below are two lists of some different types of specialised plant and animal cells for you to compare with the generalised cells you have drawn in part (a). Draw diagrams of each of the cells in the lists. Label the diagrams and write down any special features of these cells that link them to the functions they carry out.

   You may need to look in other parts of this book. Add more types to the lists if you can find them.

Plant cells	Animal cells
palisade mesophyll cell	red blood cell
root hair cell	white blood cell
guard cell	nerve cells
phloem cell	cells lining trachea
xylem cell	cells with microvilli (small intestine)
	sperm cell

1.4 Biological molecules

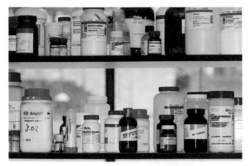

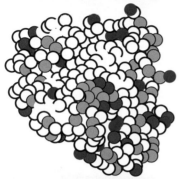

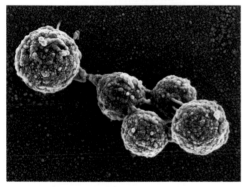

What living organisms are made of

If the human body is broken down into its basic elements, it is made up mainly of the following atoms:

- oxygen 65%
- carbon 18%
- hydrogen 10%
- nitrogen 3%
- calcium 1.5%
- phosphorus 1.2%.

Other elements, including potassium, sulfur and iron, contribute to the remaining 1.3% of the mass of the human body. The proportions of elements present in the cells of all living organisms, ranging from bacteria to whales and giant redwoods, are much the same. This is because, at the biochemical level, all life forms are made from the same groups of compounds – carbohydrates, proteins, lipids, nucleic acids – with mineral ions and water.

All of the elements listed are abundant in rocks and other non-living materials. How can they be bonded together to form the thousands of compounds that are only found in living organisms? There are many other abundant elements in the rocks of planet Earth, such as silicon and aluminium, so why do these play no part in the formation of biological compounds? How do the chemicals in the bottles on the shelf become organised into a ribonuclease molecule (shown as the model) and finally become part of a working human cell (in the third image)? How do these biological molecules work together to provide the special features we recognise in living organisms?

■ The large molecules in living organisms

Carbon atoms can link to each other and are at the heart of most biological molecules. The term 'organic' (carbon containing) is used to describe these large biological molecules. Table 1.3 summarises information about three important types of large molecule of living organisms. They all contain the elements carbon (C), hydrogen (H) and oxygen (O). The table summarises the basic units (building blocks) of each kind of molecule and gives some examples.

Table 1.3 The molecular units found in carbohydrates, proteins and lipids and some examples of each.

Class of compound	Elements present	Basic units	Examples
carbohydrates	carbon (C) hydrogen (H) oxygen (O)	simple sugars (monosaccharides) e.g. glucose, fructose	dissacharides (e.g. maltose, sucrose, lactose)
			polysaccharides (e.g. starch, glycogen, cellulose)
proteins	carbon (C) hydrogen (H) oxygen (O) nitrogen (N) sulfur (S)	amino acids	proteins (e.g. enzymes, haemoglobin in red blood cells, myoglobin in muscles)
lipids (fats and oils)	carbon (C) hydrogen (H) oxygen (O)	fatty acids and glycerol	lipids (e.g. oleic acid in olive oil)

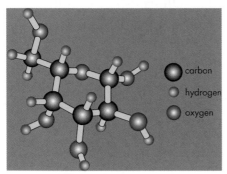

carbon
hydrogen
oxygen

Figure 1.10 In this model of a glucose molecule, the sticks between the atoms (C, H and O) represent chemical bonds. The bonds between the carbon atoms were all originally made using energy from the Sun.

STUDY TIP

Check the important role played by glucose in respiration (Section 2.3) and photosynthesis (Section 2.1).

STUDY TIP

Maltose can be used as an example of digestion – see human digestion (in Section 2.2) and in seed germination (Section 5.2).

STUDY TIP

Make sure you understand why it is important for the plant that starch has no osmotic effect (see osmosis in Section 1.5).

Carbohydrates

Carbohydrates contain the elements carbon, hydrogen and oxygen, in the ratio 1C : 2H : 1O. You can see these proportions in the formula for glucose: $C_6H_{12}O_6$. Simple sugars can join together to form larger molecules, such as disaccharides and polysaccharides; some examples are described below.

Glucose ($C_6H_{12}O_6$) – a monosaccharide

Glucose is an important simple sugar. It is a monosaccharide ('mono' = 'one' and 'saccharide' = 'sugar') and contains six carbon atoms. In plants, glucose molecules are formed by photosynthesis. Humans obtain glucose from digested food. Glucose is used for respiration by all living cells, when it is broken down to carbon dioxide and water in aerobic respiration. The energy in the bonds between the carbon atoms in the glucose molecule is released by respiration, and then used to do work in the respiring cell.

Maltose and sucrose – two disaccharides

Maltose is made up of two glucose molecules joined together. When starch is digested it breaks down into maltose. **Sucrose** is made up of a glucose and a fructose molecule joined together. (Fructose is also a monosaccharide.)

Starch and glycogen – two polysaccharides

Both of these polysaccharides are made up of many glucose units. Glucose molecules can be stored inside cells in one of these large molecules, rather than being used straight away.

In **plant cells**, glucose is stored as a polysaccharide (polymer) called **starch**. Millions of glucose molecules are linked up in a three-dimensional network to create a solid granule of starch inside the plant cells. This is an efficient way of storing the glucose in a small space. It has the added advantage that the starch granule is insoluble in water. This means that the stored sugar cannot be easily washed away and has no osmotic effect within the cell. The starch present in plant foods forms a major part of the human diet, providing an important source of energy.

In **animal cells** glucose is stored as a polysaccharide called **glycogen**. Like starch, glycogen is insoluble. Glycogen forms smaller granules than starch. This means it can quickly be broken down again to glucose when needed. Glucose that is absorbed into your blood from a digested meal is temporarily stored in your liver as glycogen. These reserves of glycogen are released between meals, whenever the concentration of glucose in the blood begins to fall below normal.

STUDY TIP

Make a link to insulin and regulation of blood glucose in Section 4.5.

STUDY TIP

Look in Section 2.2 for more details about the human diet.

Make a link to xylem and its functions in Section 3.4.

Cellulose – another polysaccharide

Plant cell walls are composed of a polysaccharide called **cellulose**. Cellulose is made up of interlinked chains of thousands of glucose molecules. These chains become twisted together to form strong fibres that wrap around the plant cell, providing it with support.

Much of the total mass of a plant is composed of cellulose and this makes an important contribution to dietary fibre in the human diet. We do not have enzymes that can digest cellulose, so it is not broken down in our guts. Many animals that feed on plants (herbivores) have symbiotic bacteria in their intestines and these bacteria digest the plant cellulose for them.

In woody plants the cellulose cell walls of some tissues (such as xylem vessels) become reinforced by another, stiffer substance called lignin. The paper of this book and your exercise book is made from wood pulp, so your ink is being drawn into the cellulose and lignin fibres of trees. Globally, the biomass of plants is greater than the biomass of everything else, and thus the most abundant organic substance on the planet is cellulose.

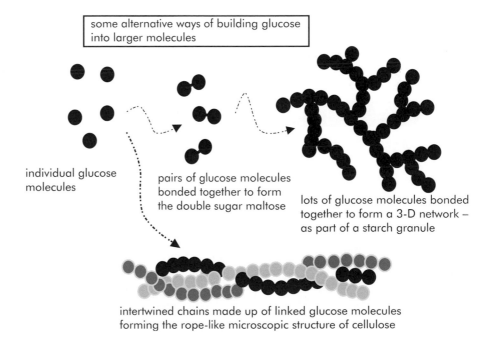

some alternative ways of building glucose into larger molecules

individual glucose molecules

pairs of glucose molecules bonded together to form the double sugar maltose

lots of glucose molecules bonded together to form a 3-D network – as part of a starch granule

intertwined chains made up of linked glucose molecules forming the rope-like microscopic structure of cellulose

Figure 1.11 Building glucose units into larger carbohydrate molecules.

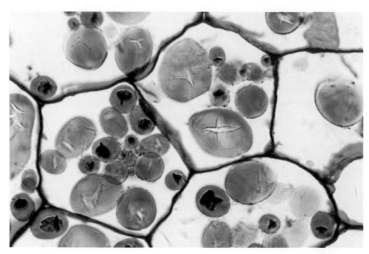

Figure 1.12 Potato tuber cells stained to show starch grains, seen with the light mocroscope.

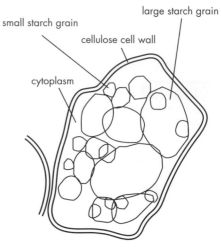

Figure 1.13 Diagram showing starch granules inside a potato cell, similar to the image in Figure 1.11.

Figure 1.14 Cellulose fibres (from cell wall of an alga), as revealed with an electron microscope.

Testing for glucose and for starch

You can carry out simple tests in the laboratory to detect glucose and starch. These are described in more detail in the following practical activities. To summarise briefly:

■ test for **glucose** using **Benedict's solution** – heat the solution and a reddish colour develops if glucose is present
■ test for **starch** using **iodine solution** – a blue–black colour develops if starch is present.

■ Practical activity – testing for glucose in a solution

1 Place 5 cm³ of the sample solution to be tested in a test tube. Add several drops of Benedict's solution.
2 Place the test tube(s) in a beaker of boiling water for 5 minutes.
3 Observe the colour of the solution in the test tube, as shown in Figure 1.15.

Pale blue (no change in colour of the Benedict's solution) shows no glucose is present. A colour ranging from yellow through orange to brick red shows glucose is present. Red indicates more glucose than the yellow or orange colours. Sometimes a precipitate forms at the bottom of the tube.

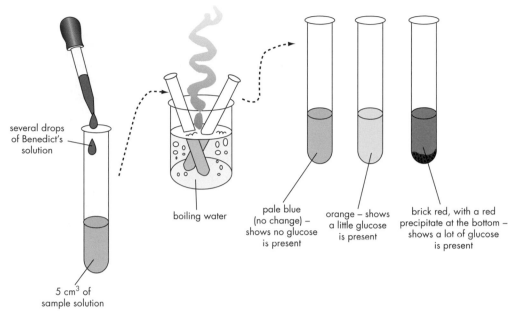

Figure 1.15 Testing for glucose in a solution.

▉ Practical activity – testing for starch in a solution

1 Place 5 cm³ of the sample solution to be tested in a test tube. Add several drops of iodine solution.

2 Observe the colour of the solution in the test tube.

Pale yellow (no change in colour of the iodine solution) shows no starch is present. A blue–black colour shows starch is present.

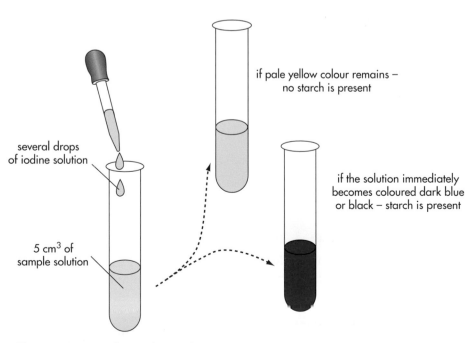

Figure 1.16 Testing for starch in a solution.

Proteins

All proteins are composed of a chain of **amino acids**. There are many thousands of different kinds of proteins found in a living cell. Yet all are made using just 20 different amino acids. The extraordinary variety of proteins is possible because each different protein is made of a unique sequence of amino acids, linked together in the chain.

A small protein may be made of a chain containing a hundred amino acids. Larger proteins might contain two or three hundred amino acids. When a protein is made, the chain is folded and twisted to give the molecule a specific overall shape. The shape of a protein determines its biological properties. A protein may, for example, be an enzyme (e.g. amylase), part of a skin flake (e.g. keratin), a carrier of oxygen (e.g. haemoglobin) or a structural substance (e.g. collagen in bones and ligaments).

Green plants make amino acids by using the nitrate ions that they absorb from the soil. They use the nitrates to provide the element nitrogen, which forms an essential part of each amino acid. Animals, fungi and bacteria get their amino acids by digesting the proteins of the organisms they eat.

Amino acids are not stored in the body of humans, so to remain healthy an adult person must eat a certain amount (proportional to body weight) of protein in their diet every day in order to replace proteins lost from the body by processes such as shedding skin flakes.

STUDY TIP

You can find more details about the nitrogen cycle in Section 6.3.

STUDY TIP

Look in Section 2.2 for more details about the human diet.

STUDY TIP

You can find out more about protein structure and protein synthesis in Section 5.4.

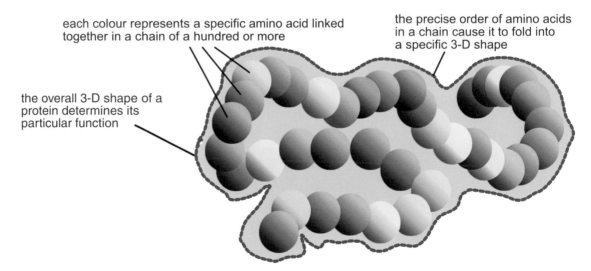

each colour represents a specific amino acid linked together in a chain of a hundred or more

the precise order of amino acids in a chain cause it to fold into a specific 3-D shape

the overall 3-D shape of a protein determines its particular function

Figure 1.17 Building amino acids into a protein.

PRACTICAL

Remember to wear eye protection.

■ Practical activity – testing for protein

1 Place 5 cm³ of the sample solution to be tested in a test tube. Add several drops of Biuret reagent.
2 Observe the colour of the solution in the test tube.

Pale blue (no change from the initial mixture of the reagent and the sample) indicates no protein is present. A pink / violet colour indicates that protein is present in the sample.

Lipids (fats and oils)

Lipids are made from a molecule called glycerol plus three attached fatty acids. The fatty acids can vary, giving different types of lipid. Some lipids are liquid (at normal temperatures) and these are described as oils, whereas others are solid and are described as fats. Lipids are substances that do not mix with water – we describe them as 'hydrophobic' ('hydro' = 'water' and 'phobic' = 'dislike or fear').

Cell membranes are composed largely of lipids, a factor that helps the cell to control the materials that can cross the membrane to enter or leave the cell.

Another important use of lipids is as a means of storing chemical energy in the body. In plants and animals, lipids provide a very efficient means of storing energy, which can then be used when food is scarce. During times when the diet of an animal contains excess energy, adipose (fat) cells swell up with additional lipid, and shrink again when the diet contains too little energy. When the human body needs to use stored energy, the lipids in the adipose tissue are broken down to glycerol and fatty acids. The fatty acids are transported in the blood from adipose tissue to respiring tissues in the body.

In humans, these adipose cells are found just under the skin. This layer of adipose tissue provides insulation for the human body, helping to conserve body heat. It also helps to protect the body from damage by acting as a cushion or 'shock absorber'.

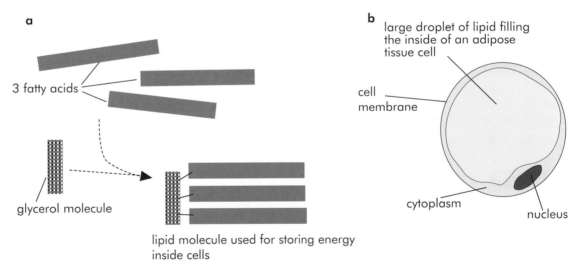

Figure 1.18 a Structure of a lipid molecule. **b** Lipid in an 'adipose' tissue cell.

■ Practical activity – testing for fat (or oil)

Emulsion test
1 Grind up some food particles (with a dry mortar and pestle).
2 Add a level teaspoonful to a test tube.
3 Add 3 cm³ ethanol to the test tube.
4 Stopper the test tube with a bung and shake it well.
5 Leave the test tube in a test tube rack until the particles have settled. This allows the lipids to dissolve in the ethanol.
6 Strain the ethanol solution off into another test tube. (Leave the food debris behind.)

7 Add 3 cm^3 of water to the ethanol solution.

8 Observe the appearance of the mixture in the test tube.

A clear solution means no lipids are present in the food sample. A milky emulsion (cloudy white) indicates that lipids are present in the food sample.

Grease spot test

Simply rub a dry, solid piece of food, such as a piece of nut, against a piece of paper. If the food contains lipids, then the place where the paper is rubbed becomes more translucent (lets light shine through) than the rest of the paper.

Buoyancy test

Pure oil (such as olive oil) is less dense than water, so forms a layer that floats on top of the water – something easy to see if water and oil are mixed in a test tube.

Water

Water is the most abundant molecule in all living cells. A human living cell contains 70% water by mass and a lettuce leaf cell more than 90% water (mostly in the vacuole). Water is, therefore, an important constituent of cytoplasm. Substances found in living organisms are dissolved in water inside the cytoplasm, between cells and inside transport systems. Water can act as a lubricant, for example in peristalsis.

Water forms the basis of the liquids for transport – for example, the blood, which is pumped around the body in vessels by a muscular organ such as the heart in animals. In plants, in xylem vessels, the water molecules form an unbroken column with each molecule clinging to its neighbours by cohesion. This means that a pull at one end of the column is transmitted along the whole column and this is how water can be pulled up xylem vessels.

Water also participates in many biochemical reactions – large molecules (macromolecules) are broken down by the addition of water (hydrolysis) or are joined together by the removal of water (condensation). In addition, the respiration of sugars (see Section 2.3) results in the formation of water, and water plays a central part in the reaction with carbon dioxide during the production of sugar in photosynthesis.

The growth of plants requires lots of water, much of which forms the cell sap in the plant cell vacuole. A swollen vacuole pushes the cytoplasm against the cell wall, making the cell turgid, and giving support to plant tissues.

STUDY TIP

Make links to Section 3.6 for transport in humans and Section 3.4 for transport in plants.

STUDY TIP

Check the role of water in digestion of food (Section 2.2) and germination of seeds (Section 5.2) then look in Section 2.3 for reactions of respiration and Section 2.1 for photosynthesis.

STUDY TIP

Make a link with Section 1.5 to see how water helps make plant cells turgid.

Enzymes

Chemical reactions inside living organisms are known as **metabolic reactions**. Nearly every metabolic reaction is catalysed by an enzyme, so enzymes are often described as **biological catalysts**. Enzymes speed up the rate of a metabolic reaction – it might otherwise be so slow that it does not occur (at the temperature inside living cells). As with other catalysts, enzymes are not permanently changed by the reaction. Enzymes are proteins and some of their properties are linked to their protein structure.

We can represent a series of enzyme reactions as follows:

$$A + B \xrightarrow{\text{enzyme 1}} C + D \xrightarrow{\text{enzyme 2}} E + F \dots$$

A and **B** are the substrates (the molecule or molecules being acted on or that are reacting), **C** and **D** are the products. **E** and **F** may be the products of a second, linked reaction (part of a pathway of reactions). A different enzyme catalyses each of the steps, as shown by the arrows.

Each enzyme has a particular fold on its surface (known as the **active site**). This is where the molecules involved in the reaction (the substrates) meet. The shape of the active site precisely fits one substrate and no other.

When the **substrates** meet at the active site, they react to form the **products** of the reaction. The active site of an enzyme catalyses the reaction by lining up the substrate molecules so they are in the right position to exchange atoms with each other. This means that a reaction can occur without the need, for example, for additional heating.

Once the products of the reaction have formed, they are released from the active site, and they are replaced by another set of substrate molecules. The enzyme is unchanged by the process, so the process can be repeated over and over again, sometimes hundreds of times a second.

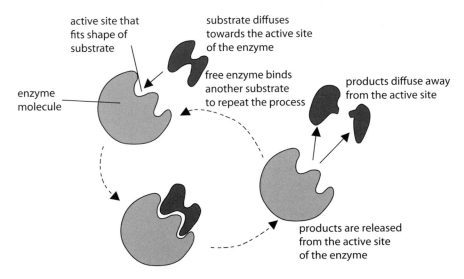

active site that fits shape of substrate

substrate diffuses towards the active site of the enzyme

enzyme molecule

free enzyme binds another substrate to repeat the process

products diffuse away from the active site

products are released from the active site of the enzyme

enzyme–substrate complex where catalysis of reaction leads to the formation of the products

Figure 1.19 Diagram of an enzyme, showing the relationship between substrate, products and active site.

The product of one enzyme-catalysed reaction inside a cell usually becomes the substrate for another enzyme reaction, and so on. Substances (metabolites) are modified in different ways by a series of enzymes, as they pass along a **metabolic pathway**. Examples of metabolic pathways include photosynthesis and cellular respiration – even though we summarise these processes as simple equations, the actual pathways for both processes involve a series of reactions.

If the overall shape of an enzyme changes for some reason, this might mean that the precise fit between active site and substrate is lost. When something causes an enzyme to become inactive in this way, we say it has been **denatured**. **High temperatures** or **extreme pH** conditions can cause this denaturation.

Enzymes have particular conditions in which they work best, known as their **optimum** conditions. Outside their optimum range, enzymes work less well or not at all.

Temperature and enzyme activity

At low temperatures, such as near the freezing point of water, the movement of enzyme and substrate molecules is slow. The molecules have less kinetic energy and it takes longer for substrate molecules to travel to the active site of the enzyme. Consequently, enzyme reaction rates are slower at lower temperatures, but increase as the temperature rises.

Most enzymes in an organism have an **optimum** working temperature in that organism. For humans it is 37 °C and for plants this is often lower, say 15 °C. Some extremes of enzyme function have been recorded, including working at very low temperatures and also very high temperatures.

At temperatures above the optimum for an enzyme, the shape of the enzyme may become distorted. The distortion may affect the shape of the active site, so that it no longer fits the substrate properly. This means that the enzyme no longer catalyses the reaction, because it has been **denatured**. The denaturation process is not reversible. A denatured enzyme molecule remains permanently distorted, even if the temperature is lowered again.

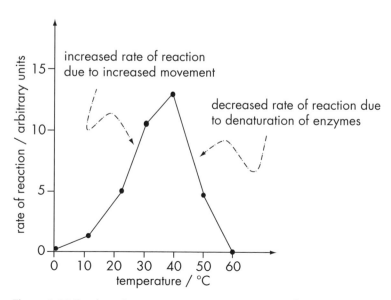

Figure 1.20 Graph to show enzyme activity over a range of temperatures.

■ Practical activity – investigating the effect of temperature on amylase activity

Amylase is involved in the digestion of starch. When starch is present, iodine solution is blue–black in colour, but when no starch is present, the iodine solution is yellow. The method measures the time taken for all of the starch to be digested. When no blue–black colour is seen, the starch has disappeared (it has been digested).

1 Take a dimple tile (cavity tile) and add a drop of iodine solution to each dimple (cavity). This is used to test for the presence of starch.

2 Place some amylase solution in one test tube and 5 cm³ of starch solution in a second test tube.

3 Place both test tubes in a beaker of water of known temperature for five minutes. This allows the contents of the tubes to be at the same temperature as the water in the beaker. The temperature of the water in the beaker is monitored with a thermometer.

4 Transfer a few drops of amylase solution to the test tube with starch. This is now the reaction mixture.

5 At 1-minute intervals, transfer a drop from the reaction mixture to a drop of iodine solution in one of the dimples on the tile. Use a clean dropper or glass rod each time and work round the iodine drops on the tile in sequence. Observe the colour of the drop of iodine solution each time the reaction mixture is added.

6 Note the time taken for the starch to disappear from the reaction mixture. You know this when the drop of iodine solution remains yellow and does not turn blue–black.

7 Repeat steps 2 to 6 but use a different temperature in the water bath (beaker with water). You can add hot water to get higher temperatures and you can add ice to get lower temperatures.

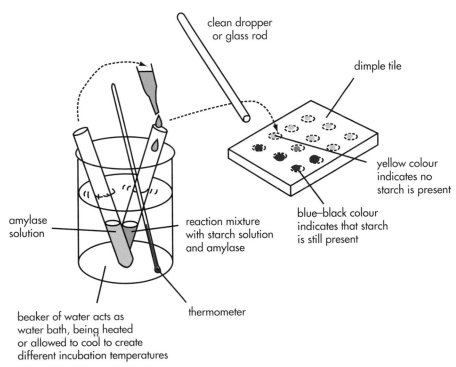

Figure 1.21 Investigating the effect of temperature on amylase activity.

pH and enzyme activity

Inside the living cell, the pH of the cytoplasm is neutral (pH = 7). Most enzymes inside the cell work well at this pH – that is, they have an optimum of pH 7. For enzymes secreted (released) outside the cell, such as those released into the intestines of a human being, the pH optimum may vary. In the stomach, where the protease pepsin is found, the pH = 2 (very acid), yet in the mouth, where salivary amylase is secreted, the saliva is pH 7.5 (slightly alkaline). In order to function, different enzymes have different optimum pHs suited to where they work and what they do. Another enzyme (arginase) works in the liver cells, acting on alkaline chemicals.

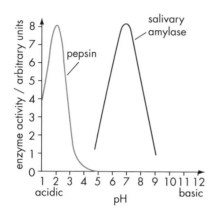

Figure 1.22 Activity of pepsin and salivary amylase enzymes over different pH ranges.

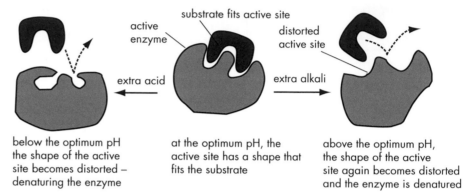

below the optimum pH the shape of the active site becomes distorted – denaturing the enzyme

at the optimum pH, the active site has a shape that fits the substrate

above the optimum pH, the shape of the active site again becomes distorted and the enzyme is denatured

Figure 1.23 How pH changes can distort the active site and affect the activity of an enzyme. When the enzyme is denatured, the substrate does not fit the active site.

PRACTICAL

This allows you to see the way changes in pH affect the activity of an enzyme. Where else in this book do you learn about coagulation of milk as a result of pH changes? (Check in Section 7.2.)

Remember to wear eye protection.

■ Practical activity – investigating the effect of pH on the activity of the enzyme rennet

Rennet is an enzyme that coagulates proteins in milk (makes them go solid). The time taken for the milk to coagulate in the tube gives a measure of the rate of the reaction.

1 Take 5 test tubes, A to E, as shown in Figure 1.24. Each tube contains 10 cm^3 of milk.
2 Add a few drops of dilute HCl to tubes A and B to acidify the milk. Check the pH by dipping a strip of universal indicator paper into the milk. Tube A should be pH 3 and tube B should be pH 6.
3 Add a few drops of dilute KOH to tubes D and E to make the milk alkaline. Check the pH by dipping a strip of universal indicator paper into the milk. Tube D should be pH 8 and tube E should be pH 11.
4 Add nothing to the milk in tube C so that the pH is about neutral (pH 7).
5 When the pH has been adjusted, place all the test tubes in a water bath for 5 minutes, to equilibrate (reach the temperature of the water bath).
6 Add 1 cm^3 of rennet to each test tube and note the time.
7 At regular intervals, tilt each test tube to check whether the milk has coagulated (gone solid). Note the time when coagulation occurs. At some pH values, coagulation does not occur, so a time limit should be set to the observations.

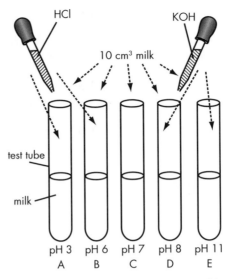

Figure 1.24 Investigating the effect of pH on the activity of the enzyme rennet.

STUDY QUESTIONS

1 a) i)   Arrange the following structures in order of size, starting with the smallest:

               cell; organelle; molecule; atom; macromolecule

       ii) Which **three** elements are *always* present in carbohydrates, proteins and lipids?
       iii) Which **two** additional elements are found in proteins?
    b) i)   Name the building blocks that are used in making proteins.
       ii) List **three** examples of proteins and say where these examples are found in cells.
    c) i)   List **three** examples of large carbohydrate molecules made from glucose.
       ii) State where you would find each of these large carbohydrate molecules and the function that they have in cells.

2 a) How do you prepare a food sample to do a practical test to find out what it contains?
    b) Draw up a table to summarise how you test a food sample for each of the following molecules: starch, glucose, protein, fat.
        In your table, include the name of the reagent (or how you do the test), the colour or appearance at the start, the colour or appearance if the molecule you are testing for is present, whether you need to heat your mixture during the test and how long you would leave it before recording your results.

3 a) Explain what is meant by the term **biological catalyst**.
    b) Make a list of ways that a biological catalyst differs from a catalyst used in chemical reactions, and ways that biological catalysts and chemical catalysts are similar.

1.5 Movement of substances into and out of cells

Watching tiny particles move

In 1827, the botanist Robert Brown was looking at tiny pollen grains suspended in a drop of water, using a simple microscope. He saw that the tiny particles seemed to be jiggling about, but he was unable to explain why.

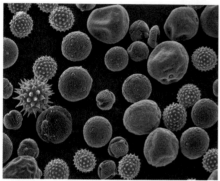

We now know that he was the first person to describe the effects of the random movement of molecules in a fluid. The water molecules were jostling about, bumping into the pollen grains and nudging them this way and that.

Imagine that you make a similar slide, but put all your pollen grains at one end of the slide. As time passes, the pollen grains would spread out so that they become evenly distributed throughout the drop of water. On your slide, what you have observed is diffusion. The pollen grains have moved down a concentration gradient, from where there were a lot of pollen grains to the rest of the water drop where there were none. Even when the particles are spread out evenly, they don't stop moving, but for every particle moving one way, another moves in the opposite direction. You can see this same effect in the bottles. A few crystals of a coloured substance (potassium permanganate) have been dropped into the first one and gradually the colour diffuses through the water so that it becomes coloured throughout.

So how is this process important in living organisms? And how do substances enter and leave the cells of living organisms?

■ Molecules and movement

In gases and liquids all molecules move freely and randomly, in no particular direction. This movement is the result of the kinetic energy the particles possess. If the molecules are warm they have more kinetic energy and move faster than if they are cold.

Suppose you add a few drops of a strongly coloured liquid to a beaker of water and watch what happens. You can see the colour gradually spreads through the water until it is even throughout. Each molecule of the coloured substances is moving (at random) among the water molecules, but the overall movement of the coloured molecules results in the colour becoming evenly distributed. The net overall movement of the coloured molecules is from a region of higher concentration (where the drops were added) to a region of lower concentration (where originally there was no colour). We call this difference in concentration a **concentration gradient**.

You can observe a similar effect if you place a cube of transparent jelly (such as agar jelly) on a dish with a coloured liquid. Gradually you can see the colour spreading through the jelly, until the whole cube of jelly becomes coloured. You can use an even simpler everyday example and watch the effect of adding a tea bag to a glass of hot water. You soon see the net effect of the dark coloured molecules in the tea spreading out into the water.

Living cells can be considered as being between the jelly and liquid described above. Molecules move (or are transported) from cell to cell for different reasons. The **cell membrane** acts as the boundary for the cell contents. The cell membrane also acts as a barrier for substances moving into and out of cells. Some small molecules are able to move through the cell membrane, as though it has pores or small holes in it. Other molecules are relatively large and cannot pass through the membrane in this simple way, so different mechanisms are involved in helping these molecules pass through the barrier, into and out of cells. The cell membrane also has the job of preventing some molecules entering or leaving the cell. In this way, the cell has some control over what is inside it.

Movement of substances into and out of cells can be by **diffusion**, **osmosis** and **active transport**. Each of these mechanisms is considered in more detail below.

Diffusion

Diffusion is the net movement of molecules (or ions) from a region with a higher concentration to a region with a lower concentration (of that molecule).

The descriptions above (coloured liquid, jelly, teabag) present a simple view of diffusion using non-living materials. Examples of diffusion in living organisms include:

- in plants – oxygen gas and carbon dioxide gas diffuse into and out of leaves through the stomata. The direction of diffusion depends on the relative concentrations of each gas inside and outside the leaf.
- in plants – the movement of water vapour molecules out of leaves through the stomata into the air outside during **transpiration.**
- in humans – the movement of oxygen from air in the alveoli (in the lungs) into the blood in the lungs, and movement of carbon dioxide from the blood in the opposite direction into the alveoli in the lungs.
- in humans – movement of dissolved substances (such as sugars and amino acids) from the blood into cells.

Osmosis

Osmosis is a special case of diffusion. The term **osmosis** refers to the movement of **water** molecules from a region with a higher concentration of water molecules (a more watery or more dilute solution) to a region with a lower concentration of water molecules (a less watery or more concentrated solute solution) through a **partially permeable membrane**. In living cells, the **cell membrane** acts as a partially permeable membrane, which can also be described as a **selectively permeable membrane.**

Another way of describing movement of molecules by osmosis is to use the term 'water potential'. So osmosis can be described as the movement of water from an area of high water potential (high concentration of water) to an area of lower water potential (lower concentration of water).

STUDY TIP

Look in Section 1.3 for more details about cell structures including the cell membrane.

STUDY TIP

You can find out about gas exchange in flowering plants in Sections 2.1 and 3.1.

Check in Section 3.4 to help you understand about loss of water and transpiration in plants.

You can find out about gas exchange in humans in Section 3.2 and transport in humans in Section 3.5.

Make links with respiration in Section 2.3, digestion in humans in Section 2.2 and transport in humans in Section 3.5.

The term **partially permeable** describes a membrane that effectively has gaps in it – these gaps allow small molecules (such as water) to pass through, whereas larger molecules are unable to pass through. Other mechanisms are involved in their transport across the membrane. These larger molecules can be carried through by special transport proteins that act like a 'door' in the membrane.

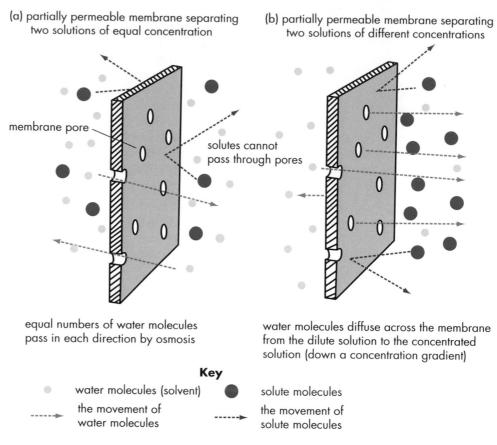

(a) partially permeable membrane separating two solutions of equal concentration

(b) partially permeable membrane separating two solutions of different concentrations

membrane pore

solutes cannot pass through pores

equal numbers of water molecules pass in each direction by osmosis

water molecules diffuse across the membrane from the dilute solution to the concentrated solution (down a concentration gradient)

Key

○ water molecules (solvent)   ● solute molecules

- - - → the movement of water molecules   - - - → the movement of solute molecules

Figure 1.25 Illustration of osmosis.

The diagram above:

(a) shows movement of water molecules through a partially permeable membrane when the concentration (of solute) is the same on each side. Water can move through both ways but there is no net movement. Solute molecules cannot move through because they are too large, so stay on the same side of the membrane.

(b) shows net movement of water through a partially permeable membrane when the concentration (of solute) is higher on one side. The net movement of water is from the region where there are more water molecules to the region where there are fewer water molecules, as in osmosis. When the sides are equal there is no NET movement, as in (a).

Examples of osmosis in living organisms include:

- in plants – entry of water into root hair cells from the soil water around the root hair cell and nearby soil particles. The soil water is the solution with a high concentration of water molecules (a dilute solution). The cell sap inside the root hair cell is a less dilute solution (because of the sugars and salts dissolved in the cell sap). The cell membrane acts as the partially permeable membrane.

STUDY TIP

Check in Section 3.4 for details about water uptake by the roots of plants.

STUDY TIP
Look in Sections 4.2 and 4.3 to help you understand homeostasis and osmoregulation in humans – why it is important and how the correct water balance is maintained.

- in animal cells (such as a red blood cell) – if too much water enters a cell by osmosis, the cell swells up and it may burst because there is no cell wall (as in plants) to restrain the cell membrane. If surrounded by a more concentrated solution, water moves out by osmosis and the cell shrinks (Figure 1.28).

In the human body, blood and other fluids are in close contact with all cells. If you drink a lot of fluid your blood becomes more dilute, so there are ways of adjusting the level of water in the blood, otherwise too much water would pass into cells by osmosis. Conversely, if you sweat a lot or lack water for other reasons, you become dehydrated. Your blood is less watery (more concentrated) and this would draw water out of the nearby cells by osmosis. It is very important to keep a correct osmotic balance inside your body. This is described as **osmoregulation**.

PRACTICAL

Dialysis (Visking) tubing provides a useful model of a partially (selectively) permeable membrane, as it shows some properties similar to that of the cell membrane in living cells.

■ Practical activity – demonstration of osmosis using dialysis (Visking) tubing

Dialysis (Visking) tubing is partially permeable. Water molecules can pass through because they are small but sucrose molecules cannot.

1 Take a length of dialysis tubing. Knot it at one end to seal it.
2 Fill the dialysis tubing with strong sucrose solution.
3 Insert a glass tube through the rubber bung, then fit the end of the bung into the open end of the dialysis tubing. Tie the tubing to the bung with fuse wire.
4 Place the dialysis tubing in a beaker with water, as shown in diagram A (Figure 1.26).
5 Observe the level of the solution in the glass tube.

After several minutes, the meniscus rises up the glass tube, as shown in diagram B. Water has entered the dialysis tubing by osmosis.

If the dialysis tubing is filled with water and placed in a beaker with sucrose solution (as shown in diagram C), the level of the meniscus falls. This is because water has moved out of the dialysis tubing into the sugar solution, by osmosis.

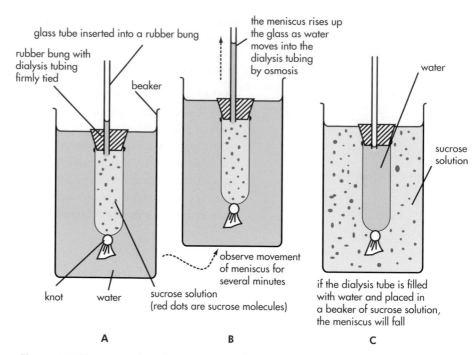

Figure 1.26 Demonstration of osmosis, using dialysis (Visking) tubing.

Active transport

Active transport describes the way that some molecules (or ions) can be moved across a membrane against the concentration gradient, from a lower concentration to a higher concentration of that molecule. This movement requires energy. The energy is provided by ATP from respiration. Sometimes a 'carrier' is required to help the molecule through the membrane. This is one way that quite large molecules can be moved through the cell membrane into a cell. You can compare this to a pump raising water to a higher level – some energy is used in this activity too.

Examples of active transport in living organisms include:

- in plants – useful mineral ions enter into root hair cells from the soil water by active transport. The soil water contains a lower concentration of the mineral ion than inside the cell, so the ions are moved into the cell against a concentration gradient.
- in humans – in the kidney, glucose is actively transported back into the blood from the glomerular filtrate in the kidney tubule, so that no glucose passes out of the body in the urine.

■ Comparing plant and animal cells

An important difference between plant and animal cells is that plant cells have a **cellulose cell wall** and the cell membrane lies inside this, whereas in animal cells, the cell membrane is the only boundary around the cells.

The cellulose cell wall is made of woven cellulose fibres. The gaps between the cellulose fibres are big enough to allow water and many much larger molecules to pass through. The real barrier for the movement of substances into and out of plant cells is the partially permeable cell membrane.

STUDY TIP

Look in Section 3.4 for details about how mineral ions are taken up in flowering plants.

STUDY TIP

Details about excretion in humans are given in Section 4.2.

STUDY TIP

Look in Section 1.3 for more details about cell structure.

You can isolate a few (living) plant cells by peeling off a strip from the epidermis of an onion (or from a coloured stalk such as rhubarb). Place the peel in a drop of water and examine the cells under a microscope. The events that occur in the cells are summarised in Figure 1.27.

plant cell in pure water
(a very dilute solution)

nucleus

plant cell in solution that equals
the concentration inside the cell

cytoplasm

plant cell in solution that is more
concentrated than that inside the cell

cell membrane

cell wall

cell membrane pushes against the
cell wall, as water enters the cell
by osmosis and the cell contents swell

cell membrane touches the cell wall, but does
not push against it as there is no net movement
of water into the cell by osmosis

the cell becomes flaccid, and if the concentration of
the salt solution is high, the cell membrane peels away
from the cell wall as the cell contents shrink, due to the
movement of water by osmosis out of the cell

Figure 1.27 Plant cells placed in pure water take in water (by osmosis) and become turgid, but if the cells are surrounded by a more concentrated solution (say of sugars or salts) water passes out and the cell becomes flaccid.

If you do the same thing to an animal cell, such as a red blood cell, the water continues to enter and the cell continues to increase in size and it may burst. This is because there is no structure equivalent to the cellulose cell wall to provide a counter-pressure and hold it in shape.

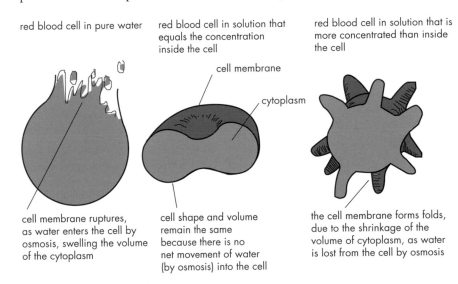

red blood cell in pure water

red blood cell in solution that
equals the concentration
inside the cell

cell membrane

cytoplasm

red blood cell in solution that is
more concentrated than inside
the cell

cell membrane ruptures,
as water enters the cell by
osmosis, swelling the volume
of the cytoplasm

cell shape and volume
remain the same
because there is no
net movement of water
(by osmosis) into the cell

the cell membrane forms folds,
due to the shrinkage of the
volume of cytoplasm, as water
is lost from the cell by osmosis

Figure 1.28 Osmosis and red blood cells – in pure water, red blood cells take in water until they burst, whereas if surrounded by a more concentrated solution (of sugars or salts) water passes out and the cell shrinks and becomes folded.

In plants, turgor makes an important contribution to the support of plant tissues. In addition, plants have hard materials like that in the walls of xylem and these also provide support. Xylem is found in the veins of the leaves and in woody stems.

You can see the effect of wilting if a plant organ, such as a lettuce leaf, has too little water. The cells of the leaf lose turgor and the lettuce leaf goes floppy or flaccid.

STUDY TIP
Make a link to Section 3.4 (Transport in flowering plants).

STUDY TIP

Make a link to Section 3.3 (Transport in living organisms).

MATHS TIP

Substitute numbers into equations using appropriate units
Useful scientific theories often generate equations which can be used to understand biological processes. When new measurements are put into an equation, the results make biological sense, supporting the original theory.

Now go to page 43 to apply this to an equation relating to the rate of diffusion.

MATHS TIP

Surface areas and volumes of cubes
Surface area of a cube
(units^2) = length x width x 6

Volume of a cube
(units^3) = length x width x height

Now go to page 41, question 1, to apply this to some calculations looking at surface area to volume ratios.

STUDY TIP

Look for examples that provide a large surface area in relation to volume and for ways in which living organisms maintain a concentration gradient, allowing more molecules to diffuse.

■Factors affecting movement of substances into and out of cells

As stated at the start of Section 1.5, molecules move (randomly) all the time as a result of their kinetic energy. The rate of net movement (by diffusion or osmosis) is influenced by various factors and it is important to link this to how molecules move into cells and are distributed to all cells inside a multicellular organism:

- **Surface area in relation to volume** – suppose you have 100 cells (think of them as a lump of jelly). You can either spread them out flat on a surface, so that the cells are in a layer just one cell thick, or you can keep them in a spherical lump. You can see that any molecule that is going to diffuse into these cells gets to the inside of the cells faster in the single layer than in the sphere. Many organisms show adaptations that increase the surface area available for exchange of materials in relation to the volume to be supplied.
- **Distance** – the greater the length of the journey travelled by diffusing molecules (from a region of higher to a region of lower concentration), the slower the rate of diffusion (the longer the journey takes). This is the main reason why single-celled organisms can receive the oxygen they require by simple diffusion from the water surrounding them – the distance from the outside world to the centre of the cell is only a few micrometres. As soon as multicellular organisms grew in size to more than a mm or so thick, then the rate of diffusion was too slow to allow them to acquire the oxygen they need fast enough over this distance. This is why specialised exchange organs (like gills and lungs and villi) evolved, along with transport systems to take substances where they are needed in the body (see Section 3.5). Diffusion still occurs, but only over tiny distances, such as between the inside and outside of a blood capillary.
- **Temperature** – the random movement of molecules is a result of their kinetic energy. At higher temperatures molecules have more energy than at lower temperatures, so their movement is faster at higher temperatures. This means that the rate of net movement (by diffusion or osmosis) is greater at higher temperatures.
- **The difference in concentration of molecules** (the concentration gradient) – this is another important factor in determining the rate of movement of substances. If there is a big difference the net movement is faster than if there is very little difference (even though the molecules continue to move all the time). Again, you can find examples in living organisms to help you understand this. In the lungs, the blood is flowing so constantly takes away the oxygen it has collected from the lungs, so more can diffuse through from the alveoli. In a leaf, on a windy day, the water vapour outside the stomata is blown away, so more water vapour diffuses from the leaf air spaces to the air outside the leaf.

■ Practical activity – investigating osmosis in potato tissue

1 Take a large potato. Use a cork borer to cut and remove several cylinders of tissue from the potato. The cylinders all have the same diameter. This is shown in diagram A (Figure 1.29).
2 Cut the cylinders to an identical length then weigh them and record their mass.
3 Prepare a series of salt solutions (NaCl), each of different concentration, ranging from 0% to 10% NaCl (for 0% NaCl, use distilled water.) Fill a set of boiling tubes with each of these solutions.
4 Place one cylinder of potato tissue in each of the different NaCl solutions and leave for about 30 minutes or longer (diagram B). (You can leave them for up to 24 hours and then it may be advisable to leave the boiling tubes in a refrigerator if room temperature is too warm.)
5 After 30 minutes (or the time you choose in 4), remove the potato cylinders. Blot them (with a paper towel) and weigh them again.
6 Calculate the percentage change in mass for each of the potato cylinders.
7 Plot a graph of salt concentration against percentage change in mass of the potato cylinders.

In water and the more dilute solutions of NaCl, the potato cylinders take in water by osmosis and their mass increases. When the potato cylinders are in the more concentrated solutions of NaCl, water is drawn out by osmosis and their mass decreases. Somewhere in the middle, the concentration of the NaCl solution is about the same as the concentration (of cell sap) inside the potato cells. At this concentration, there is no net movement of water in or out and the mass of the potato cylinder stays the same. You can find this point from your graph.

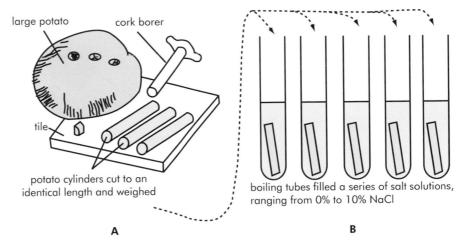

large potato   cork borer

tile

potato cylinders cut to an identical length and weighed

boiling tubes filled a series of salt solutions, ranging from 0% to 10% NaCl

A   B

Figure 1.29 Investigating osmosis in potato tissue.

▪ Practical activity – investigating the diffusion of red pigment out of cells of beetroot

This practical activity demonstrates the importance of the cell membrane both as a barrier in a living cell and to keep substances inside the cell. In beetroot, the red pigment molecule is too large to diffuse out of the cell when the cell membrane is intact. However, when the cell has been killed by heating, the cell membrane is also destroyed and the colour passes out into the water.

1 Use a knife to cut a block of tissue from a fresh (uncooked) beetroot.
2 Cut the block into two equal-sized pieces of beetroot and wash thoroughly in water.
3 Place one piece of beetroot in water in beaker A and one piece in water in beaker B.
4 Heat the water in beaker A to 90 °C for 1 minute. Keep beaker B at room temperature.
5 Allow the water in beaker A to return to room temperature and observe the colour in both beakers.

In beaker A, you can observe the diffusion of the red colour of the pigment in beetroot cells into the water. Heating to above 45 °C is known to damage plant cell membranes.

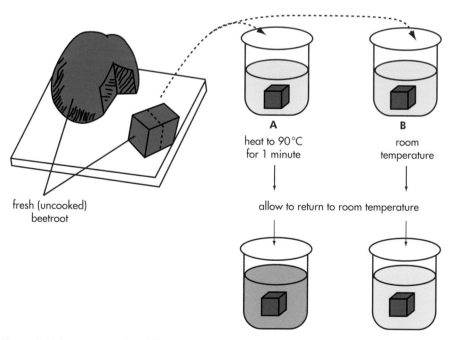

Figure 1.30 Investigating the diffusion of red pigment out of living cells and non-living cells.

STUDY QUESTIONS

These questions should help you understand the different factors that affect movement of substances into and out of cells.

1 Surface area to volume ratio

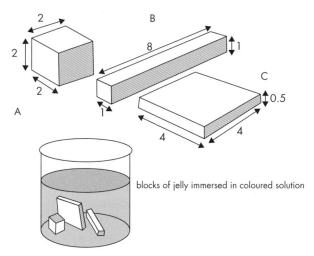

Dimensions are in cm but diagram not to correct scale.

blocks of jelly immersed in coloured solution

Look at the blocks of jelly and work out which block would become coloured inside most quickly (i.e. which has the shortest distance to the middle?). Use these steps to help you work out your answer.

a) Calculate the volume of each block. Draw a table for your answer to this part and for parts **b)** and **c)**.

b) Calculate the area of each side then add them together to get the total surface area of each block.

c) If these blocks represented part of a living organism, which has the largest surface area available for exchange of materials? Which has the largest surface area in relation to its volume?

d) Give some examples of structures in living organisms that show a large surface area that helps exchange of materials into and out of the organism.

2 Temperature

Explain why the rate of diffusion (say of a red dye dropped into a beaker of water) is faster in a warm temperature than in a cold temperature. In your explanation, include the words **random movement** and **kinetic energy** of molecules.

3 Concentration gradient

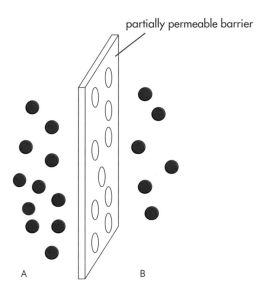

partially permeable barrier

a) i) At the start, which way do molecules move through the barrier?

ii) Explain your answer to part **i)**.

iii) What happens when there is an equal number of molecules on each side – do they continue to move through the holes?

iv) Suppose molecules on side B were continually being swept away, what would happen to the net movement of molecules through the barrier?

b) Give some examples from living systems that show how maintaining a concentration gradient helps to increase the rate of diffusion of molecules through a surface. Include plants and animals in your examples.

Summary

I am confident that:

✓ I know that all living organisms require nutrition, that they respire and excrete their waste, that they respond to their surroundings and move, that they control their internal conditions, and are able to reproduce, grow and develop.

✓ I understand how the characteristics listed above are carried out by all living organisms, but sometimes in different ways by plants, animals and other living organisms.

✓ I understand that plants, animals, fungi and protoctists are eukaryotic organisms and that bacteria are prokaryotic organisms.

✓ I understand the meaning of the terms eukaryotic and prokaryotic.

✓ I know that plants and animals are multicellular organisms but that they differ in their method of obtaining food and I can list differences in the structure of their cells.

✓ I can describe features of fungi, bacteria and protoctists, and know why they are not included with plants and animals.

✓ I can name examples of all the main groups and can describe their structure.

✓ I understand why, in some classifications, viruses are not included as living organisms.

✓ I can describe some examples of pathogens from the different groups of fungi, bacteria, protoctists and viruses.

✓ I can draw and label a diagram of a plant cell and of an animal cell and use this to show the features they contain.

✓ I can describe the functions of the structures shown in the cell diagrams.

✓ I know ways that a plant cell differs from an animal cell, in both structures and functions.

✓ I know that carbohydrates, proteins and lipids (fats and oils) contain the elements carbon, hydrogen and oxygen and that proteins also contain nitrogen and sometimes sulfur.

✓ I can describe how smaller basic units are built up into larger molecules and that starch and glycogen are made from simple sugar units, that proteins are made from amino acids and that lipids are made from fatty acids and glycerol.

✓ I understand that enzymes are proteins and that they act as catalysts in metabolic reactions.

✓ I understand that enzymes are affected by temperature and pH, and can describe how to do experiments to illustrate the effect of temperature on the activity of an enzyme.

✓ I can describe how to test a food sample for the presence of glucose, starch, proteins and fats (oils).

✓ I know definitions for the terms diffusion, osmosis and active transport and understand how each of these contributes to the movement of substances into and out of cells.

✓ I can describe simple experiments that illustrate diffusion and osmosis in living and non-living systems.

✓ I understand that different factors can affect the rate of movement of substances into and out of cells.

✓ I can describe how surface area to volume ratio, distance, temperature and concentration gradient can affect the rate of movement of substances into and out of cells, and describe examples that illustrate these effects.

MATHS SKILLS

Substitute numbers into equations using appropriate units

Useful scientific theories often generate equations which can be used to understand biological processes. When new measurements are put into an equation, the results make biological sense, supporting the original theory.

The rate of diffusion of a substance (J) across a membrane =

$$\frac{\text{difference in concentration between the two sides of the membrane (C)} \times \text{area of membrane (A)}}{\text{thickness of the membrane (D)}}$$

Does the rate of diffusion (J) increase or decrease in each of the following situations?

a)   Increase in C
b)   Decrease in A
c)   Decrease in D
d)   Increase in A

MATHS SKILLS

Surface areas and volumes of cubes

Surface area of a cube / units^2 = length × width × 6
Volume of a cube / units^3 = length × width × height

Now try Question 1 on page 41

MATHS SKILLS

Determine the slope and intercept of a linear graph

The slope (gradient) of a straight line on a graph shows how the magnitude of the *y*-variable changes in relation to the *x*-variable. An intercept (where a straight line passes through an axis) on a graph shows what happens when the magnitude of one variable = 0.

The graph shows the result of an experiment to determine the effects of different sugar solutions on the mass of cylinders of beetroot.

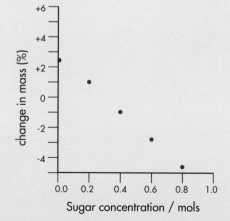

a)   Estimate the gradient of a best fit line for the graph, giving your answer as % change / mol dm^3.
b)   Use the graph to find out the sugar concentration at which no change occurs.
c)   Explain why no change occurs at the concentration given in (b).

Example of student response with expert's comments

■ Using and interpreting data

1 Catalase is an enzyme found in many plant and animal cells. It catalyses the breakdown of hydrogen peroxide, a toxic by-product of metabolism, into oxygen and water.

Students investigated the effect of temperature on the activity of catalase. They measured the oxygen given off at different temperatures by catalase from potato tissue and by catalase from liver tissue, as shown in the table.

a) Plot a line graph on a graph grid 11 cm × 11 cm to show the results of their investigation. Use a ruler to join the points with straight lines. (6)

b) The temperature at which an enzyme works best is called the optimum temperature. How does the optimum temperature for liver catalase differ from that for potato catalase? (2)

c) Both enzymes show a decrease in activity at temperatures higher than the optimum. Explain the reason for this. (3)

(Total = 11 marks)

Temp / °C	Rate of oxygen production / cm³ per minute	
	potato	liver
5	0	4
15	16	8
25	17	21
35	11	46
45	2	44
55	0	37

Student response Total 10/11	**Expert comments and tips for success**
a) 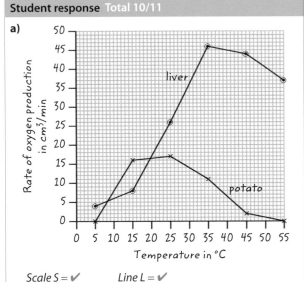 Scale S = ✔ Line L = ✔ Axes (correct way round) A1 = ✔ Axes (labelled + units) A2 = ✔ Points accurate P = O Key K = ✔	S: scale chosen uses sensible intervals that are easy to use when plotting points, and when reading values from graph. Here, 1 small square = 1 °C on horizontal axis and 1 cm³ per minute on vertical axis. L: points joined with neat, straight lines drawn with a ruler. Always plot points and draw lines with a sharp HB pencil, in case you make a mistake. Remember to take with you into the examination room a ruler, a sharp pencil, a sharpener and an eraser. P: one mistake in plotting (25 °C, 21 cm³ per minute, for liver) so no mark. K: graphs for liver and potato are labelled.
b) The optimum for the <u>potato catalase is 25°C. The optimum for the liver catalase is 35°C</u> ✔ — much <u>higher</u> ✔.	The plotted points should be joined with straight lines, in which case the optimum temperature can be taken from either the table or the graph.
c) At high temperatures the <u>active site on the enzyme molecule begins to lose its shape</u> ✔ so it <u>can't bind to the substrate</u>. ✔ The enzyme is <u>denatured</u>. ✔	Full marks.

Applying principles

1 A student carried out an investigation into the processes that take place inside bean seeds at the start of germination. The beans were divided into three groups:

Group X   Beans soaked in water for 48 hours
Group Y   Beans soaked in water for 48 hours then boiled for 15 minutes and cooled
Group Z   Beans left in their original dry state

Three Petri dishes containing agar (a jelly) and starch were prepared. They were marked X, Y and Z. Beans from each group were cut in half and placed, cut side down, onto the agar jelly. The plates were kept in an incubator at 25 °C.

After 24 hours, the student removed the beans but used pins to show where the beans had been on the surface of the agar jelly. The student poured iodine solution over the agar jelly, gently rinsing it off after 2 minutes. The results are shown in the diagram.

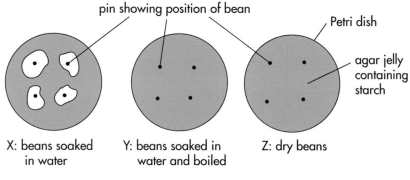

pin showing position of bean
Petri dish
agar jelly containing starch

X: beans soaked in water   Y: beans soaked in water and boiled   Z: dry beans

Key ▨ blue–black colour   ☐ red–brown colour

a) **i)**   Describe the distribution of starch in dishes X, Y and Z.   *(2)*
   ii) Suggest an explanation for the distribution of starch in dishes X and Y.   *(3)*
   iii) Suggest an explanation for the distribution of starch in dish Z.   *(2)*
b) The student noticed that the beans soaked in water for 48 hours had begun to germinate, but the dry beans had not. Give **one** factor, other than water, which is necessary for beans to germinate. *(1)*

(Total = 8 marks)

Student response Total 7/8	Expert comments and tips for success
a) i) <u>In X most of the dish contains starch, apart from the areas that were underneath the beans</u>. ✔ These have no starch. <u>In Y and Z there is starch in the whole of the dish</u>. ✔	The answer is clear and concise.
ii) In X the beans produced <u>something like amylase</u> ✔ which <u>broke down the starch</u>. ✔ When they were cooked this <u>didn't work anymore</u>. ✔	In this type of question you are not expected to have studied the investigation yourself. The marks are awarded for the way you apply your knowledge of biology to a new situation. So recognising that the breakdown of starch might be due to something 'like' amylase is enough for a mark.
iii) Unlike X, the dry beans didn't produce any amylase ✔ to break down the starch. ✔	Full marks.
b) Light is necessary for germination. O	Light is necessary for plant growth but not for germination.

Exam-style questions

1 Living organisms show a variety of features and these features are used to classify them into different groups.

a) This group are eukaryotic. They cannot carry out photosynthesis and feed by absorbing the products of extracellular digestion.

Which one of the following matches this description? [1]

  A bacteria

  B fungi

  C protoctists

  D viruses

b) This group have cells with no nucleus or mitochondria. Their cytoplasm contains DNA in a circular chromosome.

Which one of the following matches this description? [1]

  A bacteria

  B fungi

  C protoctists

  D viruses

c) Which one of the following is a carbohydrate molecule formed from glucose units that may be stored in animal cells and some fungi? [1]

  A cellulose

  B glycerol

  C glycogen

  D starch

d) i) Name one group of single-celled organism involved in the conversion of milk to yoghurt.

  ii) Describe the fermentation reaction and explain why the texture of the milk changes as it becomes yoghurt.

e) i) Name a single-celled organism involved in making bread.

  ii) Describe the fermentation reaction and explain why baked bread has holes in it.

[Total = 11]

2 **a)** The features listed in the table are found in cells of flowering plants or animals cells or both. Copy and complete the table by placing ticks in the appropriate boxes if the feature is present in that type of cell.

Feature	Plant cell	Animal cell
contains cytoplasm		
has a cell wall made of cellulose		
contains DNA in the nucleus		
has a cell membrane		
stores glycogen in the cell		
never has chloroplasts		
usually has large vacuoles		

[7]

b) Cells of fungi and bacteria show features that differ from flowering plants and animals in a number of ways.

  i) How do the cell walls of bacteria and fungi differ from those of plant and animal cells?
  1. bacteria
  2. fungi [2]

  ii) How do bacteria differ in the way DNA is contained in the cell? [1]

[Total = 10]

3 The table lists some statements and descriptions. Each of these refers to molecules found in the cells of living organisms. The first box lists some large molecules and the second box lists the basic units (or building blocks) for these large molecules.

Box 1 – large molecule

cellulose; glycogen; lipid; proteins; starch

Box 2 – basic units (building blocks)

amino acids; fatty acids + glycerol; glucose

Copy and complete the table by choosing the correct large molecule and basic unit for each statement. Some have been done for you. For one table row you are asked to provide a suitable description.

Statement/description	Name of large molecule	Basic unit/building block
found in cell walls of plants		
some of these are enzymes in plants and animals		
	haemoglobin	
large molecule, stores energy in liver and muscles		glucose
carbohydrate stored in seeds and broken down when seed starts to germinate		
does not mix with water and forms part of cell membrane		

[Total = 6]

4 Some observations are listed below. They can be explained by referring to the way substances move into and out of cells.

a) Explain how mineral ions can accumulate inside plant root cells even though the concentration of mineral ions in the soil water is very low. [3]

b) A gardener had some tomato plants ready to plant in his polytunnel. He left them on the beach and a wave of seawater completely covered them. Later he saw that the plants had wilted and become floppy. Explain this observation. [3]

c) Some red blood cells were placed in a few drops of water and observed under the microscope. Soon the blood cells seemed to break up and were no longer whole cells. Explain this observation. [3]

[Total = 9]

EXTEND AND CHALLENGE

1 Using enzymes in industrial processes

In living organisms, enzymes act as catalysts; they allow metabolic reactions to take place within a narrow temperature range, such as human body temperature or other lower temperatures in plants and microorganisms.

Enzymes also have many industrial applications. These include use of enzymes in processes in the food industry. Often the enzymes used come from microorganisms, including fungi and bacteria. Traditionally, fermentations have been carried out with whole microorganisms – for example, yeast is used in the production of bread, beer and wine, and *Lactobacillus* is used in the production of yoghurt and cheese. (For more details, see Section 7.2.)

Here we ask you to find out information (using books or the internet) about enzymes that have been extracted from their source organism and are used in different processes in the food industry.

a) The list below gives some examples of processes that are carried out by enzymes.

1 preparation of milk suitable for people with lactose intolerance
2 increasing the yield of juice when crushing and pressing fruit
3 clarification of beer, wine and fruit juices
4 tenderisation of meat
5 fermentation of milk to produce 'vegetarian' cheese
6 production of high-fructose corn syrups (HFCS) from maize (corn)

For each process, find out more information under the following headings.

- the name of the enzyme used
- the reaction that is catalysed by the enzyme
- the source of the enzyme (usually from fungi or bacteria)
- any other information of interest linked to the process, including the benefit to people and the commercial advantage of using the enzyme.

b) Inside living cells, enzymes are often held in some of the cells' organelles. For example, enzymes linked to photosynthesis are found in the chloroplasts and some of the enzymes linked to respiration are found in the mitochondria. In many industrial processes, enzymes are often held on an insoluble material and are described as 'immobilised enzymes'.

Find out more about some examples of immobilised enzymes and suggest why this is an advantage in the industrial process.

2 Using bioinformatics as a tool in classification

Bioinformatics is a rapidly growing area of biology. The power of modern computers has been harnessed to create huge databases of information.

Originally, classification and naming of organisms was based mainly on comparisons of the physical anatomy or other visible features of the organisms. It was thought that, the more physical features shared by two groups of organisms, the more likely they are to be closely related.

Modern biologists now routinely compare the molecules in organisms, such as the individual amino acids present in a protein or the sequence of bases (known by the letters G, C, A and T) inside certain genes in their DNA. Biologists can then use this information as a basis for classification.

Living species that are closely related to one another, in an evolutionary sense, are likely to have a more recent ancestor in common than those more distantly related. If you have to go back a long time to find a common ancestor then there has been more time for random changes in the DNA of the two species to accumulate and the two species are likely to have become more different from each other.

This approach can be used to calculate how long ago the two species shared a common ancestor.
The information can then be used to construct a tree-like diagram illustrating the relationships of a group of different species – whether they are closely related or only distantly related.

As an example, a study was made of similarities and differences in the haemoglobin molecule (a protein) in eight mammals, listed in the box.

opossum; fox; human; bear; rhesus monkey; dog; capuchin monkey; gorilla

Further information about this study is given below. To answer the questions, you may have to refer to other parts of this book. You are also encouraged to use outside sources of information.

a) Use books or the internet to find images or descriptions of these eight mammals. Use information about their physical features and try to arrange the mammals in groups so that those with most similarities are in a group together, separated from others showing more differences. Give reasons to support the way you have made the groups.

b) Haemoglobin is a protein, made up of chains of different amino acids. The table summarises information about the structure of haemoglobin found in each of these eight mammals. Part of the haemoglobin molecule was analysed to work out the sequence of amino acids in each species. The figures in each cell of the table show the percentage similarity in amino acid sequences for each pair of species linked to that cell.

On the basis of the information given in this haemoglobin data, answer these questions.

1 Which **two** mammals are most closely related to a human?

2 Which mammal is most closely related to a dog?

3 Which mammal is most distantly related to all the other mammals?

4 Is there any other information in the table that is of interest or surprises you in any way?

c) i) Use the haemoglobin data in the table and try to arrange the mammals in groups so that those with most similarities are in a group together, separated from others showing more differences.

   ii) Compare the grouping you have drawn up on the basis of the haemoglobin data with the grouping you devised from images or physical features of the mammals. How closely do your groupings match?

d) i) When doing a study of this sort, suggest why is it important that scientists compare many different genes.

   ii) Suggest what is meant by 'random changes in the DNA'. What environmental factors might be responsible for causing such changes in the DNA?

e) Find out about two more examples that show how analysis of molecular sequences (in amino acids or in DNA) have been used to study classification or evolution of species in flowering plants and in animals.

Opossum	Fox	Human	Bear	Rhesus monkey	Dog	Capuchin monkey	
72%	92%	99%	90%	97%	90%	94%	Gorilla
	73%	71%	73%	**	73%	72%	Opossum
		91%	95%	90%	99%	92%	Fox
			90%	96%	90%	95%	Human
				90%	95%	91%	Bear
					89%	**	Rhesus monkey
						91%	Dog

**information not available

Source of information: www.radford.edu/jkell/molecular_phylogeny.pdf

2 Nutrition and respiration

TO THINK ABOUT . . .

Write down a food chain to include something you have eaten today (or would like to eat). Where did the energy in the food come from and how are you going to use that energy in your body? What do plants do with the energy contained in the food materials they build up?

Living organisms and their need for energy

The smallest mammals in the world are the bumblebee bat and the pygmy shrew. Each weighs less than 2 grams. The bat lives in the warmth of Thailand and the insects it eats provide only just enough energy to keep its body warm. The pygmy shrew (shown in the photograph) lives in southern Europe and to keep warm it has to eat up to twice its own body weight in insects every day. The lower limit for size in mammals is determined by the need to eat sufficient food to be respired to keep warm.

All the energy required by life on Earth is harvested from the Sun, by plants, during photosynthesis, when they lock the energy of the rays of light into the bonds between the carbon atoms of sugar molecules. Eventually these complex organic molecules are passed up the food chain, for example from plant to insect to these small mammals. And as the sugar molecules are respired in the cells of these tiny mammals, the released energy (originally from the Sun) warms their bodies, keeping them alive.

We need plants and plants need the Sun and we are all locked together in mutual dependency. So what are the different strategies used by flowering plants and by humans to obtain their food? And how do organisms break food down in respiration to release energy for life?

2.1 Nutrition in flowering plants

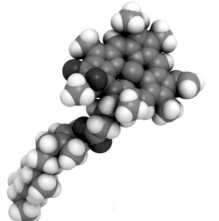

A view of plants on planet Earth

The true colour picture of planet Earth, made by NASA, shows three main colours: white for snow and ice, blue for the oceans and green for most of the land surface. The only truly biological colour visible on Earth from space, is that of the green pigment chlorophyll used by all photosynthetic organisms.

Chlorophyll absorbs the red and blue parts of the Sun's light rays, converting the light energy into chemical energy which is then used to build the bonds for making sugars from carbon dioxide and water. The model represents a chlorophyll molecule. All the green you can see in the NASA photo is light being reflected from *inside* plant leaf cells. That is where chloroplasts use the red and blue rays, but reflect the green rays. So plants make their own food using solar energy, and all other forms of life ultimately depend upon this food energy for life. On land, it is only the deserts and high mountains, where there is little liquid water, that are not green. If only we could find a way to get plants growing in these difficult regions!

■ The reactions of photosynthesis

All life on Earth depends on plants. Green plants convert the energy from sunlight into a chemical form. Plants use trapped solar energy to make sugars (carbohydrates) in a process called **photosynthesis**. Green plants are described as **autotrophs** because they can make complex molecules using simple molecules and energy from the Sun.

The process of photosynthesis occurs inside **chloroplasts**. Each cell in the photosynthetic tissue of a plant contains many green chloroplasts. Inside each chloroplast, carbon dioxide is combined with water to make glucose, using trapped energy from the Sun. The light energy from the Sun is trapped by the green pigment **chlorophyll**. This process involves the splitting of water, resulting in oxygen as a waste product.

This equation summarises the process of photosynthesis.

$$\text{carbon dioxide} + \text{water} \xrightarrow{\text{sunlight trapped by chlorophyll}} \text{glucose} + \text{oxygen}$$

$$6CO_2 + 6H_2O \xrightarrow{\hspace{4cm}} C_6H_{12}O_6 + 6O_2$$

When a ray of light is absorbed by a chloroplast, the chlorophyll traps its energy, using the power to split water molecules. Enzymes catalyse the photosynthetic reactions, using carbon dioxide and leading to the formation of glucose.

> **STUDY TIP**
> Look at Section 6.2 and write a concise explanation of why **autotrophs** are also known as **producers** in ecological terms.

> **STUDY TIP**
> Look at Figure 1.5 and Section 1.3 and make a list of the ways in which the structure of a palisade mesophyll cell is adapted for photosynthesis.

The glucose formed inside the chloroplasts in photosynthesis may be used by the plant in several ways. When it is first made, glucose is stored temporarily inside the leaf cells as granules of starch. (Note that when doing experiments to investigate photosynthesis, a leaf can be tested for starch to show that photosynthesis has occurred – see practical activities on page 23.)

Most of the sugar is then transported to the rest of the plant (in the form of sucrose), where it may be used in a variety of ways including:

- respiration inside plant cells to provide **energy** for chemical work
- as a building block of cell walls, made of **cellulose**, a polymer of glucose
- be converted into **lipids** (fats and oils) as a form of long-term energy storage
- be converted into **amino acids**, using minerals such as nitrates absorbed from the soil. The amino acids are used to synthesise **proteins**.

▪ Practical activity – experiment to show the evolution of oxygen from a water plant during photosynthesis

1 Take a bundle of shoots of a water plant, such as *Elodea* or *Cabomba*. Place the shoots in a beaker of water under an inverted conical funnel, as shown in Figure 2.1.
2 Fill a test tube with water. Keep your thumb over the end of the test tube while you invert it over the end of the funnel. The funnel directs rising bubbles of gas from the water plant into the test tube.
3 Leave the apparatus in normal daylight at room temperature for a few days.
4 Observe the bubbles of gas that come from the ends of the shoots of the water plant and collect in the test tube. As gas collects, the water is displaced from the test tube.
5 Carry out a glowing splint flame test on the gas collected in the test tube. This test can be used to show that oxygen has collected in the test tube.

▪ Practical activity – experiments to show the production of starch during photosynthesis

(a) The requirement for light

1 Take a plant in a pot (a 'potted plant') and place it in a dark cupboard for 2 days, so that its leaves do not contain starch (described as a 'destarched plant').
2 Bring the destarched plant out of the cupboard into the light. Mask one or more of the leaves with a light-proof card, held on with paper clips, as shown in Figure 2.2. Leave the plant in daylight for several hours.
3 Remove the masked leaf from the plant and test it for starch, as follows:
step 1 – dip the leaf in boiling water. This kills the leaf and stops all metabolic reactions in the leaf
step 2 – dip the leaf in boiling alcohol (ethanol) for about 2 minutes. The alcohol removes the chlorophyll (green colour).

STUDY TIP

Refer to Section 3.4 for more details about transport in the phloem.

PRACTICAL

This practical activity is simple to set up and shows how a gas is given off by a water plant during photosynthesis. You can compare evolution of gas bubbles from the plant in the light and in the dark. Provided you collect a good sample of gas, you should be able to show that it is oxygen, using the glowing splint test.

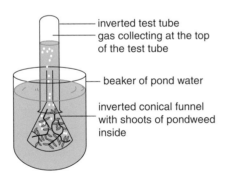

inverted test tube
gas collecting at the top of the test tube

beaker of pond water

inverted conical funnel with shoots of pondweed inside

Figure 2.1 Experiment to show the evolution of oxygen from a water plant during photosynthesis.

PRACTICAL

This group of practical activities shows the production of starch during photosynthesis and factors required for photosynthesis to occur. It is important that you know how to test a leaf for starch.

Remember to wear eye protection.

Note – when using alcohol, you must ensure there is no naked flame to avoid the alcohol catching fire. As a safety precaution, you should turn off any Bunsen burners and place the tube containing alcohol in a beaker of hot water to boil it.

step 3 – dip the leaf again into hot water for about 2 seconds. This softens the leaf

step 4 – spread the leaf on a white tile (or on a Petri dish)

step 5 – cover the leaf with iodine solution

If starch is present, the leaf goes black (or blue-black). If no starch is present, the leaf becomes a yellowish colour (same as the iodine solution).

Expected results are shown in the diagram.

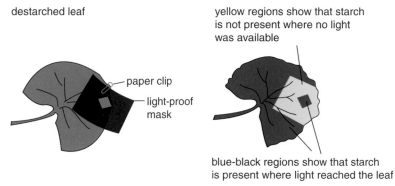

Figure 2.2 Experiment to show that light is necessary for the production of starch during photosynthesis.

(b) The requirement for chlorophyll

Use a potted plant with variegated leaves. A variegated leaf is one that appears green in part of the leaf where chlorophyll is present but is white or yellowish in other parts of the leaf where there is no chlorophyll.

1. Take a potted plant with variegated leaves and place it in a dark cupboard for two days, so that its leaves do not contain starch (described as a 'destarched plant').
2. Bring the destarched plant out of the cupboard into the light. Leave the plant in daylight for several hours.
3. Remove one leaf from the plant and make a drawing to show which parts of the leaf are green and which parts are white. Test the leaf for starch as follows:

   steps 1–3 – as above. **Note precautions to be taken when using alcohol**.

   step 4 – spread the leaf on a white tile (or on a Petri dish)

   step 5 – cover the leaf with iodine solution

   If starch is present, the leaf goes black (or blue-black). If no starch is present, the leaf becomes a yellowish colour (same as the iodine solution).

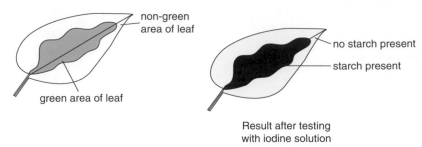

Figure 2.3 Experiment to show that chlorophyll is necessary for the production of starch during photosynthesis.

Expected results are shown in the diagram. Compare the results with the leaf you drew at the start. You should see that the parts of the leaf that were green contain starch but the parts that were white do not contain starch.

(c) The requirement for carbon dioxide

This experiment can be modified to show the need for carbon dioxide in photosynthesis. Use a plant with green leaves.

1 Take a potted plant and place it in a dark cupboard for 2 days, so that its leaves do not contain starch (described as a 'destarched plant').
2 Bring the destarched plant out of the cupboard into the light. Enclose one leaf on the destarched potted plant in a tube or transparent bag containing potassium hydroxide. This absorbs carbon dioxide so there is no carbon dioxide in the air around the leaf. Leave the plant in daylight for several hours. *Note that potassium hydroxide is corrosive. The concentration of the potassium hydroxide solution should not exceed 2M.*
3 Remove this leaf from the plant and remove a second leaf that has been in the air. Test both leaves for starch as follows:
Follow *steps 1 – 5* in previous experiment to test the leaves for starch. **Note precautions to be taken when using alcohol**.
Compare the results for the two leaves. If starch is present, the leaf goes black (or blue-black). If no starch is present, the leaf becomes a yellowish colour (same as the iodine solution).
Expected results are that there is no starch in the leaf from the container without carbon dioxide but there is starch in the leaf that was in the air.

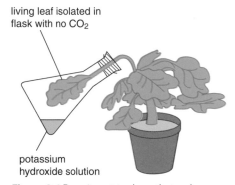

Figure 2.4 Experiment to show that carbon dioxide is necessary for the production of starch during photosynthesis.

■ What affects the rate of photosynthesis?

Plants compete vigorously with one another – they need to gain access to the light required to power photosynthesis. Nevertheless, even if a plant leaf does get plenty of light, its rate of photosynthesis may be limited by another factor. In any situation, there is always some factor that provides a limit to the maximum rate of photosynthesis that can be achieved. This is called the **rate-limiting factor**.

The rate-limiting factor for photosynthesis is likely to be one of the following:
- light intensity
- carbon dioxide concentration
- temperature.

Light intensity

At night, when there is no light, there is no photosynthesis. As dawn approaches and a little light becomes available, photosynthesis begins to occur. As the light level increases, the rate of photosynthesis also increases. However, as the light intensity increases further, there comes a point at which there is no corresponding increase in the rate of photosynthesis. This is due to the fact that something else is setting the maximum rate of photosynthesis – another factor has become rate-limiting.

Temperature

Generally enzymes work better at warmer temperatures. If a plant leaf is cold, the slow activity of the photosynthetic enzymes, and the other molecules involved in the reactions, set a limit on the maximum rate of photosynthesis – even when lots of light is available. At the optimum temperature, the photosynthetic enzymes and substrate molecules move faster and meet more frequently to carry out the reactions of photosynthesis. At higher temperatures, say above 45 °C, photosynthetic enzymes may become denatured and the rate of photosynthesis falls. But a few plants (such as some cacti) are known to carry out photosynthesis at temperatures as high as 55 °C.

Carbon dioxide

Only a tiny fraction of the air is composed of carbon dioxide gas (0.04%). This gas is the source of the carbon atoms plants need to make glucose. During the day, a leaf has to exchange a large volume of air to provide sufficient carbon dioxide to the photosynthetic cells of the leaf. If there is plenty of light and warmth available to a leaf, but not enough carbon dioxide, then the carbon dioxide concentration becomes the rate-limiting factor in photosynthesis. Growers often boost the carbon dioxide levels in the air inside glasshouses and polytunnels to improve the growth of their plants.

■ Practical activity – investigating the effect of different factors on the rate of photosynthesis

Aquatic plants, such as *Elodea* or *Cabomba*, provide suitable plant material for these experiments.

1 Place a shoot of the water plant upside down in a boiling tube, beaker or measuring cylinder, as shown in the diagram. Attach a paperclip to the tip of the shoot of the water plant to prevent the plant floating to the top of the tube.

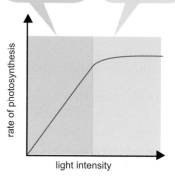

light intensity is the limiting factor as every time it is increased there is a corresponding increase in rate of photosynthesis

some other factor is limiting the rate of photosynthesis (temperature or carbon dioxide concentration?)

rate of photosynthesis

light intensity

Figure 2.5 How light intensity affects the rate of photosynthesis, until another factor limits the rate.

STUDY TIP

Look at Section 1.4 concerning enzymes and temperature. Draw a simple graph to show the expected effects of temperature on the rate of photosynthesis.

STUDY TIP

Make a link with Section 3.1 for more details on gas exchange.

STUDY TIP

Look at Section 7.1 and make a list of the ways in which a polytunnel can improve crop production.

STUDY TIP

Make reference to Section 6.3 for links to the carbon cycle.

PRACTICAL

This practical activity is a simple extension of the practical activity on page 52 and helps you to gain an understanding of how different factors can affect the rate of photosynthesis.

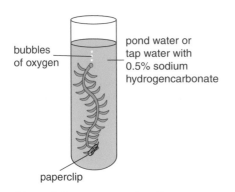

Figure 2.6 Investigating the effects of different factors on the rate of photosynthesis.

2 Fill the tube with tap water containing 0.5% sodium hydrogencarbonate. This provides a source of carbon dioxide for photosynthesis.
3 Place the tube in a water bath, such as a glass beaker, to protect it from becoming warmed by the light source.
4 Place the whole apparatus with the water plant in the light and observe a stream of bubbles being released from the cut end of the shoot.

The rate of production of bubbles is assumed to be an indication of the rate of photosynthesis. Using this apparatus, you can measure the effect of different factors:
- light intensity – vary distance from the light source
- carbon dioxide – vary the concentration of sodium hydrogencarbonate in the water
- temperature – change the temperature of the water bath that contains the tube with the water plant.

Adaptations of plant leaves for photosynthesis

In the photograph (Figure 2.7) note how these thin young tree leaves are held horizontal by their stalks and veins. This helps them to catch as much light as possible. The folds in the leaves stiffen the blades and help to channel rain water off the leaves.

The photomicrograph in Figure 2.8 shows part of a leaf from a plum tree, cut through at the midrib. The thickness of this leaf is only 1 mm. (The red stained areas show some pigments in the leaf.) You can identify other cells and match them with the diagram of a vertical section through a typical leaf (Figure 2.9). Notice the large numbers of chloroplasts in the palisade mesophyll cells, where most of the light is intercepted. This is where most of the sugar is made in photosynthesis. Each mesophyll cell is close to a leaf vein. The xylem vessels bring water and mineral ions from the soil, and the phloem vessels transport the sugars (as sucrose) and amino acids made by the mesophyll cells away to the rest of the plant.

Figure 2.7 Veins in a leaf help hold the leaves in a suitable position to capture light.

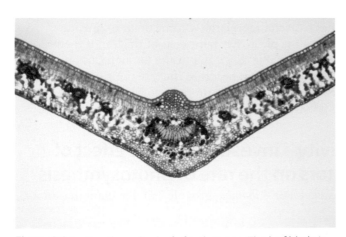

Figure 2.8 Section through a leaf of a plum tree. The leaf blade is less than 1 mm in thickness. This photomicrograph passes through the midrib (vein along the centre of a leaf). Other cells shown can be identified by comparison with the diagram in Figure 2.9.

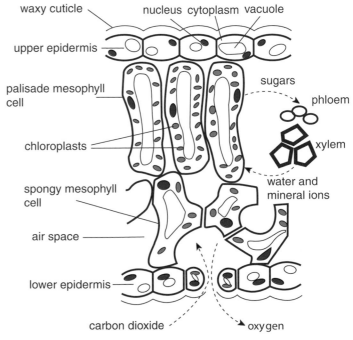

Figure 2.9 Vertical section through a typical leaf during the day.

These are some of the features that help a plant leaf carry out photosynthesis:

■ **Large surface area** – exposing a large surface area to the Sun helps to ensure that lots of light is collected.
■ The leaf is **thin**, making it lighter and easier to support and giving a short internal distance for penetration of light and transport of materials (including carbon dioxide and oxygen as gases in the air spaces).
■ **Waterproof waxy cuticle** – the leaf surface is coated with a waxy cuticle. This is transparent, so allows light to reach the chloroplasts in the mesophyll cells. It also prevents direct evaporation of water from the leaf cells to the atmosphere. The surface is often folded so that rain water runs off easily. The cuticle also helps protect the leaf from infection by fungal spores.
■ **Mesophyll cells** – the photosynthetic tissues are composed of two layers:
   ■ the **palisade mesophyll** cells are packed closely together, just under the **upper epidermis**. Most photosynthesis occurs in this layer as most light rays are captured here.
   ■ below this layer is the **spongy mesophyll**, with large interconnected **air spaces** between the cells. These air spaces allow free diffusion of carbon dioxide and oxygen as gases between the stomata and mesophyll cells.
■ **Leaf veins** – these form a network within the leaf and are made up of xylem and phloem. The **xylem** brings **water and mineral ions** up from the soil to the leaf tissues, to be used by mesophyll cells. In addition, the thickened walls of the xylem vessels help to **hold the leaf flat and horizontal**, enabling it to collect the light. The largest leaf vein is the **midrib**, which runs into the **leaf stalk**.
■ **Lower epidermis** – the lower surface of the leaf is covered in tiny pores called **stomata**. These pores are open during the day and this allows efficient diffusion of carbon dioxide into the leaf, for use by mesophyll cells for photosynthesis. The waste oxygen produced by the mesophyll cells diffuses out of the leaf during the day. At night the stomata are closed and gas exchange between the leaf and atmosphere stops. This helps the plant conserve water.

■ Mineral ions and plant growth

The glucose made in a plant leaf by photosynthesis is a carbohydrate and contains the elements carbon, hydrogen and oxygen. However, more than a dozen other elements are required by plants to synthesise other essential substances.

These other elements are the **mineral ions** (also called mineral salts or mineral nutrients). They are absorbed from the soil by the roots of the plant. The mineral ions are dissolved in the water that is drawn up the xylem, from the plant roots to the leaves. In the mesophyll cells, the mineral ions are used in the conversion of carbohydrates into proteins, chlorophyll, DNA and other important molecules.

A fertile soil should have a good supply of a range of mineral ions, but if the soil is infertile, there is likely to be a shortage of one or more of the necessary mineral ions. Plants growing in such habitats may have stunted growth and show symptoms that indicate a lack of one or more particular mineral ions. This is one reason why farmers examine crops for signs of a shortage of mineral ions, and add fertilisers to encourage rapid plant growth.

Two examples of important mineral ions are given below, though from your knowledge of biological molecules in plants, you can probably work out others that are needed for the large molecules that make up plants.

STUDY TIP
Look at Section 3.4 to make sure that you understand how water gets from the soil to the cells in a plant leaf.

STUDY TIP
Check Section 3.1 on exchange of gases in a flowering plant and review how internal leaf structure helps with exchange of gases.

STUDY TIP
Section 1.5 explains how active transport is involved in the absorption of mineral ions.

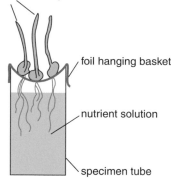

Figure 2.10 Investigating the requirement for different mineral ions for growth of plant seedlings.

Nitrate ions

Nitrates provide a source of the element nitrogen. In the plant, the element nitrogen is used to make amino acids. The **amino acids** become joined together to make the **proteins** of the plant. Plants usually need to absorb quite a lot of this mineral ion so that they can make the proteins they require. Plants affected by a shortage of nitrates scarcely grow at all – their leaves are often small and purple.

Magnesium ions

The element **magnesium** is a critical part of the **chlorophyll** molecule, found in chloroplasts. The magnesium ion is found at the heart of the molecule, and if there is not enough the plant cells cannot synthesise chlorophyll. Leaves of plants affected by a shortage of magnesium are yellow, rather than green, and the growth of the plant is poor.

■ Practical activity – investigating the requirement for different mineral ions for growth of plant seedlings

1 Set up a series of tubes similar to the one in Figure 2.10. Each tube should contain a different solution, as follows:
   tube 1 – standard nutrient solution, with all the mineral ions ('complete solution')
   tube 2 – complete solution but **lacking** in **one** necessary mineral ion
   tubes 3, 4, 5, etc. – complete solution but each lacks a different necessary mineral ion (one at a time), for example minus nitrate, minus phosphate, minus magnesium etc.
   final tube – distilled water (lacks all mineral ions).

2 Make an aluminium foil 'hanging basket' over the top of each tube. Make holes in the aluminium foil so that roots of seedlings can pass through.

3 Place a few young wheat seedlings (about 3 days old) in the hanging basket. Use forceps to pass the roots through the foil so that the roots project through the foil into the solution in the tube. A group of 5-day old seedlings is shown in Figure 2.10.

4 Place the tubes in the light and top up the solution in the tubes every 2 days with the appropriate solution.

5 Observe the growth in the different solutions. Measure the height of the seedlings in the different solutions. To determine the mass, remove the seedlings from the baskets and weigh them. Calculate the mean fresh mass of a seedling for each solution.

6 Plot the data on a bar chart to compare the changes in height and mass after a certain time and determine the relative importance of individual mineral ions.

If the seedlings are placed in continuous light from a bright fluorescent light, you can see differences in height and mass of the seedlings in as little as 7 days.

STUDY QUESTIONS

1 A student wanted to study the effect of light intensity on the rate of photosynthesis in some tree leaves.

She collected three leaves and placed one in each of three boiling tubes, set up as shown in tube C in the diagram.

She treated the tubes in different ways:
- one tube was wrapped in light-proof aluminium foil (no light)
- another tube was wrapped in a single layer of tracing paper (dim light)
- the final tube was left unwrapped (bright light).

She left the tubes in the light for 30 minutes. At the start, the hydrogencarbonate indicator in all tubes was orange, as shown in tube C. After 30 minutes, the colours had changed as shown in the diagram.

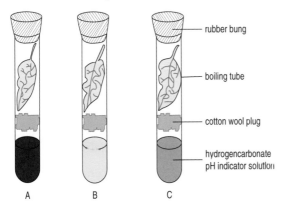

Note that hydrogencarbonate indicator is used to detect changes in pH, as shown in the colour chart.

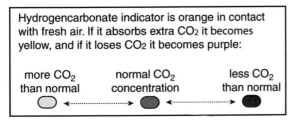

Hydrogencarbonate indicator is orange in contact with fresh air. If it absorbs extra CO_2 it becomes yellow, and if it loses CO_2 it becomes purple:

| more CO_2 than normal | normal CO_2 concentration | less CO_2 than normal |

a) Which of the tubes had no light? Explain how you worked out your answer.
b) Which of the tubes had bright light? Explain how you worked out your answer.
c) Using your own knowledge and information in this experiment, decide on whether each of the following statements is true or false. In each case, give a reason to support your answer.

- The leaves should have been selected from different trees.
- The leaf kept in the dark was an experimental control.
- Plant leaves respire only in the dark.
- Plant leaves carry out photosynthesis only in the daytime.

2 Some students set up an investigation into the effects of mineral ion deficiencies on the growth of wheat seedlings. They used the technique described on page 58.

The wheat seedlings were allowed to grow in the different nutrient solutions for 14 days. The students then weighed the individual wheat seedlings in each of the solutions. Their results are shown in the table.

Mass of individual seedlings / g					
Complete solution	Solution minus nitrate	Solution minus phosphate	Solution minus magnesium	Solution minus potassium	Distilled water (no nutrients)
2.5	0.3	0.5	1.4	1.3	0.1
2.1	0.2	0.4	1.0	1.2	0.05
1.9	0.2	0.4	1.3	0.9	0.2
2.8	0.4	0.2	1.2	1.1	0.05

a) Calculate a mean value for each column of figures.
b) Which single mineral ion appears to be the most important for seedling growth?
c) The students used wheat seeds from the same seed packet.
   i) Suggest why the students used seeds from the same packet for their experiment.
   ii) Suggest another factor the students could control in their experiment.
d) Why do you think there is some seedling growth in the distilled water even though there is no supply of mineral ions?
e) Use other sources of information and find out the specific roles of phosphate ions and potassium ions in plants.
f) Describe and explain the roles of microorganisms in supplying nitrates to plant roots (use Section 6.3 to help you).

2.2 Human nutrition

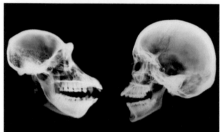

Getting value out of our human diet

Using a controlled fire to cook food was an important milestone in the evolution of humans. Cooking makes the food easier to digest, tastier, and more energy can be extracted from it. Cooked food is also more likely to be free from potentially harmful bacteria. The extra energy obtained means that less time has to be spent gathering the food, and there is no need to spend half the day chewing tough raw food, like chimpanzees and gorillas. As a consequence, modern humans (*Homo sapiens*) have evolved shorter digestive systems, smaller jaws and bigger brains than our ape relatives.

So what do we eat? What do we need in our food? And how do we process it in our bodies to make it become part of our cells?

■ The human diet and its requirements

Humans are **heterotrophs**, like other animals. This means that we take in food containing large complex molecules. The human body then processes these molecules, uses energy contained in them and uses parts of some of these molecules as building blocks for our human cells. The body also discards parts of some molecules if they cannot be used or have become waste products from the chemical activities (metabolism) of body cells. Humans are **omnivores** and so can eat food from both plant and animal sources.

The term **diet** means the foods and nutrients that a human body consumes. Often the term is used in a popular sense to refer to so-called 'slimming' diets, designed to help a person lose weight. In other cases it is used for special diets that people may wish to adopt in different circumstances, for example, for sports activities or high-altitude trekking. Special diets may also be adopted for medical reasons – to exclude foods that a person should avoid or include foods required to help the body cope with particular nutritional stresses.

In this section we look at the components of a healthy balanced human diet. We also look at some variations in diet that might occur when there are different demands on the body. We do not discuss the range of foods across the world, nor do we consider specifically vegetarian or vegan diets in which, for various reasons, people may avoid food derived from animal origin.

Components of the diet

Look back to Section 1.4 on biological molecules to remind yourself about the molecules that are required to make up the cells of the body. All these molecules are taken in as food and form parts of the diet.

Table 2.1 lists the necessary components of the human diet, together with some examples of food sources that supply good quantities of these components. The table also summarises some important functions of different components. Three examples of vitamins are given and two examples of mineral ions. Other vitamins and mineral ions are also required in the body.

STUDY TIP

Look at Section 1.4 to remind yourself about the range of biological molecules found in living organisms and why they are important.

Table 2.1 Components of a human diet and their functions in the body.

Component	Foods with abundant supply of the component	Functions in the body
carbohydrate	rice, pasta, potatoes, yams, maize (sweetcorn), fruits (e.g. bananas, grapes), honey	energy released (respiration) from sugars; energy stored as glycogen (in the muscles and in the liver)
protein	meat, fish, eggs, some dairy products (for example cheese), pulses and other legumes (e.g. beans, lentils)	important for growth and repair of the body; forms much of the cell cytoplasm; structures like muscles, tendons and bones; enzymes; some hormones (e.g. insulin); haemoglobin (in red blood cells); antibodies in blood plasma (Note that the proteins in food provide the amino acids that are the building blocks for the various molecules and structures listed above.)
lipid (fat and oil)	fatty foods (including meats); oily foods (some fish); oil (e.g. from olives, rapeseed, sunflower, maize, groundnuts)	stores energy as droplets in cells; part of cell membranes; insulation (heat loss); nerve insulation; some hormones (e.g. testosterone)
vitamins	A – meat (e.g. liver); fish liver oil; eggs; dairy products; some green vegetables; carrots	forms a pigment important in vision (in retina of eye)
	C – citrus and other fruits, some green vegetables	healthy skin; helps wounds to heal (lack of vitamin C leads to scurvy)
	D – fish liver oil, dairy products (butter) [also made by action of sunlight on skin]	bones (lack of vitamin D leads to brittle bones and sometimes to the condition known as 'rickets')
mineral ions	calcium – dairy products (milk, cheese, yoghurt)	bones, teeth
	iron – red meats, some green vegetables	haemoglobin (in red blood cells)
water	drinks (water, tea, coffee, juices), juicy fruits (oranges, apples, pineapples, grapes)	about 80% of cytoplasm is made of water; body fluids (e.g. blood, sweat, urine)
dietary fibre (roughage)	plant material (e.g. vegetables) contains indigestible fibres	helps with peristalsis, provides bulk (for gut contents and faeces), helps prevent constipation

MATHS TIP

Changing the subject of an equation

An equation is an expression containing a mixture of letters and numbers, and an equals sign. The components of an equation can be rearranged and still make sense.

For example, the equation for the area of a triangle is:
length (L) $\times$ ½ height (h) = Area
If you are given the area and h, L can be found by rearranging the equation:
L = area / ½ h

Now go to page 77 to apply this to some calculations involving body mass index (in humans).

The summary in Table 2.1 gives the overall range of requirements and, for a **balanced diet**, they should be consumed in appropriate proportions. The bulk of the diet is generally made up from carbohydrate foods with varying amounts of lipid and protein. Mineral ions and vitamins are essential but required in much smaller quantities.

If people eat a variety of foods, containing all these components, they are likely to have a balanced diet. The actual requirements for individual people vary at different times of life and in different situations. If certain components are absent or insufficient, people may suffer from a variety of conditions (sometimes known as deficiency diseases), or malnutrition may occur. On the other hand, an excess of food (particularly carbohydrate or lipid) can lead to obesity or other health problems.

We now look at a few examples, to expand on information in the table. Growing children require relatively more protein in their diet than a mature adult, because protein is necessary for growth (of new cells), and building up the muscles and other structures in the body. Similarly, adequate supplies of calcium are required when children are growing to ensure correct development of bones and teeth.

Figure 2.11 This collection of foods shows (from the top) various breads (mainly carbohydrate); eggs, cheese and meat (that provide protein); fruits and vegetables. A range of similar foods can contribute to a balanced diet.

Inadequate quantities of any of the components listed in the table may lead to poor development or poor maintenance of the body. Lack of iron may result in anaemia because of insufficient haemoglobin to carry oxygen in the blood to the cells where it is needed. Low levels of protein for young children leads to poor growth and development.

However, excess intake of certain foods, particularly carbohydrates and lipids, may lead to people becoming overweight or obese. In some cases this is linked to problems with the heart and circulatory system.

Energy requirements and how these may vary

Look at the bar chart in Figure 2.12. It shows average energy requirements per day for different people of different ages, when doing a similar low level of activity.

Note that the graph gives the energy values as 'kJ' (kilojoules) and you may be more familiar with the idea of how many *calories* different foods contain. Both units are ways of measuring energy content, but in science we generally use kJ rather than kcal.

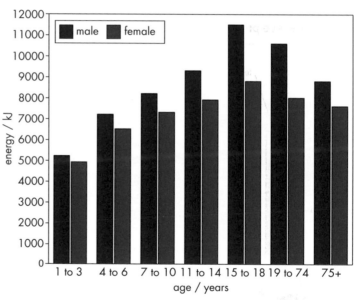

Figure 2.12 Average energy requirement per day for males and females of different ages (low level of activity).

From the graph you should be able to see the following:

■ At different ages (from child to adult, through to elderly) people require different amounts of energy. The requirement increases through childhood and is highest for adolescents, in the age range 15 to 18 years. The requirement for energy is just a little less through most of adult life, but begins to be lower for older people, from about 75 years.

■ Comparison of males and females at the same age shows that, at all ages, males require more energy than females.

Next, we look at energy requirements for different levels of activity. Table 2.2 shows data for males, at four different ages, carrying out different tasks: sedentary (e.g. mainly sitting, using a computer); low activity level (e.g. some walking around doing everyday activities); active level (e.g. some vigorous activity such as running or cycling).

Figure 2.13 Energy requirements vary, depending on the level of activity.

Table 2.2 Energy requirements of males, in different age groups, according to activity levels.

Age / years (males)	Energy / kJ per day		
	sedentary	low activity	high activity
8 to 9	6 280	7 330	8 370
17 to 18	10 260	12 140	13 820
31 to 32	9 840	10 890	12 140
75+	8 370	9 210	10 470

The trend is as you would expect – the higher the level of activity, the more energy the body requires to be able to carry out this level of activity. For females the trend is similar, but energy requirements are lower for all three categories.

Pregnant women require a higher energy intake than women of the same age who are not pregnant and carrying out similar activities. The extra energy is needed partly for the higher metabolism in the woman's body, synthesising materials for the developing baby, and partly because of the increased weight of the mother's body.

PRACTICAL

This practical activity provides a simple illustration of how scientists can work out the energy values of different foods. Even though this procedure is not 100% accurate, it gives a useful opportunity for you to understand the principle of the experimental approach.

Remember to wear eye protection.

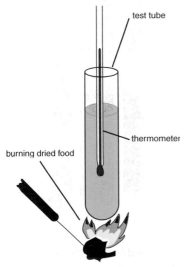

test tube

thermometer

burning dried food

Figure 2.14 Experiment to determine the energy content in a food sample.

MATHS TIP

Changing the subject of an equation
An equation is an expression containing a mixture of letters and numbers, and an equals sign. The components of an equation can be rearranged and still make sense.

Now go to page 77 to apply this to some calculations from an experiment involving heat released by a combusted food sample.

■ Practical activity – experiment to determine the energy content of a food sample

1 Place 20 cm³ of water in a test tube or boiling tube. Use a thermometer in the test tube to take the temperature of the water at the start, as shown in Figure 2.14.
2 Weigh a small portion of the food sample to be tested. (Do not use nuts or nut products.) Then hold a burning portion of this food sample under the bottom of the tube so that the flame touches the tube.
3 Note the increase in temperature of the water in the tube.
4 Calculate the energy content in the food, using the guidance below.

The increase in temperature in the water is due to the energy released by the food as it burns. Use the following steps to calculate the energy content of the food sample:

■ we know that 4.2 joules of energy is required to raise the temperature of 1 cm³ of water by 1°C
■ suppose the increase in temperature of the water is y °C
■ from this, you can calculate how much energy has been released into 20 cm³ of water in the tube:

heat energy released from food sample $= 4.2 \times y \times 20$ joules

■ this gives an estimate of the total number of joules transferred from the burning food sample to the water.

The value you calculate is likely to be lower than any values you may find given in data about the food being tested. This is because not all the heat energy in the food sample is transferred to the water. Some heat energy is lost to the surroundings.

■ What happens to the food in the body?

Humans take in food containing large complex molecules. Before making any use of these molecules, the human body must **digest** them and break them down into smaller molecules. For example, starch is broken down into sugar molecules and proteins are broken down into amino acids. The body can then use energy contained in the molecules and convert parts of the digested molecules into substances that become our human cells.

The **alimentary canal** is effectively a tube, going through the body from the **mouth** to the **anus**. As the food passes along the alimentary canal (also referred to as the gut) various things happen to it.

Figure 2.15 shows the relevant structures of the alimentary canal and summarises the processes that take place as the food progresses. Some of these processes are mainly mechanical – as a means of moving the food along the gut. Other processes are chemical and some are part of the metabolism of the human. These various processes are now considered in more detail, step by step.

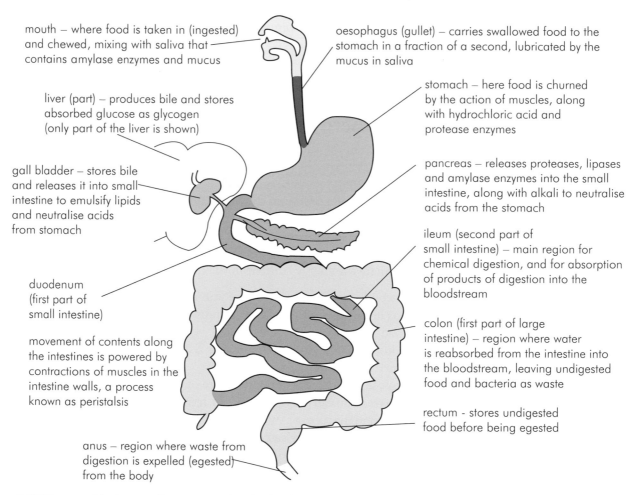

mouth – where food is taken in (ingested) and chewed, mixing with saliva that contains amylase enzymes and mucus

oesophagus (gullet) – carries swallowed food to the stomach in a fraction of a second, lubricated by the mucus in saliva

liver (part) – produces bile and stores absorbed glucose as glycogen (only part of the liver is shown)

stomach – here food is churned by the action of muscles, along with hydrochloric acid and protease enzymes

gall bladder – stores bile and releases it into small intestine to emulsify lipids and neutralise acids from stomach

pancreas – releases proteases, lipases and amylase enzymes into the small intestine, along with alkali to neutralise acids from the stomach

duodenum (first part of small intestine)

ileum (second part of small intestine) – main region for chemical digestion, and for absorption of products of digestion into the bloodstream

movement of contents along the intestines is powered by contractions of muscles in the intestine walls, a process known as peristalsis

colon (first part of large intestine) – region where water is reabsorbed from the intestine into the bloodstream, leaving undigested food and bacteria as waste

rectum - stores undigested food before being egested

anus – region where waste from digestion is expelled (egested) from the body

Figure 2.15 Diagram of the human alimentary canal, with related structures, and some functions of its different parts.

Ingestion and passage of food through the alimentary canal

Food is taken into the mouth. This is known as **ingestion** and is the first stage of processing food as it passes along the alimentary canal.

The food is moved around inside the mouth, helped by the tongue. The tongue also helps line up the food to be chopped and chewed by the action of teeth. The crushed food is mixed with mucus in the saliva. This results in the food being formed into a ball known as a **bolus**, which helps with the swallowing process. The amylase in the saliva starts the digestion of starch.

Movement of food along the **oesophagus**, and indeed through the **small intestine**, is helped by **peristalsis**. This describes the combined action of muscles in the gut walls – circular and longitudinal (see Figure 2.16). Movement of food in this way ensures that it passes through the various processes that are required before being incorporated into the cells of the body.

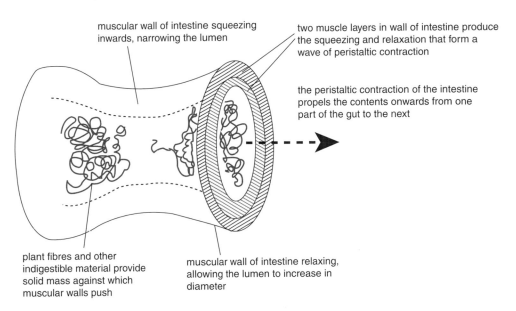

muscular wall of intestine squeezing inwards, narrowing the lumen

two muscle layers in wall of intestine produce the squeezing and relaxation that form a wave of peristaltic contraction

the peristaltic contraction of the intestine propels the contents onwards from one part of the gut to the next

plant fibres and other indigestible material provide solid mass against which muscular walls push

muscular wall of intestine relaxing, allowing the lumen to increase in diameter

Figure 2.16 How peristalsis helps to move food along the gut.

You can compare peristalsis to squeezing a marble along some fairly flexible rubber tubing. If you squeeze the tubing behind the marble (which represents the food bolus), you can push the marble forwards. This is similar to circular muscles contracting behind the food bolus. Then you have to imagine the longitudinal muscles contracting and effectively shortening the tubing, so that the marble has moved forward to a new position. If you put a drop of washing-up liquid inside the tubing, this helps lubricate the tubing and it is very much easier to push the marble along. In a similar way, mucus from the walls of the gut helps to lubricate the passage of the food bolus. The fibre that is taken in with food (as dietary fibre or roughage) is important to give some bulk to food and help with peristalsis.

In the **stomach**, the muscular walls churn the food to ensure that it is mixed well with the digestive enzymes. The hydrochloric acid is important to provide a suitable low pH for the activity of protease enzymes. The acid also kills bacteria and other pathogens that may have been taken in with the food.

The **small intestine** is sometimes considered in two parts – the first part that leaves the stomach is called the **duodenum** and the rest is called the **ileum**. Bile enters the duodenum from the gall bladder. Bile 'emulsifies' lipids, breaking them into smaller droplets and thus helping lipases to break down the large lipid molecules. **Bile** is alkaline, which is important for neutralising the acid materials from the stomach, thus providing a suitable pH for digestive enzymes in the rest of the small intestine. Digestive enzymes are secreted from the pancreas into the duodenum. These include amylases, lipases and proteases.

In the **small intestine** the food is again churned by the action of muscles in the walls. This keeps the food constantly in contact with the **villi** (see Figure 2.17. The villi give an enormous surface area inside the gut. The villi have two roles: firstly, as the site for secretion of digestive enzymes (amylases, lipases and proteases) and secondly, as the site for absorption of the digested food (smaller molecules).

The small intestine is very long – approximately 7 m in an adult body – so it is coiled many times to fit inside the abdomen, below the diaphragm. Movement of food through the small intestine is again by peristalsis.

By the time the food reaches the **colon** of the **large intestine**, most of the useful nutrients from the food have been absorbed. Large quantities of water are absorbed back into the body from the large intestine. The waste material is known as **faeces**. This is stored in the **rectum** and passes out of the body at the anus (**egestion**).

Digestion

The carbohydrates, proteins and lipids in our diet are large, complex molecules. They are too large to pass through the walls of the small intestine for distribution by the circulatory system around the body.

Digestion is the process that breaks down large complex molecules into smaller simpler molecules. These are mainly chemical processes, catalysed by enzymes. The mechanical action of teeth grinding and chewing the food helps to provide a larger surface area for the action of digestive enzymes on the food molecules.

Large molecules that need to be broken down (digested) into their smaller units include:

- **carbohydrates** such as starch, which is broken down in two stages into simple sugars (first to maltose, then maltose to glucose)

- **proteins** – broken down into smaller molecules (first to peptides, then to amino acids)

- **lipids** – broken down into smaller molecules (glycerol and fatty acids).

The different enzymes involved in the digestion of these food molecules are summarised in Table 2.3.

Some smaller components of food do not need to be digested. Mineral ions are small and can be absorbed from the gut into cells of the body. Vitamins are larger molecules, but can be absorbed from the gut without being changed.

STUDY TIP

Check back to Section 1.4 on biological molecules for more detail about the structure of carbohydrate, protein and lipid molecules and the units that they are made up of.

STUDY TIP

Make a link to Section 1.4 for information about enzymes.

Table 2.3 A summary of chemical digestion in the gut.

Large molecule	Enzyme	Products	Where in gut
starch	amylase	maltose (+ glucose)	mouth and small intestine
maltose	maltase	glucose	small intestine
proteins	pepsin	peptides	stomach (needs low pH)
peptides	protease	amino acids	small intestine
lipids	lipase	fatty acids + glycerol	small intestine (need to be emulsified by action of bile)

Note that pepsin (which acts on proteins in the stomach) requires a low pH (acid conditions) for optimum activity. All the other digestive enzymes work best in a pH around neutral. The acidity in the stomach is neutralised by the alkaline pancreatic juice to give a suitable pH for digestion in the small intestine.

Absorption

Absorption is the stage at which the small molecules, produced by the action of the enzymes in the gut, are taken into the body. Up to this stage, the food has remained in the tube that makes up the alimentary canal (gut) and has not yet entered any cells of the body. Absorption occurs through the walls of the small intestine, which are adapted to allow such absorption to occur efficiently.

Inside the small intestine, there are many finger-like projections called **villi** (see Figure 2.17). (Note – villus = singular, villi = plural.) These villi have several important functions in the digestion and absorption of food:

- They provide a very large surface area inside the intestine.
- They move constantly, helping to keep the food moving and mixing it with enzymes.
- Enzymes are produced from their walls (see page 66).
- They allow absorption of digested molecules into the blood capillaries and lacteals (part of the lymphatic system).
- They are richly supplied with blood (in the capillaries) so can absorb digested food and carry it away (initially to the liver, through the hepatic portal vein).

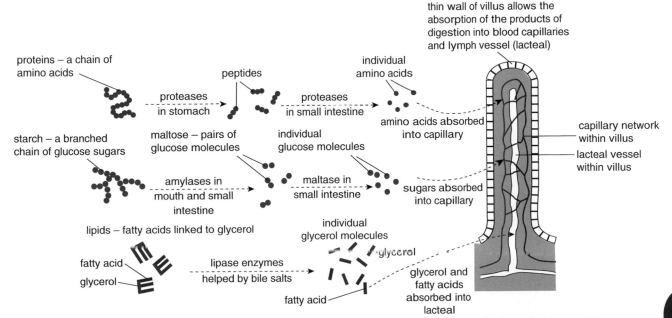

Figure 2.17 Diagram to illustrate the structure of a villus and events that take place leading up to absorption of digested food.

Inside the small intestine, as the large food molecules are broken down, there is a high concentration of the smaller digested molecules, such as sugars and amino acids. These can pass through the thin walls of the small intestine into the network of blood capillaries inside the villi.

Most digested molecules (including sugars and amino acids) pass through by diffusion, down the diffusion gradient (from the high concentration in the small intestine to the lower concentration in the blood capillaries). The blood keeps moving away, removing the absorbed molecules and maintaining the diffusion gradient. In some cases, absorption is assisted by active transport. The blood, containing the absorbed nutrients from the digested food, goes first to the liver via the hepatic portal vein.

Fatty acids and glycerol, from the digestion of lipids, take a different route into the lacteals (part of the lymphatic system). Later, these vessels empty their contents into the blood system, so digested lipids are eventually transported to cells in the body where they are needed.

> **STUDY TIP**
> Look back at Section 1.5 to remind yourself about concentration gradients.

Assimilation

We have now, at last, reached the stage that shows us what eating or taking in food is all about. **Assimilation** is the process by which the absorbed food molecules are built into molecules that form the cells of the human body.

We will look at just a few examples. Glucose is available for respiration and release of energy. Some glucose is built up into glycogen, which can be stored in the liver or muscles, to be available for energy release when required. Amino acids are the building blocks of proteins, such as enzymes and other body structures, including bones, ligaments and muscles. Fatty acids and glycerol combine again to form lipids. Some are laid down as fat deposits (which may provide protection, insulation or reserves of energy to be used when required), others contribute to cell membranes or the formation of certain hormones. Further metabolic reactions involving absorbed nutrients lead to the formation of DNA and many other complex molecules, contained in your hair, teeth, hormones, blood plasma and any other body structures you can think of.

Egestion

We are nearly at the end of the story. The undigested materials (such as plant fibres) that remain in the gut pass into the large intestine. During its journey through the gut, water was added to the ingested food in various digestive juices. These juices contained secretions from different glands. This water is absorbed back from the large intestine into the bloodstream. The remains, consisting of undigested materials and *E. coli* bacteria are **egested** (passed out) from the body, through the anus, as **faeces**.

STUDY QUESTIONS

1 a) Different digestive juices are produced as food moves through the gut, as follows:
   - saliva
   - gastric juice
   - pancreatic juice
   - intestinal juice.

   i) For each of these digestive juices, list the enzymes they contain, the food molecules that these enzymes act on and the products of the enzyme action.

   ii) State where each of these enzyme reactions takes place.

   b) List any other important functions carried out by these digestive juices. In your answer, include reference to other substances included in the juice, in addition to the enzymes.

   c) List mechanical or physical processes that contribute to the digestion of food. State where the processes occur in the gut and how they help.

2 Scientists have invented a tiny camera that can be swallowed as a capsule. It then passes through the whole length of the alimentary canal, taking pictures on the way. The pictures can be viewed on a screen, outside the body.

   Imagine you are that camera. Describe what you would see on your journey, what it would feel like and what is happening to the food you swallowed at the same time. You can be creative in the language you use or the style you adopt, but make sure the biology is correct. Write your account in a way that someone who is not a science specialist could understand.

2.3 Respiration

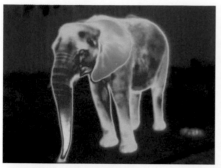

Respiration and keeping warm

These images are thermograms – photographs showing the heat given off from the surface of an elephant, a human and an arum flower. All this heat comes from inside the cells of these organisms, released as part of the process of cellular respiration.

For the elephant and the human, the heat helps to maintain a constant warm body temperature (they are homeothermic). Some animals, such as reptiles, are poikilothermic – their temperature is that of their environment. The arum flower uses heat to attract insects during a cool night for pollination (but plants are not considered to be homeothermic). Being homeothermic comes at a cost – about 80% of the respiration in your body is used to generate the heat you need to keep warm. As long as you get enough fuel (food) to respire, you can continue to survive when it is cold.

STUDY TIP

Use the Greek origin of the words **poikilothermic** and **homeothermic** to help you understand them and remember their meaning (poikilo = varied; homeo = alike, similar; thermos = heat).

STUDY TIP

Look at Section 6.2 to find out about the role of producers in the energy flow of an ecosystem. You could list reasons why only a fraction of the energy in one trophic level gets passed to the next level.

■ Living organisms require energy

The processes of life require energy to make them occur. Moving a muscle, pumping a molecule across a cell membrane, building a large molecule from smaller ones, or just keeping warm may seem very different activities. Yet each of these processes requires energy and the body uses the same energy-rich compound (called **ATP**) to make them happen. Cellular respiration involves the breaking of bonds within molecules, such as glucose and fatty acids, which releases the energy needed to make ATP. The energy in the glucose and fatty acids originally came from the Sun, having been made in producer organisms, and passed along food chains as part of the bodies of organisms.

Cellular respiration

The chemical reactions in living organisms (known as metabolism) can be divided into those that release energy and those that require energy. All living organisms carry out **cellular respiration**, in which substances such as glucose, are broken down and release energy. This energy is then carried as the chemical substance **ATP** (adenosine triphosphate) to wherever it is needed in the cell for useful work to be done.

Most ATP produced by respiration is made in cell organelles called **mitochondria**. Inside these organelles reactions take place that use oxygen to help break down glucose and produce waste carbon dioxide and water.

Aerobic and anaerobic respiration

Glucose can be respired in different ways, depending upon whether oxygen is available or not.

■ **Aerobic respiration** – the glucose is completely broken down into carbon dioxide and water. Oxygen is involved in the reaction and lots of energy is released.

glucose + oxygen → carbon dioxide + water + lots of energy (as ATP)

$$C_6H_{12}O_6 + 6O_2 \rightarrow 6CO_2 + 6H_2O + \text{lots of energy (as ATP)}$$

■ **Anaerobic respiration** – the glucose is only partly broken down, and oxygen is not used. Anaerobic respiration allows a little energy to be released from the glucose, but much less than with aerobic respiration. However, this process does generate some energy, even when there is a shortage of oxygen.

There are two ways in which anaerobic respiration of glucose occurs, depending on the type of organism involved.

In **animals**, anaerobic respiration of glucose produces **lactic acid** as a waste product.

glucose → lactic acid + a little energy (as ATP)

$$C_6H_{12}O_6 \rightarrow 2C_3H_6O_3 + \text{a little energy (as ATP)}$$

In **plants and fungi**, anaerobic respiration of glucose produces **ethanol** (**alcohol**) and **carbon dioxide** as waste products.

glucose → alcohol + carbon dioxide + a little energy (as ATP)

$$C_6H_{12}O_6 \rightarrow 2C_2H_5OH + 2CO_2 + \text{a little energy (as ATP)}$$

Most living organisms benefit from being able to respire both aerobically and anaerobically. This means they can respond to the changing circumstances in which they find themselves. For most organisms, anaerobic respiration is a short-term solution to a temporary shortage of oxygen. However, some organisms (including certain bacteria) can only respire anaerobically and cannot live in the presence of oxygen.

Here we look at some examples that show how organisms respire both aerobically and anaerobically.

When an athlete is exercising very hard, the oxygen supply may be insufficient to support aerobic respiration. Despite the best effort of the athlete's lungs and circulation, the supply might not be enough for the work the muscles are doing. In this case, some of the energy for the exercise may be obtained from glucose by anaerobic respiration. The lactic acid formed as a consequence represents the 'oxygen debt'. This debt is 'repaid' by the athlete, who continues to breathe hard after the exercise has finished. This allows the lactic acid to be reprocessed to glucose.

Air spaces in the soil usually allow plant root cells to respire aerobically. However, if the soil becomes flooded, the air spaces become filled with water, so the plant root cells may have to respire anaerobically. If the flooding goes on too long, the alcohol produced can poison the roots, leading to the death of the plant.

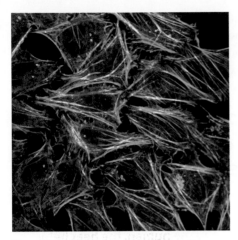

Figure 2.18 These human cells have been stained so that the nuclei appear purple and the hundreds of mitochondria appear red. The mitochondria are the sites within the cell where ATP is produced during aerobic respiration.

STUDY TIP

Look at Section 3.6 and list the changes that occur inside the body of an exercising athlete that would help to reduce the size of their oxygen debt.

PRACTICAL

This practical activity gives you a way of showing that carbon dioxide is given off by living organisms. Different living material can be used, provided it fits in the tubes. You should make sure you are familiar with the colour changes in the hydrogencarbonate indicator and why they take place in relation to changes in the amount of carbon dioxide in the tubes. You have opportunities to use hydrogencarbonate indicator in practical activities in other parts of this book.

Remember to wear eye protection.

When yeast cells (a fungus) are mixed into dough for making bread, they soon use up all of the oxygen. During the proving stage of bread making, the yeast cells respire the sugars in the dough anaerobically to produce ethanol and carbon dioxide. The trapped carbon dioxide makes the holes in the bread and this helps the bread to rise. The alcohol evaporates off during the baking stage.

■ Practical activity – experiment to demonstrate the evolution of carbon dioxide by living material

Hydrogencarbonate indicator changes colour depending on the pH of the solution. When more carbon dioxide is dissolved in the solution, it becomes more acidic. When less carbon dioxide is dissolved in the solution, it becomes more alkaline. The colour changes are shown in the chart in Figure 2.19.

1 Set up three boiling tubes (A, B and C) as shown in the diagram.
2 Each tube contains some hydrogencarbonate indicator solution that has been standing in air (or air has been bubbled through it). At the start, the colour of the indicator in each tube is orange, as shown in tube C.
3 Place material in each tube, as follows:
    tube A – dead material (for example seeds that have been boiled to kill them)
    tube B – living material (for example soaked seeds that are germinating or living maggots)
    tube C – no material (or something inert such as more cotton wool or a small piece of paper tissue).
4 Place the bung in each tube. After three hours observe the colours in each tube.

After three hours, the expected colours in each tube are as shown in the diagram.

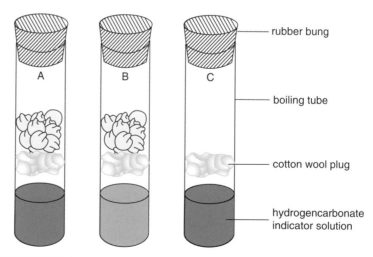

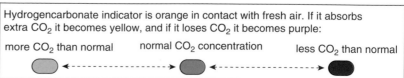

Hydrogencarbonate indicator is orange in contact with fresh air. If it absorbs extra CO_2 it becomes yellow, and if it loses CO_2 it becomes purple:

more CO_2 than normal   normal CO_2 concentration   less CO_2 than normal

Figure 2.19 Experiment to demonstrate the production of carbon dioxide by living material.

tube A – orange colour, showing no change in the carbon dioxide concentration. This shows that dead material does not give off carbon dioxide.

tube B – yellow colour, indicating that the tube contains more carbon dioxide than at the start. This demonstrates that living material produces carbon dioxide.

tube C – orange colour, showing no change in the carbon dioxide concentration. This is as expected as there is no living material in the tube.

■ Practical activity – demonstration of the production of carbon dioxide by yeast in anaerobic conditions

1 Set up the apparatus as shown in Figure 2.20. Place a yeast and glucose (or sucrose) solution in a specimen tube (or boiling tube) and cover it with a layer of paraffin oil or cooking oil. This prevents oxygen diffusing into the yeast + glucose mixture from the air above.

2 Insert the rubber bung (with delivery tube) to form an air tight seal to the specimen tube. Pass the end of the delivery tube into a second specimen tube containing limewater. The delivery tube is drawn out into a fine nozzle. The limewater is clear at the start.

3 Observe the bubbles of gas that pass into the limewater and watch for it to turn milky.

Limewater turns milky in the presence of carbon dioxide. This shows that carbon dioxide is given off by the yeast during respiration. Oxygen has been kept out of the yeast + glucose mixture by the layer of oil on the surface.

If diazine green is added to the yeast mixture, you can see when the respiration becomes anaerobic. Diazine green is blue in the presence of oxygen but turns pink if there is no oxygen.

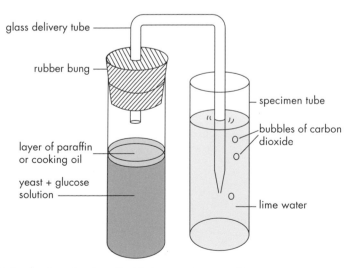

Figure 2.20 Production of carbon dioxide by yeast in anaerobic respiration.

■ Practical activity – measurement of the rate of respiration by living material

1 Set up the apparatus as shown in Figure 2.21. The potassium hydroxide absorbs carbon dioxide released by the living material. The cotton wool plug provides a support for the living material and prevents it making contact with corrosive chemicals (for example the KOH). It also allows circulation of the air in the specimen tube. *Note that potassium hydroxide is corrosive. The concentration of the potassium hydroxide solution should not exceed 2M.*

2 Place the living material (for example respiring maggots) on the cotton wool plug.

3 Insert the rubber bung (with delivery tube) to form an airtight seal to the specimen tube. Pass the end of the delivery tube into a second specimen tube containing water, coloured with ink or a food dye. The colour helps you see the level of liquid (meniscus) in the tube. Make sure the delivery tube goes below the level of the liquid.

4 Observe the height of the liquid in the delivery tube at the start and over a period of time.

As the living material respires, it takes in oxygen and gives out carbon dioxide. The carbon dioxide is absorbed by the KOH, so this lowers the pressure of gas in the apparatus. The result is that the meniscus (liquid level) rises up the delivery tube.

The distance moved by the meniscus in a certain time gives a measure of the rate of respiration of the living material. You can use this apparatus to make comparisons of the rate of respiration of the same material under different conditions or of different living material.

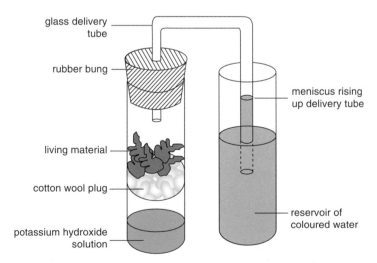

Figure 2.21 Measurement of the rate of respiration by living material.

This practical activity helps you understand that heat is given off by living material. Sometimes the 'dead' seeds are treated with a disinfectant to kill microorganisms that might grow on them. Can you suggest why?

■ Practical activity – demonstration that heat is given off by living material during respiration

1. Set up the apparatus as shown in Figure 2.22 using two vacuum flasks, A and B.
2. In flask A place some seeds that have been killed by boiling, then cooled. In flask B place some seeds that have been soaked and are germinating.
3. In each flask, place a plug of cotton wool in the neck and insert a thermometer through the cotton wool so that they are held in place by the cotton wool. Air can pass through the cotton wool plug into and out of the flask.
4. Invert both flasks and record the temperature at the start and over the next 3 to 4 days.

The thermometer in flask B shows a higher temperature. This demonstrates that the living seeds have given off heat.

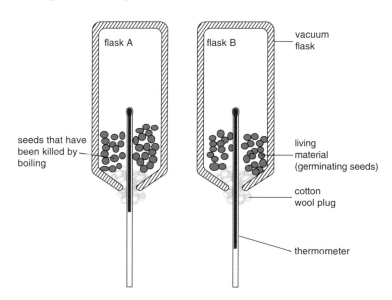

Figure 2.22 Demonstration that heat is given off by living material during respiration.

STUDY QUESTION

1. Copy and complete the table. For each statement about respiration, tick the aerobic and / or anaerobic boxes in the table, depending on whether the statement applies to one form of respiration, the other form, *or both*.

Statement	Aerobic respiration	Anaerobic respiration
Animals and plants can carry out the process		
The process results in the formation of ATP		
The process results in the formation of carbon dioxide by yeast		
The process results in the production of heat		
The process results in the production of lactic acid in animals		
The process requires oxygen to be available		
The process involves the formation of ethanol in plants		
The process can only be used for a temporary period of time in animals		

Summary

I am confident that:

✓ I can describe the process of photosynthesis and understand how green plants convert solar energy to chemical energy and build up complex carbohydrates from carbon dioxide.

✓ I know the word equation and balanced chemical equation for photosynthesis.

✓ I am familiar with experiments that support the equation for photosynthesis and show that oxygen is evolved, carbon dioxide is used, starch is made and that light and chlorophyll are necessary.

✓ I understand how carbon dioxide concentration, light intensity and temperature can affect the rate of photosynthesis.

✓ I know some examples (nitrate and magnesium) which illustrate that plants require mineral ions.

✓ I know the sources and functions of the components of a balanced diet, including carbohydrates, proteins, lipids, vitamins (A, C and D), mineral ions, water and dietary fibre.

✓ I understand how energy requirements vary in different people in different situations.

✓ I can describe the structures of the alimentary canal and know their functions in relation to the digestion of food and absorption of the products of digestion.

✓ I understand the importance of different digestive juices and the part played by certain enzymes in the digestion of food.

✓ I know how to determine the energy content in a sample of food.

✓ I know that respiration releases energy in all living organisms.

✓ I understand that the energy produced from respiration is carried as ATP and that ATP provides energy for living cells.

✓ I understand the difference between aerobic and anaerobic respiration and can write a word equation for both (and a balanced chemical equation for aerobic respiration).

✓ I know how to demonstrate that carbon dioxide and heat are given off by living organisms.

MATHS SKILLS

Arithmetic means
The arithmetic mean of a set of data us used to describe the overall result and allows different sets of data to be compared.

To find the arithmetic mean, add all the values in a set together, and divide by the number of individual values.

Now try Question 2 a), page 59.

MATHS SKILLS

Changing the subject of an equation
An equation is an expression containing a mixture of letters and numbers, and an equals sign. The components of an equation can be rearranged and still make sense.

For example the equation for the area of a triangle is:
> length (L) × ½ height (h) = Area

If you are given the area and h, L can be found by rearranging the equation:
> L = area / ½ h

The human body mass index (BMI) can be found using the equation below. Note that body mass is measured in kg and height is measured in metres:

$$BMI = \frac{body\ mass}{(height)^2}$$

Units of BMI are kg / m^2 and a value between 18 and 25 is considered optimal.

a)   Person A weighs 59 kg and is 1.88 metres tall. Is his BMI healthy or not?
b)   Person B has a BMI of 24 and weighs 84 kg. What is his height?
c)   Person C has a BMI of 33 and is 1.68 metres tall. What is his weight?

(c)   93 kg
(b)   1.87 metres
(a)   BMI too low

MATHS SKILLS

Changing the subject of an equation
An equation is an expression containing a mixture of letters and numbers, and an equals sign. The components of an equation can be rearranged and still make sense.

The heat released by a combusted food sample heating a container of water
> = temp rise in °C × volume of water in cm^3 × 4.2 joules

a)   If the volume of water is 20 cm^3 and the temp rise is 15 °C, find the heat released
b)   If the energy released was 300 joules and the temp rise was 4 °C, find the volume of water to the nearest whole cm^3.

(b)   18 cm^3
(a)   1260 joules

Example of student response with expert's comments

■ Practical activities

1 Sam investigated the energy content of different snacks. To determine the energy content of potato crisps, he used the apparatus shown. He weighed one of the crisps. He set light to it with a Bunsen burner and quickly held it underneath the test tube. He recorded the rise in temperature of the water in the test tube and used this to calculate the energy released when one crisp was burnt.

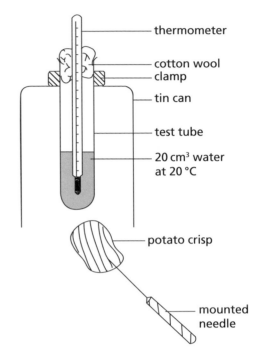

a) Sam used some special features in his apparatus to help him obtain more accurate results. These are listed below. For each feature, suggest how it would help him obtain more accurate results:
   i)   He used a test tube with thinner glass walls than usual. *(1)*
   ii) He surrounded the test tube of water with a tin can. *(1)*
   iii) He placed cotton wool in the opening of the test tube. *(1)*
b) Suggest **two** other sources of error that Sam did not take into account. *(2)*
c) Suggest **one** way that Sam could improve the reliability of his results. *(1)*

(Total = 6 marks)

Student response Total 5/6	Expert comments and tips for success
a) i) There would be less glass to heat up ✔ so the water would get hotter.	The student has understood that the experiment will be more accurate if heat loss can be reduced, and that if less glass has to be heated up more heat energy will be transferred to the water and the recorded temperature will be higher.
ii) Less heat lost to the air ✔ so more heating of the water and the temperature rises.	The student has understood that the can reduces heat loss to the surrounding air.
iii) The cotton wool traps the heat. ✔	The student has said just enough for the mark.
b) Part of the crisp might not have burnt, ✔ or left soot O, so the value for the heat produced would be too low.	Only 1 mark here because soot is left when something does not burn completely, i.e. it is the same marking point.
c) He could repeat the experiment several times and calculate the mean result. ✔	For the results to be reliable they should be repeatable and show little variation. Experiments should be repeated a minimum of three times to check that there is very little variation between the outcomes. The mean result can then be calculated.

■ Extended writing

General advice

- Before starting to write, underline the key words in the question, and think carefully about what is being asked.
- Make a brief list of ideas for your answer. This list will help you organise your answer sensibly and logically, and it also might help to jog your memory so that you remember to include all the relevant points.
- The mark allocation for extended writing questions is usually 6 marks. This gives a guide as to how long you should spend on this question in relation to the rest of the paper – probably about 6 minutes.
- Often a single word is not enough for a mark but requires further description or elaboration in some way.
- The question may refer to two different parts of a topic — say, features or processes to compare. Make sure you cover both aspects, or you will not be able to gain full marks.
- Sometimes you may decide that it is useful to draw a diagram to help with your answer. If you do this, make sure the diagram is labelled or annotated with notes so that it can be understood and makes links with the rest of your written answer.

1 Lipids (fats) are an important source of energy in the human diet. Describe how lipids are digested in the small intestine. *(6)*

Student response

In the small intestine, pancreatic juice breaks down lipid to fatty acids and glycerol. The enzymes in the small intestine need a neutral pH to work best so bile, which is alkaline, pours onto the food and neutralises the acid from the stomach. Bile also breaks up the fat into tiny droplets to speed up digestion.

Mark scheme

The mark scheme shows how marks are awarded.

(1) digestion involves the breakdown of large food molecules to smaller ones

(2) (digestion / breakdown) by enzymes

(3) lipase

(4) (lipase) from pancreas / pancreatic juice / pancreatic duct

(5) (breaks down) lipid to fatty acid + glycerol

(6) bile

(7) from gall bladder / liver

(8) (bile) alkaline / neutralises acid from stomach / provides correct pH for lipase action

(9) (bile) emulsifies lipids / converts lipids to small droplets / equivalent

(10) so large surface area for lipase to act

Total: 6

Student response Total 5/6	Expert comments and tips for success
In the small intestine, <u>pancreatic juice breaks down lipid</u> ✔	Although lipase is not mentioned, 'breaks down lipid' together with 'pancreatic juice' is just enough for the award of mark (4). Enzymes not mentioned, so mark (2) not awarded.
to fatty acids and glycerol. ✔	The student gains mark (5). Adding the words 'by lipase' would have gained mark (3). In answering questions about enzyme action, like this one, always try to include the name of the enzyme involved and a word equation for the reaction.
The enzymes O in the small intestine need a neutral pH to work best, so bile, ✔ which is alkaline, ✔ pours onto the food and neutralises the acid from the stomach.	'Enzymes' is not linked to digestion or breakdown so not enough for mark (2). The student gained mark (6) for 'bile' and mark (8) for 'alkaline'. 'Neutralises acid from the stomach' is another way of getting mark (8).
Bile also <u>breaks up the fat into tiny droplets</u> ✔ to speed up digestion.	'Breaks up the fat into tiny droplets' is equivalent to 'emulsifies lipids' for mark (9). Try to use the correct scientific term, lipid, although you will not usually be penalised for using the word 'fat' instead. A reference to 'large surface area' here would have gained mark (10).

Exam-style questions

1 A student used a shoot of pondweed to investigate the effects of light on photosynthesis.

She immersed the pondweed in a boiling tube containing pond water to which some sodium hydrogencarbonate had been added. She then placed this boiling tube in a large beaker containing cool tap water. When ready, she illuminated the apparatus with a bright lamp.

The student then watched the bubbles coming from the shoot of pondweed. She counted the number of bubbles in 1 minute and recorded this as the bubble rate. She recorded the bubble rate three times.

She then changed the light intensity by moving the position of the lamp and took three more readings of the bubble rate. This process was repeated with the lamp at different distances from the apparatus. She recorded her results as reading 1, reading 2 and reading 3. These results are shown in the table.

| Distance of lamp from tube / cm | Bubble rate / bubbles per minute | | | |
	Reading 1	Reading 2	Reading 3	Mean
10	36	36	36	36
20	24	34	20	
30	13	14	15	14
40	9	7	8	8

a) i) Calculate the missing mean value in the table. [1]

ii) Identify an anomalous result in the table. [1]

b) Explain why the student used each of the following:

i) sodium hydrogencarbonate solution. [1]

ii) the beaker of cool tap water. [1]

c) What do the results tell you about the effect of light intensity on the rate of photosynthesis? [2]

d) Suggest how the student could modify the experiment to investigate the effects of temperature on the rate of photosynthesis in pondweed. [3]

e) i) Write out a balanced chemical equation to summarise the process of photosynthesis. [3]

ii) Choose **one** factor that may affect the rate of photosynthesis and describe a situation when this factor may become a rate-limiting factor [2]

[Total = 14]

2 Visking (dialysis) tubing is an artificial partially permeable membrane, which allows small molecules (such as sugars) to diffuse through, but does not allow larger molecules (such as starch) to pass through.

The diagram shows how Visking tubing can be used as a model of digestion.

Beakers containing distilled water, with knotted lengths of Visking tubing floating in them

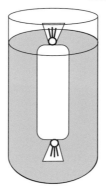 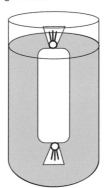

beaker A	beaker B
length of Visking tubing containing starch solution	length of Visking tubing containing starch solution mixed with the enzyme amylase

Some students set up the model as shown in the diagram. The Visking tubing in beaker A contained starch solution only. The Visking tubing in beaker B contained starch solution mixed with the enzyme amylase. They placed each length of Visking tubing in a beaker of water. After 3 hours, the water in each beaker was tested to see if it contained starch or glucose.

a) i) To test for starch, the students poured some of the liquid in the beaker into a test tube and added a few drops of iodine solution to it. What colour would they expect if starch is present? [1]

ii) Describe a test they could do to find out if glucose is present. [2]

b) i) Draw a table that you could use for the expected results to show whether starch and glucose are present or absent in each beaker. Predict the results you would expect for the starch and glucose tests and write your answers in the table you have drawn. [3]

ii) Explain the reasoning for your predictions. [3]

[Total = 9]

3 Plant root cells need to carry out respiration. In a good, well-drained soil, they obtain the gas they require from the air spaces in the soil around the roots.

When it rains, water fills these air spaces, preventing the roots from having access to the gas in the air spaces. The soil is described as being waterlogged. Many plants cannot tolerate having their roots in waterlogged soil and it appears that their root cells have been killed.

a) i) Name the gas used for respiration by root cells in a good well-drained soil. [1]

   ii) Write a balanced equation for respiration carried out by the root cells in waterlogged soils. [3]

b) Suggest **two** ways in which cells in the root would use the energy released by respiration. [2]

c) Some plants are able to grow in waterlogged soils. These plants often have large interlinked air spaces inside the stems and roots. Suggest how this might explain their tolerance to waterlogged soils. [3]

[Total = 9]

EXTEND AND CHALLENGE

1 The leaves of plants found in different habitats around the world often show specific adaptations to the climatic conditions in which they live. The intensity of light in tropical regions is greater than that found in regions towards the poles and the lushest vegetation occurs where there is a lot of light and a lot of rain.

On a single tree, the leaves on the sunniest side grow larger and thicker than those on the shadier side. All plant tissues respire, so a large leaf respires more than a smaller one. However, a large leaf can convert more light energy into sugar when light is available. The trade-off for a tree making large leaves on one side and smaller leaves on the other side is that this arrangement is an efficient use of resources.

A study was made of carbon dioxide uptake and carbon dioxide release in a sun-adapted leaf and a shade-adapted leaf of the same tree, at different light intensities. The graph shows the results.

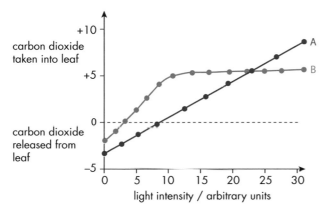

a) What process generates carbon dioxide within the tissues of a leaf?

b) Explain why both leaves export carbon dioxide from their leaves at certain light intensities.

c) On the basis of the data, which leaf is likely to be the larger one? Give reasons for your answer.

d) At a certain light intensity, the rate at which photosynthesis occurs just balances the rate of respiration. This is called the compensation point. Identify the compensation point for each leaf.

e) The table shows some features of leaves. Decide whether each feature is likely to be true of a shade leaf or a leaf from the light side of a tree. Copy and complete the table. Place a tick in the appropriate box and also give a reason to support your answer.

Feature	Shade leaf	Light leaf	Reason
Thickest leaf cuticle			
Fewer stomata			
Two layers of palisade mesophyll cells			
Larger air spaces between spongy mesophyll cells			
More numerous chloroplasts in each cell			

f) Suggest why many trees in temperate regions shed their leaves in autumn.

2 Most of us take it for granted that, provided we have a healthy balanced diet, our digestive system does its job properly.

However, there are a number of diseases that can disrupt the normal functioning of our intestines. One of these is coeliac disease. This condition occurs when the lining of the small intestine becomes allergic to a gluten protein called gliadin. This protein is found in wheat. If someone suffering from coeliac disease eats food containing gluten proteins, the lining of their small intestine becomes inflamed and the villi that should be present become much shorter or may even be absent.

For a person who knows they have coeliac disease, the only treatment is to have a gluten-free diet. The disease is now so common that many large supermarkets have special sections selling items suitable for gluten-free diets.

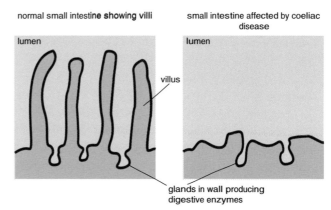

a) List some foods that contain wheat products.

b) Draw a diagram summarising the internal structure of a normal villus and explain how a villus is adapted for its roles in the alimentary canal.

c) People with coeliac disease usually lose weight and lack energy. Use the information in your diagram and your own understanding of the role of villi to give a biological explanation of these symptoms.

d) The incidence of diabetes in adults is rising along with the incidence of obesity. As indicated above, some supermarkets have special sections selling foods for particular diets. One example is diets for people with diabetes. What differences would you expect in a foodstuff intended for a diabetic and someone who does not have diabetes? How far might there be a link between diabetes and obesity? Suggest ways that modern supermarkets might be encouraging poor eating habits among their customers.

e) Another example of a controlled diet is for people with lactose intolerance. Find out the biological basis for lactose intolerance. In which regions of the world is lactose intolerance found? Which foods may (and may not) be eaten by people who are lactose intolerant and how do they ensure they have a balanced diet?

3 Look at passages 1, 2 and 3 and use your own knowledge to answer the questions that follow. You may need to use information from other sources to help you.

Passage 1

The crucian carp is a fish that can survive for many days entirely without oxygen. It lives in fresh water in cold areas where the water freezes over during the winter. However, the fish gets on with its life, but slowly, relying on anaerobic respiration to supply the energy needs of its muscles. Remarkably the muscles in the fish convert the lactic acid, produced as waste in anaerobic respiration, into ethanol. This ethanol rapidly diffuses from its gills into the water surrounding the fish. Crucian carp are known to grow more slowly when they are without dissolved oxygen in the water.

Passage 2

Another remarkable fish is the Antarctic ice fish, which lives in seawater at −1.5 °C. The water does not freeze because of the salt in it. These fish respire aerobically, like you and me, but they do not have any red blood cells to help carry oxygen around their body.

Passage 3

When an organ is removed from a person's body for transplantation into another person at a distant location, the organ is often transported to the recipient in a chilled container with ice.

a) In passage 1, why does the oxygen content of water covered in ice eventually get reduced to zero?

b) Why does the crucian carp move slowly during the winter?

c) Why is the growth rate of the crucian carp reduced in anaerobic conditions?

d) In passage 2, why does the water not freeze, even though the temperature is below zero?

e) How does the Antarctic ice fish get enough oxygen to its cells without the help of red blood cells?

f) Suggest why organs for transplant are transported in chilled containers.

g) The examples in the passages above all refer to respiration in cold situations. Find out about organisms living in other situations that are low in oxygen. For example, many invertebrate animals that live in the mud at the bottom of freshwater ponds, show adaptations that enable them to live there. Describe some of these adaptations or collect images to show them.

h) What other situations can you think of involving people going to areas with lower than normal oxygen? How do the people manage to obtain enough oxygen for respiration in these situations?

i) Check in other Sections of this book to remind yourself of situations in which living organisms survive (or suffer) in situations with low oxygen.

3
Movement of substances in living organisms

TO THINK ABOUT . . .

How are the needs of cells provided for – first in the exchange surfaces of living organisms, then by internal transport systems? As a start, make three lists under these headings:

1 things a cell needs to take in and what it must get rid of
2 exchange surfaces (in multicellular plants and animals) and the special features they may have
3 transport systems within the body, what they carry and how materials are kept moving along them.

Supplying the needs of cells

Living cells are in a state of dynamic activity, continually exchanging materials with their surroundings and responding to chemical signals. Nutrients and gases are taken in. Waste products are released.

All of these materials must cross the cell membrane – the partially permeable boundary of the cell. The surface area of the cell membrane sets an upper limit to the rate at which materials can cross. For single cells, the distance from the cell surface to the inside of the cell is short and materials can diffuse in and out efficiently.

But in large multicellular organisms the distances are too great for diffusion alone to meet their needs – for exchange of materials with the environment and transport to all the cells of the body. Multicellular organisms have exchange organs, such as lungs, gills, leaves and roots, which help exchange of materials with the environment outside the organism. Inside, there are mass transport systems and further adaptations that help with the exchange of materials where they are needed in the cells. The photographs show how a part of the human body is adapted to increase absorption of digested food in the gut. The villi (first and second images, shown in red and purple) already increase surface area massively. Yet these specialised cells that line the small intestine have tiny folds called microvilli (bottom image) and these densely packed folds increase the surface area available for absorption a further 1000 times. The microvilli are shown in blue and this image was obtained with an electron microscope. This helps give you an idea of the relative scale of these three images

3.1 Gas exchange in flowering plants

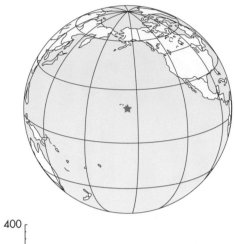

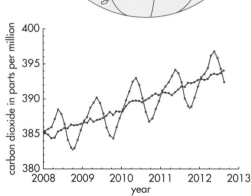

Monitoring the carbon dioxide concentration in the air

There is a weather station in Hawaii, marked with a green star on the globe. For just over 50 years the carbon dioxide concentration of the air has been monitored at this weather station. It is in a very useful position (globally) for this monitoring because the air moving around the northern and southern hemispheres swirl together here so that the readings provide an average for the world.

The graph shows recent data over a 5-year period. The monthly readings for carbon dioxide concentration are shown in red. You can see that they seem to fall and rise each year. In the northern hemisphere plants take up more carbon dioxide in the summer than they do in the winter. Can you think of reasons why plants do this? What other trend can you see in the data?

■ Processes that use and produce gases in plants

There are two important processes that occur in plants and which use and produce gases:

■ **respiration** – uses oxygen and gives off carbon dioxide. Respiration occurs all the time, in all the living cells in a plant

■ **photosynthesis** – uses carbon dioxide and gives off oxygen. Photosynthesis occurs only when it is light and only in plant cells that contain chloroplasts.

How the leaf is adapted for gas exchange

In Section 1.5 you study the features of gas exchange surfaces that allow diffusion of materials through the cell membrane (which is the boundary to the cell), between the cell and the outside atmosphere.

Now we look at how far the structure of a leaf helps gas exchange take place, between the leaf of a flowering plant and the surrounding air. We look first at the whole leaf and then consider details of its internal structure.

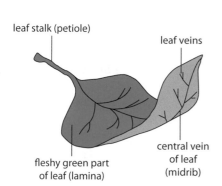

Figure 3.1 Features of a whole leaf.

leaf stalk (petiole)

leaf veins

central vein of leaf (midrib)

fleshy green part of leaf (lamina)

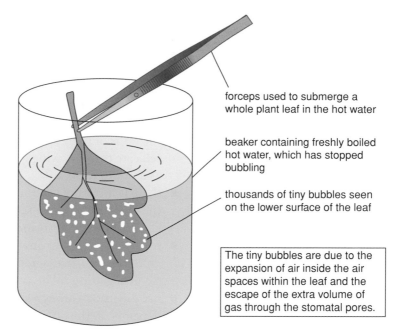

forceps used to submerge a whole plant leaf in the hot water

beaker containing freshly boiled hot water, which has stopped bubbling

thousands of tiny bubbles seen on the lower surface of the leaf

The tiny bubbles are due to the expansion of air inside the air spaces within the leaf and the escape of the extra volume of gas through the stomatal pores.

Figure 3.2 Dip a leaf into hot water and watch the air bubbling out of the stomata. This shows you where the stomata are on the lower surface of the leaf.

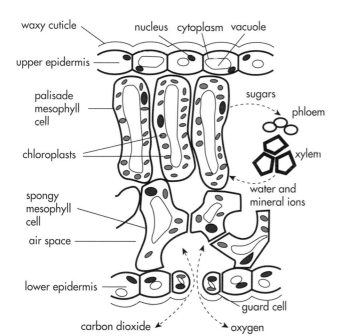

waxy cuticle
nucleus cytoplasm vacuole
upper epidermis
palisade mesophyll cell
chloroplasts
spongy mesophyll cell
air space
lower epidermis
guard cell
sugars
phloem
xylem
water and mineral ions
carbon dioxide
oxygen

Figure 3.3 Vertical section of a leaf showing internal structure. The arrows show movement of substances, as given. Note that the arrows passing through the stoma (between the guard cells) show the movement of carbon dioxide and oxygen, though the net movement of each gas varies at different times of day (or with different light intensities).

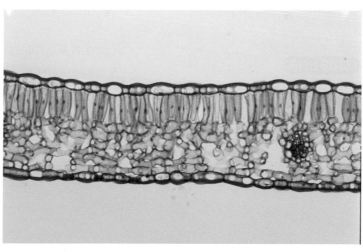

Figure 3.4 Vertical section of leaf of hellebore, as seen under a light microscope. Compare structures shown in this photomicrograph with the diagram in Figure 3.3.

Here are features important for gas exchange:

1 For the whole leaf (see Figures 3.1 and 3.2):
   - large flat surface exposed to the air, giving a large surface area in relation to volume
   - many stomata (holes or pores), usually more on the underside. When open, the stomata allow movement of gases into and out of the air spaces inside the leaf
   - leaf is thin, so the internal distance for movement (diffusion) of gases is relatively short.

2 For the internal structure, as seen in the vertical section of the leaf (Figure 3.3)

- many stomata (more details on page 89), which open in sunlight
- air spaces inside the stoma surrounded by loosely packed mesophyll cells and this is where gases diffuse into or out of the living cells
- palisade mesophyll cells, the site of most photosynthesis, are in close contact with the air spaces, thus allowing gases used and produced by photosynthesis to be exchanged with air in the air spaces
- the walls of both the spongy mesophyll and the palisade mesophyll cells are thin, allowing gases to pass through the cell membrane into and out of the cell
- there is moisture in the air inside the air spaces and the surfaces of the cells inside the air space are moist, so this helps gases dissolve and allows diffusion into and out of the cells
- respiration is going on in all of these cells, so their close contact with air in the air space is important for exchange of gases (obtaining oxygen, getting rid of carbon dioxide)
- net diffusion through stomata – which gases diffuse in which direction depends on the diffusion gradient for each gas. This in turn depends on the rates of respiration and photosynthesis, and on the intensity of light. The net diffusion of either gas (oxygen or carbon dioxide) can be:
   either into the leaf to the air spaces (then to the cells)
   or out of the leaf from the cells to the air space then out to the air outside
- the stomata must be open if any gas exchange is to take place.

Stomata and role of the guard cells

Each stoma (plural = stomata) is made up of two guard cells with a slight gap between them (Figures 3.5 and 3.6). The guard cells are unusual in that they contain chloroplasts whereas usually the cells on the epidermis do not contain chloroplasts. The guard cells can change in shape so that the gap (stomatal pore) can increase in size or close completely. When the guard cells take in water by osmosis, they become turgid and their shape is curved. This opens up a gap between them. When they are flaccid, the gap between them closes. The gap is the stoma and it is open in daylight (usually) and closed at night.

STUDY TIP

Check in Section 1.5 to make sure you understand the terms **turgid** and **flaccid**, **osmosis** and **active transport**.

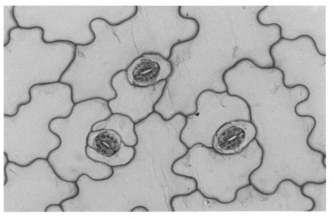

Figure 3.5 Surface view of lower epidermis of leaf of *Kalanchoe* showing stomata and guard cells. The photomicrograph was taken using a light microscope and the stomatal pores are about 20 μm long.

stomatal pore is open during the daytime because the guard cells are turgid – swollen with water absorbed from the neighbouring epidermal cells

stomatal pore is closed in the dark because the guard cells are flaccid – with water being drawn by osmosis into neighbouring epidermal cells. The sagging walls of the guard cells press together

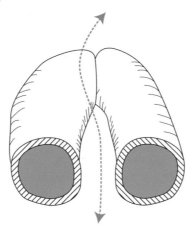
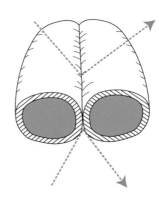

gases and water vapour can diffuse through the open stomatal pore

no gas exchange occurs when the stoma is closed

Figure 3.6 How a stoma opens and closes.

STUDY TIP

Make a link with Section 3.4 and the part played by stomata in the loss of water from a leaf by transpiration.

Events inside the air space

The balance of gases inside the air space is crucial for the direction of diffusion of gases – in other words, which gas is going in and which is going out, and when.

Let's start with respiration, which goes on all the time in living cells. During the night, photosynthesis does not occur and the stomata of the leaf are closed. The air trapped in the air spaces of the leaf contains about 20% oxygen. This is enough for the respiration of the mesophyll cells in the leaf overnight, while it is dark. During the night, the concentration of oxygen in the air spaces inside the leaf falls and the concentration of carbon dioxide rises.

Next we consider what happens during the early morning, when it is beginning to get light. Respiration continues but photosynthesis starts to occur. The same mesophyll cells now use the carbon dioxide in the air space and give out oxygen. The stomata now open and this allows diffusion of gases between the air outside the leaf and the air spaces inside the leaf. As the mesophyll cells take up more carbon dioxide, the concentration of carbon dioxide in the air space falls below that of the air outside (usually about 0.04%).

The net effect is that carbon dioxide diffuses in through the open stomata from the air outside into the air space. It is then available for the mesophyll cells to use for photosynthesis. At the same time, the oxygen produced is likely to be at a higher concentration inside the air space, so the net effect is for oxygen to diffuse out through the open stomata into the air outside. This continues as long as the light is bright enough for photosynthesis to occur.

Gas exchange in other parts of a flowering plant

In this section, the emphasis has been on the events in the leaf and its role in gas exchange. However, all living cells throughout the plant carry out respiration so require oxygen and need to get rid of waste carbon dioxide. Cells containing chloroplasts also carry out photosynthesis when the light intensity is appropriate, so these processes are linked to gas exchange.

We can look at an example to illustrate gas exchange in parts of the plant other than the leaf. In roots, gas exchange takes place through the root hairs – walls are thin and allow diffusion of oxygen into the root cells (from the air spaces in the soil) and carbon dioxide is given off. If soils are flooded the air spaces in the soil fill up with water and less oxygen is available to the plant roots.

■ Practical activity – experiment to investigate the effect of light on net gas exchange from a leaf, using hydrogencarbonate indicator

Hydrogencarbonate indicator changes colour depending on the pH of the solution. When more carbon dioxide is dissolved in the solution, it becomes more acidic. When less carbon dioxide is dissolved in the solution, it becomes more alkaline. The colour changes are shown in the chart in Figure 3.7.

1 Set up three boiling tubes (A, B and C) as shown in tube C in the diagram.
2 Each tube contains some hydrogencarbonate indicator solution that has been standing in air (or air has been bubbled through it). At the start, the colour of the indicator in each tube is orange, as shown in tube C.
3 Record the colour of the hydrogencarbonate indicator in tube C at the start.
4 Place a leaf in tube A and keep this tube in the light.
5 Place a leaf in tube B and cover this tube with dark material or place it in the dark.

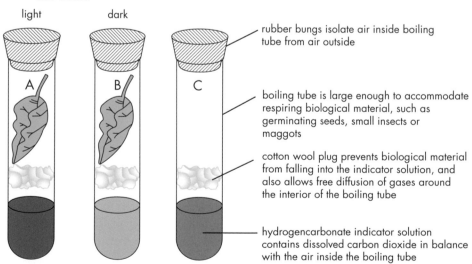

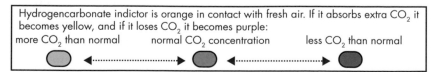

Figure 3.7 Experiment to investigate the effect of light on net gas exchange in a leaf, using hydrogencarbonate indicator. This diagram shows expected colours after about 30 minutes (see explanation on page 91).

6 After 30 minutes, observe the colours in tubes A and B.

After 30 minutes, the expected colours in tubes A and B are shown in Figure 3.7.

tube A – purple colour – this shows that there is less carbon dioxide in the tube, compared with tube C at the start. The carbon dioxide is used by the leaf as it carries out photosynthesis in the light. More carbon dioxide is taken into the leaf for photosynthesis than is released by respiration going on in the mesophyll and other cells in the leaf.

tube B – yellow colour – this shows that the tube contains more carbon dioxide than at the start. The leaf has given out carbon dioxide from respiration. The leaf does not carry out photosynthesis in the dark.

STUDY QUESTION

1 The statements in the table describe events in the leaf that are related to gas exchange.

a) Copy and complete the table. Find a suitable word (or words) from the box to match each description. Words may be used once, more than once or not at all.

> carbon dioxide; diffusion; oxygen; photosynthesis; respiration; stomata; transpiration

Statement or description	Word(s)
This process goes on in all living cells in the leaf all the time.	
These structures allow gases to pass into and out of the leaf, but they are closed at night.	
This gas is given out in bright light from cells containing chloroplasts.	
This process takes place in cells containing chloroplasts and produces oxygen.	
This gas is used when plants make sugars.	
This gas is given out in the process that releases energy from sugars.	

b) Give an explanation of how the stomata open and close in different light intensities.

c) List features of gas exchange surfaces that help them to be effective. Link these features to root hair cells and to the cells around the air spaces in a leaf.

d) At what time of day might there be no **net** diffusion into and out of a leaf? Give reason(s) to support your answer.

e) Hydrogencarbonate indicator can be used in experiments with water plants to compare levels of carbon dioxide in the water.

   i) Give the colours shown by this indicator when there is a lot of carbon dioxide in the water and when there is very little carbon dioxide in the water.

   ii) What colours would you expect if you had some pondweed in the hydrogencarbonate solution, first in the dark and secondly in bright light? Give reason(s) to support your answer.

3.2 Gas exchange in humans

Ventilation and hyperventilation

If you are shocked or scared by something and suffer a panic attack, you may end up fainting. While you are unconscious, the systems in your body restore normal functioning and you then recover.

One of the causes of fainting is breathing too hard and fast – this is called hyperventilation. Normally, of course, you do not think about how much air you process in your lungs, because the ventilation rate is controlled unconsciously. This ensures that just the right amount of oxygen enters your lungs and carbon dioxide is removed.

During hyperventilation, however, too much carbon dioxide is removed from your blood. This causes the pH of your blood to change. It is this disturbance of your blood chemistry that eventually causes your collapse. Once your breathing rate returns to normal, your blood pH is restored and you wake up.

One treatment for a panic attack is to ask the panicking person to breathe in and out of a paper bag. Can you suggest why this helps? And what happens in normal breathing to allow a supply of fresh air to reach your lungs and your body to get rid of unwanted carbon dioxide?

■ Processes that use and produce gases in humans

Human cells, like the cells of all living organisms, carry out respiration – the chemical process that releases energy, used for the activities of life. For the process of aerobic respiration, oxygen is required and carbon dioxide is produced. Section 3.2 describes how we take oxygen into the human body and how we get rid of carbon dioxide.

In Section 1.5 you have the chance to look at the way that size is important for exchange of materials between an organism and the surroundings. You should understand that the human body is far too large for the needs of all the body cells to be met by the simple diffusion of gases through the surface skin. Like all large multicellular organisms, the human body has a system that allows **exchange of gases** with the environment and a system that enables these gases (oxygen and carbon dioxide) to be **distributed** throughout the body.

> **STUDY TIP**
> Check that you know the equation for respiration (Section 2.3).

> **STUDY TIP**
> Refer to Section 1.5 to make sure you understand how size is important in relation to exchange of materials for living organisms.

■ The lungs and their role in gas exchange

The lungs, together with related structures in the thorax, provide the system for exchange of gases and distribution of them through the body in three ways:

- a large, much folded **gas exchange surface** (the alveoli in the lungs)
- a mechanism for **ventilation** (breathing), moving air in and out of the lungs, allowing the air to come into contact with the gas exchange surface
- close association of the gas exchange surface with the **blood system** to enable **distribution** of the gases around the body to and from cells.

The lungs, related structures and pathways for air

The following structures in the thorax (thoracic cavity, also known as the chest cavity) contribute to the breathing process (Figure 3.8):

- The lungs lie inside the **ribcage,** which, in addition to its function in breathing, provides physical protection to the lungs (and heart) lying inside the thoracic cavity.
- The **intercostal muscles** lie between each of the ribs. When these muscles contract, the ribs are lifted upwards and outwards. When the muscles relax, the ribs fall back to the previous position. The role of these muscles in ventilation is described below.
- The **diaphragm** lies below the lungs, forming the base of the thoracic cavity. The diaphragm is a sheet of muscle that separates the thoracic cavity from the abdominal cavity below. In its relaxed position it is dome-shaped, but when the muscles contract, the diaphragm is pulled flatter and this contributes to an increase in size of the thoracic cavity. The role of the diaphragm in ventilation is described below.
- The **pleural membranes** form a double layer, closely surrounding each lung. Between the layers there is a fluid. This allows smooth movement of the lungs as they inflate and deflate so that they can slide smoothly against the wall of the thoracic cavity.

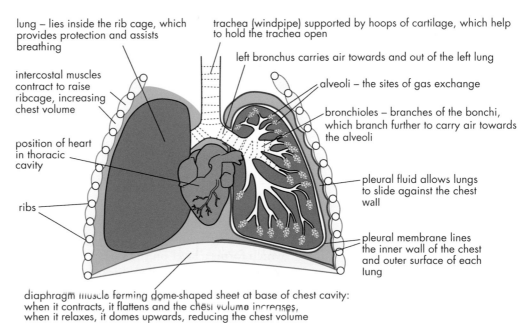

lung – lies inside the rib cage, which provides protection and assists breathing

intercostal muscles contract to raise ribcage, increasing chest volume

position of heart in thoracic cavity

ribs

trachea (windpipe) supported by hoops of cartilage, which help to hold the trachea open

left bronchus carries air towards and out of the left lung

alveoli – the sites of gas exchange

bronchioles – branches of the bonchi, which branch further to carry air towards the alveoli

pleural fluid allows lungs to slide against the chest wall

pleural membrane lines the inner wall of the chest and outer surface of each lung

diaphragm muscle forming dome-shaped sheet at base of chest cavity: when it contracts, it flattens and the chest volume increases, when it relaxes, it domes upwards, reducing the chest volume

Figure 3.8 Diagram of lungs and related structures in the thoracic (chest) cavity. The left-hand side of the diagram shows a whole lung but the surrounding ribs have been cut away. The right-hand side (left side of the body) shows structures inside the lung and some of their functions.

The heart also lies in the thoracic cavity close to the lungs but plays no part in the breathing process.

When air enters the body, it passes in through the **nose** (or mouth), where it becomes warmed and moist. The pathway for air continues down the **trachea** (windpipe), which divides into the left **bronchus** and right bronchus (plural = bronchi) when entering each lung. The trachea is strengthened by C-shaped **rings of cartilage**. These cartilage rings help to keep the airway open but are not too rigid and allow some flexibility when food passes down the oesophagus, which lies closely beside the trachea.

Inside the lungs, the bronchi divide into very many smaller branches (the **bronchioles**), which end in masses of little sacs, known as the **alveoli** (also described as 'air sacs'). The alveoli are surrounded by many blood capillaries, in close contact with the surface of the alveoli (Figure 3.9).

deoxygenated blood from pulmonary artery flows into alveolar capillaries

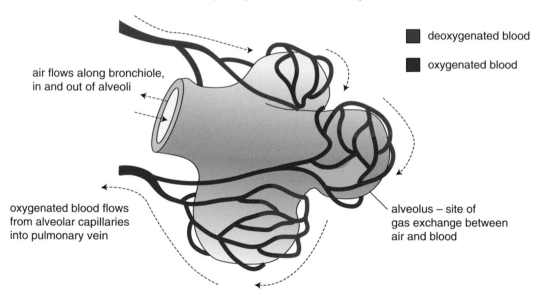

deoxygenated blood

oxygenated blood

air flows along bronchiole, in and out of alveoli

oxygenated blood flows from alveolar capillaries into pulmonary vein

alveolus – site of gas exchange between air and blood

Figure 3.9 Enlarged view of some alveoli with surrounding blood capiliaries. This is where gas exchange takes place.

Estimates vary, but if spread out, the total area of the alveoli is probably about that of a tennis court. This emphasises the enormous surface provided by the alveoli. This is the surface through which gas exchange takes place.

Along the surface lining of the bronchus and bronchioles, there are cells that have tiny hairs, known as **cilia**. Other cells in the lining produce **mucus**, which forms a layer on the surface. Particles of dust and bacteria become trapped in the mucus and this is kept moving by the beating action of the cilia. The mucus is moved up and out of the lungs, so effectively 'sweeps' the particles out of the lungs. This helps to clean the air entering the pathways in the lungs.

Air passes out from the alveoli in the lungs to the nose through the same route, in reverse.

Breathing and ventilation

Breathing is an active mechanical process – each breath brings air into the lungs and forces it out of the lungs again. This process **ventilates** the lungs, so that fresh air is brought into the alveoli and then the stale air is breathed out again. We use these terms to describe breathing in and out:

- **inhalation** (or inspiration) = breathing in
- **exhalation** (or expiration) = breathing out.

The mechanism for breathing in and out is a direct consequence of changes in the volume of the thoracic cavity. When the volume is increased, the pressure inside the lungs is reduced and becomes lower than that of the outside atmosphere. The result is that air is forced into the lungs to equalise the pressure. This makes the lungs inflate (fill up with air). Conversely, when the volume of the thoracic cavity is reduced, the pressure inside the lungs is then higher than that of the outside atmosphere, so air is forced out (breathing out) and the lungs deflate.

The changes in volume are brought about by the movement of the ribs and the diaphragm (see Figure 3.10).

For inhalation (breathing in):

- The **intercostal muscles** contract. This raises the ribcage up and outwards, thus increasing the volume inside.
- The **muscles of the diaphragm** contract. This flattens the diaphragm, thus increasing the volume inside.

For exhalation (breathing out):

- The **intercostal muscles** relax. This allows the ribcage to drop down and inwards, thus decreasing the volume inside.
- The **muscles of the diaphragm** relax. This means the diaphragm resumes its dome shape, thus decreasing the volume inside.

To summarise – for inhalation, the air is drawn in whereas for exhalation, the air is forced out.

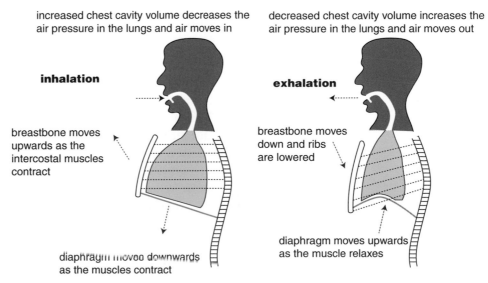

increased chest cavity volume decreases the air pressure in the lungs and air moves in

decreased chest cavity volume increases the air pressure in the lungs and air moves out

inhalation

exhalation

breastbone moves upwards as the intercostal muscles contract

breastbone moves down and ribs are lowered

diaphragm moves downwards as the muscles contract

diaphragm moves upwards as the muscle relaxes

Figure 3.10 Diagram to show how movements of the ribs and diaphragm are involved in inhalation (breathing in) and exhalation (breathing out).

Figure 3.11, showing the bell jar with balloons, gives a useful model to help you understand what is happening in the chest cavity.

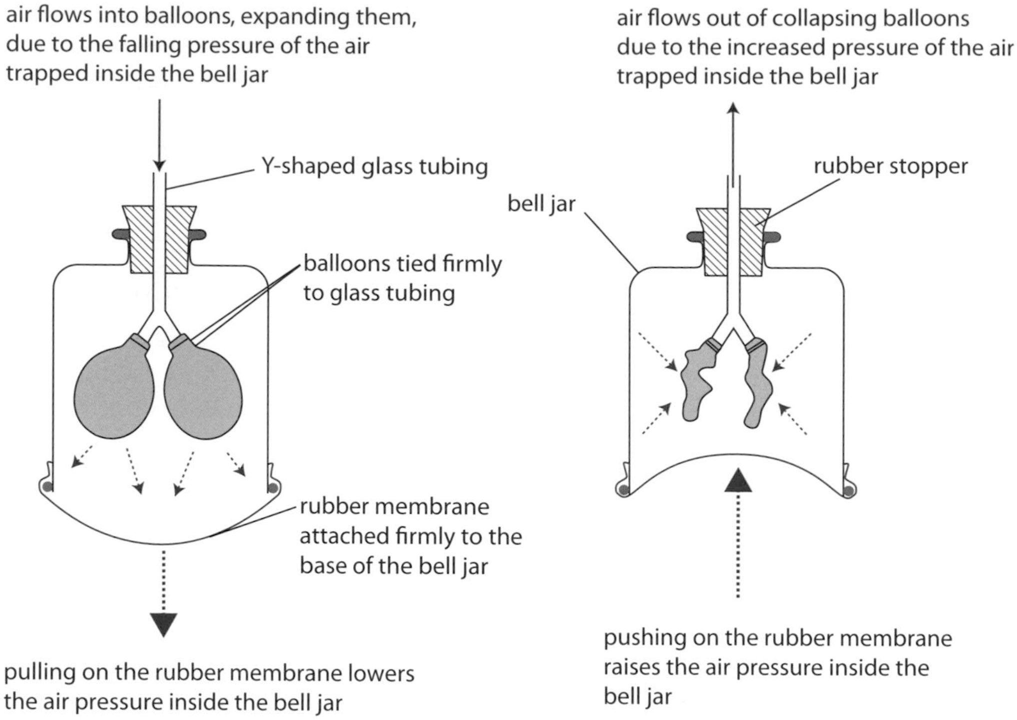

air flows into balloons, expanding them, due to the falling pressure of the air trapped inside the bell jar

air flows out of collapsing balloons due to the increased pressure of the air trapped inside the bell jar

Y-shaped glass tubing

rubber stopper

bell jar

balloons tied firmly to glass tubing

rubber membrane attached firmly to the base of the bell jar

pulling on the rubber membrane lowers the air pressure inside the bell jar

pushing on the rubber membrane raises the air pressure inside the bell jar

Figure 3.11 Model to show how movements of the diaphragm lead to changes in pressure that result in the balloons inflating or deflating. The bell jar represents the chest cavity, the balloons represent the lungs and the rubber membrane represents the diaphragm. (The ribs are not included in this model.)

The alveoli as a surface for gas exchange

In Section 1.5 you study the features of gas exchange surfaces that allow **diffusion** of materials through the cell membrane (which is the boundary to the cell), between the cell and the outside atmosphere.

Features of the alveoli in the lungs, and the associated blood capillaries, illustrate how these provide a suitable surface for exchange of gases for the requirements of the human body (Figure 3.12):

- the alveoli provide a very **large surface area**
- the alveoli walls are **thin**
- the air inside the lungs is **moist** and the alveoli have a film of liquid on their surface
- the distance between the alveoli and the blood **capillaries** is very short
- **ventilation** of the lungs during breathing helps to maintain the **concentration gradient**, so that fresh oxygen is brought into the lungs in the air we breathe in and discarded carbon dioxide is removed from the lungs in the air we breathe out
- the walls of the blood capillaries are also **thin** and the capillaries are very narrow so the distance for gases to diffuse is short
- the blood moves through the capillaries so this helps maintain a **concentration gradient** – the freshly absorbed oxygen is transported away in red blood cells and supplies of unwanted carbon dioxide are continually being brought to the alveoli surface.

STUDY TIP

You can find details about features of gas exchange surfaces in Section 1.5. Remember that they need to be thin, moist and have a means of maintaining a diffusion gradient. Make sure you apply this to the topic in this section on gas exchange in the human lung.

Figure 3.12 shows how gas exchange takes place between the air in the alveolus and gases brought in the blood in the capillaries surrounding the alveolus.

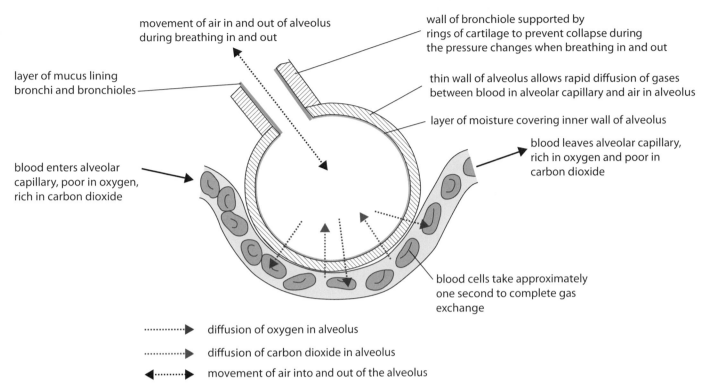

movement of air in and out of alveolus during breathing in and out

wall of bronchiole supported by rings of cartilage to prevent collapse during the pressure changes when breathing in and out

layer of mucus lining bronchi and bronchioles

thin wall of alveolus allows rapid diffusion of gases between blood in alveolar capillary and air in alveolus

layer of moisture covering inner wall of alveolus

blood enters alveolar capillary, poor in oxygen, rich in carbon dioxide

blood leaves alveolar capillary, rich in oxygen and poor in carbon dioxide

blood cells take approximately one second to complete gas exchange

⋯⋯▶ diffusion of oxygen in alveolus

⋯⋯▶ diffusion of carbon dioxide in alveolus

◀⋯⋯▶ movement of air into and out of the alveolus

Figure 3.12 The diagram illustrates the events that occur during gas exchange between air in the alveolus and blood in the surrounding blood capillaries.

STUDY TIP

It would be helpful to refer to details on blood circulation and transport of oxygen and carbon dioxide in the blood in Sections 3.5 and 3.6.

MATHS TIP

Orders of magnitude

An order of magnitude is a measure of a quantity, in powers of ten-fold. For example, 10000 is 3 orders of magnitude greater than 10.

Now go to page 128 to apply this to figures relating to carbon dioxide concentrations in different samples of air.

Oxygen in the fresh air brought into the lungs is at a higher concentration than in the blood, so diffuses from the alveolus into the blood. **Carbon dioxide** in the blood is at a higher concentration than in the air in the alveolus, so diffuses from the blood into the alveolus. It then passes out of the body in exhaled air.

Table 3.1 shows the approximate composition of inhaled and exhaled air. Compared with inhaled air, you can see that exhaled air has a higher percentage of carbon dioxide and of water vapour, but a lower percentage of oxygen. The bulk of the air is nitrogen gas but this is not used in any way by the body.

Table 3.1 Composition of inhaled and exhaled air

Gas	Inhaled air	Exhaled air
nitrogen	79%	80%
oxygen	21%	16%
carbon dioxide	0.04%	4%
water vapour	variable	saturated

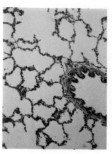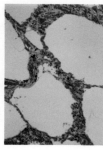

Figure 3.13 A section through alveoli in the lungs, as seen with a light microscope. The alveoli on the left are from a normal lung and show lots of small alveoli. The photomicrograph on the right shows similar tissue in the lung of a person with emphysema. The walls of many alveoli have broken down, leaving large (but fewer) air spaces. This reduces the surface area available for gas exchange.

STUDY TIP

Here you can make several links to other sections – look at information about blood and circulation in Sections 3.5 and 3.6; the importance of a large surface area for exchange of materials in Section 3.3 and carbon monoxide as an air pollutant in Section 6.4.

PRACTICAL

This experiment allows you to measure breathing rate before and after exercise. It helps you to understand the need for sufficient supplies of oxygen for cells in the body. The investigation gives a useful opportunity to collect together class results and decide how to organise these.

Smoking and its effects on the lungs and circulatory system

The harmful effects of smoking are well established. Increasing publicity has led to smoking bans in many public places in an attempt to discourage people from smoking and to protect non-smokers from the effects of 'passive' smoking.

There is plenty of evidence that smoking increases the risk of certain diseases, such as cancers, particularly lung cancer. There is also evidence that smokers suffer more ill-health in various ways compared with non-smokers. These include increased and persistent coughs, bronchitis and other lung disorders.

Here we look at some specific ways that cigarette smoke affects the lungs and circulatory system:

- **Nicotine** and **carbon monoxide** in tobacco smoke raise the heart rate and lead to higher blood pressure. These two conditions increase the risk of heart attacks or strokes.
- **Nicotine** and **carbon monoxide** also reduce the ability of the blood to carry oxygen. The carbon monoxide combines with haemoglobin so less is available to carry oxygen as oxyhaemoglobin around the body in red blood cells. This leads to the person being breathless, particularly when undertaking physical activity.
- **Tar** in the smoke damages the cilia (tiny hairs) on the surface of the bronchi and bronchioles. As described above (page 94), these cilia help to keep the mucus layer moving and help to keep the air in the lungs clean. When the cilia are damaged, there is a build up of mucus that is not removed. Bacteria that are normally removed with the mucus stay in the lungs. This leads to a greater risk of infections. Some chemicals in tar are carcinogens because they cause mutations in lung cells.

Other substances in the smoke make the alveoli less elastic and sometimes the walls begin to break down. One very large air sac has a much smaller surface area than many small ones. This reduction in surface area means the person does not obtain enough oxygen and is likely to be breathless. This condition is called emphysema and is a common condition among smokers.

■ Practical activity – investigating the effect of exercise on breathing rate

Students work in pairs for this practical investigation. Student A does the exercise and Student B records the results. They can then reverse roles and repeat the activity.

1 Student A breathes as normally as possible and raises a finger as each breath cycle begins.
2 Student B records the breathing rate (of student A) in breaths per minute, before exercise.

Figure 3.14 Investigating the effect of exercise on breathing rate.

3 Student A then carries out suitable exercise, such as running a lap of a playing field or stepping up and down a step once a second, and continues this exercise for 3 minutes.
4 Student B records the total number of breaths taken in each minute following the completion of the exercise, and regularly over the next 5 minutes.
5 The rate of breathing for each minute following the exercise is plotted on a graph.

■ Practical activity – investigating the release of carbon dioxide in breathing

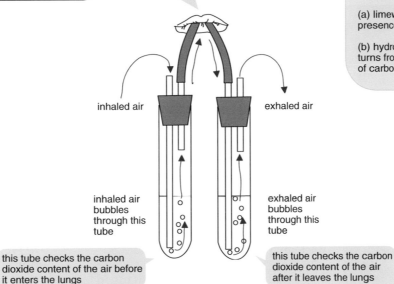

both rubber tubes are placed in mouth and *very gently* breathed through – each tube will alternately bubble as the air passes through them on route to and from the lungs

both test tubes need to be filled with an indicator for carbon dioxide:

(a) limewater – which turns milky in the presence of carbon dioxide

(b) hydrogencarbonate indicator – which turns from orange to yellow in the presence of carbon dioxide

inhaled air

exhaled air

inhaled air bubbles through this tube

exhaled air bubbles through this tube

this tube checks the carbon dioxide content of the air before it enters the lungs

this tube checks the carbon dioxide content of the air after it leaves the lungs

Figure 3.15 This apparatus can be used to show the difference in carbon dioxide content of inhaled and exhaled air.

STUDY QUESTIONS

1 Make sure you are clear about the difference between the processes of **respiration**, **breathing** and **gas exchange**.
   a) In your own words, write a definition for each term and write it in a way that you could use to explain to someone who is not studying science. Say where each process takes place in the human body.
   b) For breathing and gas exchange, write down details of how each of these processes takes place in the human body. Draw simple diagrams to help in your explanation.
2 List features of the alveoli and related structures that show how they are adapted for gas exchange.

3.3 Transport in living organisms

Why does size matter?

The amoeba is a single-celled organism. It is very small and is found crawling across surfaces in ponds where it feeds on bacteria and algae cells. Even though it is large compared to a human cheek cell, it exchanges materials successfully with the pond water that surrounds it.

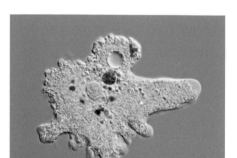

Volvox is made up of a colony of several thousand cells and it is green. A volvox is much bigger that an amoeba, but the individual cells in the colony all exchange materials directly with the pond water, as though each is a single-celled organism.

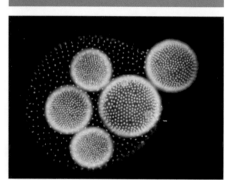

The sea anemone grows much larger than a volvox. It is big enough to kill and digest a whole fish. The walls and tentacles of the anemone are made up of millions of cells arranged in very thin sheets of tissue. Each cell in the sheet is close to the sea water and exchanges materials directly with it.

The fish is much larger than any of these organisms. So how does a fish obtain the materials it needs for its survival and distribute them to the cells inside it? In Section 3.3 we look at the specialised exchange surfaces and transport systems that the fish and other large multicellular organisms depend upon.

▮ Requirements of cells within organisms

Cells need to obtain materials from their surroundings so that they can carry out the processes of living organisms. They also need to get rid of waste materials to their surroundings.

Some examples of materials **exchanged** between a cell and its surroundings include:

- oxygen gas – taken in, for respiration (nearly all living organisms)
- carbon dioxide gas – passed out, from respiration (nearly all living organisms)
- carbon dioxide gas – taken in, for photosynthesis (green plants)
- oxygen gas – passed out, from photosynthesis (green plants)
- 'food' and other nutrients including mineral ions – taken in (all living organisms, but different ways of nutrition in plants and animals)
- waste materials – passed out (nearly all living organisms)
- water – taken in and passed out, depending on requirements (all living organisms).

A cell is enclosed by its cell membrane. Materials have to pass through this boundary when they enter or leave a cell. Sometimes the term 'surroundings' refers to the environment outside the body of the organism. Sometimes the term refers to the fluid or other material surrounding each cell inside the organism.

STUDY TIP

You need to link all the statements in this list with other sections in the book. Probably the best way is to make the links through Section 1.1 to make sure you are clear about each of the processes listed and can find the relevant references.

Inside whole organisms, there needs to be a system for distribution of these materials to different parts of the cell or to cells in different parts of the organism.

We now consider how materials are exchanged and transported, first in unicellular organisms, then in multicellular organisms.

Single-celled organisms

Yeast, *Amoeba* and *Chlorella* are all examples of single-celled (unicellular) organisms. Yeast is a fungus, *Amoeba* and *Chlorella* are protoctists. Each of these organisms is less than 0.02 mm in diameter. In fact they are so small that you cannot see individual cells with your naked eye and need a microscope to see details of their structure.

STUDY TIP

Check the names and make sure you know what a protoctist is by looking in Section 1.2 (Variety of living organisms).

STUDY TIP

Look in Section 7.2 for information about the use of yeast in making bread.

STUDY TIP

Look in Section 1.5 to make sure you understand the importance of surface area to volume ratios when considering diffusion of materials into and out of a structure.

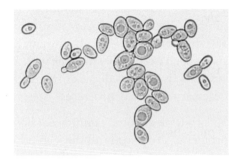

Figure 3.16 Yeast cells, seen with a light microscope. Some cells are budding (dividing asexually).

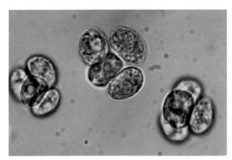

Figure 3.17 Groups of *Chlorella* cells, seen with a light microscope.

Their small size means that it is a very short distance from the outside of these single-celled organisms to the cytoplasm inside the cell where metabolic reactions are taking place. The rate of diffusion over this short distance is sufficient to supply materials as required from the surroundings and for waste materials to be lost through the cell membrane.

In addition to their small size, these single-celled organisms often have a large surface area in relation to their volume. This helps the organism exchange materials with its surroundings.

Inside the one cell of single-celled organisms distances are again short, so diffusion is adequate for movement of substances inside the cell.

Multicellular organisms

Multicellular organisms can grow to a much larger size than a single cell. In addition, certain groups of cells may specialise and are able to carry out a particular function. As examples, you can see a range of different structures in a maize plant, a palm tree, a human being, a camel or a fish. The increased complexity of the multicellular body means that it can do things that would be impossible for a single cell.

However, larger size and greater complexity increase the problems of how materials can be exchanged with the surroundings and how they can be distributed inside the organism. Dependence on diffusion alone, from cell to cell, would be far too slow. We can see how large multicellular organisms have overcome these problems by having specialised exchange surfaces.

Figure 3.18 Tip of a leaf of the liverwort *Pellia* (a simple plant), seen with a light microscope. The leaves are very thin, consisting of one layer of cells. Their simple internal structure means that materials move over very short distances.

Figure 3.19 A Candy stripe flatworm – a simple multicellular organism that relies on diffusion for transport over the short distances inside its body. The flattened body may have a length of about 12 mm and width of 5 mm.

They also have transport systems for distribution of materials inside the organism.

Some multicellular organisms do not have an internal transport system. This imposes limits on their size and on where they are able to live so that they can obtain the materials they require. Examples are a moss plant, liverwort and a flatworm (Figures 3.18 and 3.19).

As an example, a flatworm may have a body length of 12 mm and breadth of 5 mm, and the whole body is flattened so that the flatworm body is very thin. Even at this small size, the gut inside the body has many branches that reach into all parts of the body. This shape means the cells of the animal are all only a close distance from materials they need (from the surroundings and inside the organism) and diffusion is sufficient for distribution of required materials.

This size of flatworms is about the limit for an organism without an internal transport system. Diffusion alone would not work inside a maize plant, a bean plant, a camel or the human body. Here we see a combination of specialised **exchange surfaces** with a large surface area and **transport systems** that allow all cells in these larger organisms to obtain the materials necessary for living processes.

So let's look at some examples of how large organisms overcome these problems. (You study all these in other sections of the book, so check the study tips to find out more details.)

Exchange surfaces and transport systems in plants

- The **root hairs** of young roots provide an enormous **surface area**, which allows the entry of water and mineral ions.
- The **xylem** (tubes) provide a **transport system** that carries the water and mineral ions up the plant, to leaves or other parts where they are required.
- The **leaves**, with their many stomata, provide a large **surface area** through which gas exchange takes place (carbon dioxide comes in and is used in photosynthesis; oxygen comes in and is used in respiration; oxygen passes out as a waste product from photosynthesis; carbon dioxide passes out as a waste product from respiration in cells).

Exchange surfaces and transport systems in humans

- The **alveoli** in the lungs provide an enormous **surface area**, which allows exchange of gases.
- Oxygen diffuses into the blood and is then used in respiration by cells elsewhere in the body.
- Carbon dioxide (a waste product from respiration in cells) diffuses out into the air in the alveoli. (Note that if the alveoli were unravelled and flattened out, their total area would be about the size of a tennis court, and that is all fitted inside your lungs.)
- Inside the **small intestine** the **villi** provide an enormous **surface area**, which allows digested food to be absorbed into the body.
- Inside the body, the **blood circulation system** provides a **transport system** that carries materials to all cells of the body – for example, oxygen

STUDY TIP

It is worth making links to several other parts of the book so you understand the examples given. For root hairs and xylem, look in Section 3.4; for leaves and gas exchange, look in Sections 2.1 and 3.1; for alveoli in the lungs, look in Section 3.2; for the small intestine and villi, look in Section 2.2. Finally, don't forget transport in the blood circulation system in humans (Section 3.6).

STUDY TIP

It is very important that you understand about diffusion and concentration gradient, so refer to Section 1.5 for more details.

is carried from where it is absorbed from the alveoli in the lungs to muscle cells, and glucose is carried from the small intestine or liver to the muscle cells.

In the examples on page 102, diffusion is (usually) the mechanism of transfer across the surfaces where the exchanges take place. These **exchange surfaces** allow more of the materials to get into and out of the multicellular organism from the surroundings. They share some common features that help increase the rate of diffusion across them:
- they are thin – this reduces the distance across which diffusion occurs
- they are moist – this allows gases or other materials to dissolve into the water at the exchange surface, then diffuse through the surface.

In addition, a transport system close to the exchange surface removes absorbed materials quickly so that more materials diffuse across the surface. This helps to maintain a **concentration gradient** and ensure there is a difference in concentration of the substance each side of the exchange surface.

Transport systems help the materials in the organism to get closer to the cells where they are needed. Without transport systems, the large multicellular organisms with which we are familiar, could not exist.

STUDY QUESTION

1 The parts A, B and C below list some pairs of organisms or parts of organisms. You are asked to compare these organisms in relation to the way materials are exchanged with the surroundings or how materials are transported inside the organism. Sometimes they show similarities but sometimes they show differences.

To answer the questions, you may need to refer to other sections in the book, but you can use the **Study tips** to help you find the information. There are diagrams in this section and other parts of the book to help you visualise both organisms in each pair.

A A leaf and a flatworm
  a) Compare their shapes and explain how their shape helps with exchange of materials with the environment outside the organism.
  b) What gas(es) does the leaf take in? Name the process (or processes) that uses the gas(es).
  c) What gas(es) does the leaf give out? Name the process (or processes) that produces the gas(es).
  d) What gas(es) does the flatworm take in? Name the process that uses the gas(es).

  e) What gas(es) does the flatworm give out? Name the process that produces the gas(es).
B *Chlorella* and an oak tree
  These are both green organisms and carry out photosynthesis, yet they are very different in size. Describe and explain how each obtains the simple materials (carbon dioxide and water) required for photosynthesis. In your description, include reference to how the materials are transported inside the organism.
C A green flowering plant (such as a pea plant) and a human
  a) List the materials each of these organisms needs to carry out respiration in their cells. What waste materials are produced from respiration?
  b) Compare the way the pea plant and the human obtain the materials needed for respiration in their cells. In your comparison, include reference to how the materials are transported inside the organism.

3.4 Transport in flowering plants

Up to the top of the tallest tree

The tallest known tree in the world is a giant redwood tree, about 115 m tall. Scientists, like those shown in the photograph, have been able to climb to the top of the tree and look at the leaves. They found that the leaves at the top are different to those lower down – they are much smaller with fewer stomata for gas exchange.

The photographs emphasise the enormous height to the top of these trees. The force required to get the water to the leaves at the top is extreme.

Scientists suggest that the reason there have *never* been trees taller than these redwoods is because of the difficulty of supplying water to the top leaves. So how does water get to this height in a plant? Where does the water enter the plant and what is the role of leaves in transporting water to the very top of the tree?

■ Movement of water in plants

To start thinking about how water moves in a plant, you can do three simple things.

1 Take a small plant with its roots (say a pea or bean plant, or a small weed growing outside). Clean the roots gently by washing in water. Then place the plant in a beaker of water to which some red dye or food colouring has been added. You should soon see the red colouring spread up the plant and into the veins of the leaves.
2 Take the stalk of celery (or similar plant) and again place it in water with some red dye. When the red colour has moved up the stalk, make a cut across the stalk and you will see the red as a series of dots around the stalk (Figure 3.20). You can also do this with a white flower, such as a carnation, and see how the red colour travels up the stem and spreads out to give pinkish petals in the flower.
3 Place a transparent plastic bag over part of a plant shoot (or whole plant in a pot), in the sunlight. Tie the base of the bag to close it around the shoot. Watch and you will later see drops of a liquid collecting in the bag.

These three observations help establish the overall route for movement of water in a flowering plant – from the roots where water is taken in, up to the leaves where water is lost from the plant. We now look in more detail at the pathway taken by water and the mechanisms involved that help it travel through the plant, sometimes being lifted to enormous heights (as shown in the photograph of the giant redwood tree).

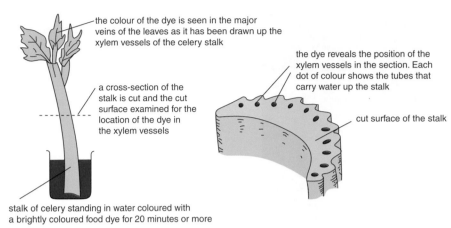

the colour of the dye is seen in the major veins of the leaves as it has been drawn up the xylem vessels of the celery stalk

a cross-section of the stalk is cut and the cut surface examined for the location of the dye in the xylem vessels

the dye reveals the position of the xylem vessels in the section. Each dot of colour shows the tubes that carry water up the stalk

cut surface of the stalk

stalk of celery standing in water coloured with a brightly coloured food dye for 20 minutes or more

Figure 3.20 Observing the movement of red dye in a celery stalk.

Absorption of water and mineral ions (salts) by root hair cells

Water in the soil surrounds the root hairs that project from the surface of plant roots (see Figure 3.21). The root hairs increase the surface area of the root in contact with the soil water, which helps the water to pass into the root by osmosis.

The cytoplasm inside the root hair cell contains lots of dissolved molecules (including sugars and some salts). This means the root hair cell has less water in it (less watery) than the soil water. The soil water passes into the root hair cell by **osmosis** and the cell membrane of the root hair cell acts as the partially permeable membrane. (The term 'selectively permeable membrane' is also used.)

STUDY TIP
You can refer to Section 1.5 for details on osmosis and Section 3.3 for general information about transport in living organisms.

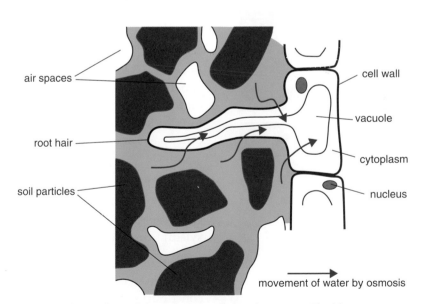

air spaces

cell wall

vacuole

root hair

cytoplasm

soil particles

nucleus

movement of water by osmosis

Figure 3.21 Root hair cell in soil showing entry of water by osmosis. The blue represents water around the soil particles and root surface. The blue arrows represent the movement of water (by osmosis) into the root hair cell.

Figure 3.22 Root hairs on a the root of a growing radish seedling.

Water passes across the root, from cell to cell, until it reaches the 'cells' making up the **xylem** in the centre of the root. These are the tubes that you may have seen as a ring of red dots when you cut across the celery stalk in red dye (see above).

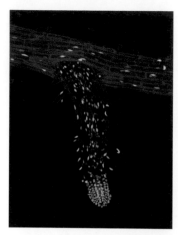

Figure 3.23 3D image of a root of thale cress (*Arabidopsis thaliana*) showing lateral root and root hairs. The diameter of the main root is approximately 150 µm.

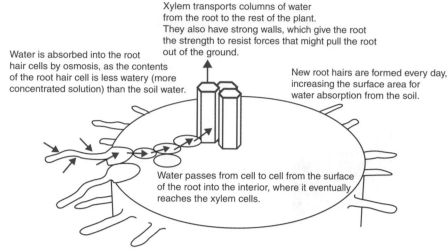

Xylem transports columns of water from the root to the rest of the plant. They also have strong walls, which give the root the strength to resist forces that might pull the root out of the ground.

Water is absorbed into the root hair cells by osmosis, as the contents of the root hair cell is less watery (more concentrated solution) than the soil water.

New root hairs are formed every day, increasing the surface area for water absorption from the soil.

Water passes from cell to cell from the surface of the root into the interior, where it eventually reaches the xylem cells.

Figure 3.24 Passage of water across cells of root from a root hair cell to the xylem.

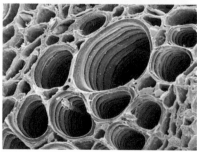

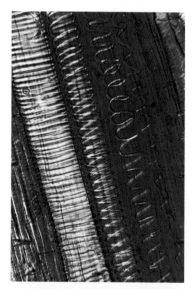

Figure 3.25 The top image is a photomicrograph of 'cells' of xylem, as seen with a light microscope. It shows rings and spirals of lignin thickening along the length of some xylem tissue. The lower image is of higher magnification, taken with an electron microscope. It gives a view across a cut surface of some xylem cells, looking inside the cells and showing the lignin thickening in the walls.

Xylem and the transport of water in the plant

The **xylem** is made up of several types of cells. One type includes dead cells and many of the end walls have broken down so that they form long vessels, with a very narrow diameter. These long xylem vessels run through the plant, from the roots, up the stems and spread out as the veins of the leaves. The walls of the xylem cells are strengthened with a material called **lignin**. Many of these xylem cells lie together and they are very strong and hard. This is what forms the wood of a tree.

As we saw in the observations of the celery stalk in Figure 3.20, water travels up the plant in these xylem 'tubes'. At this point, to help understand how water is drawn up (in the xylem) from the cells in the root, we can first consider what is happening in the leaves.

In Section 3.1, you study the structure of a leaf and that it has stomata (small holes) in the surfaces of the leaf. The stomata allow gas exchange to take place but, at the same time, water vapour is lost through the stomata into the surrounding atmosphere. This loss of water from a leaf (and from other parts of a plant above the ground) is known as **transpiration**.

As water is lost from the leaf, more is drawn from the xylem in the veins of the leaf. The water in the xylem is really a long, narrow unbroken column – or in fact lots of long unbroken columns of water, lying side by side in the various parallel xylem vessels. These columns of water run unbroken (in the xylem) from the roots to the leaves.

So, as water in the leaves transpires into the atmosphere, more is drawn up the xylem to take its place. This **transpiration pull** makes an important contribution to drawing water up through the plant, in the xylem.

Some other features of the xylem and properties of the water molecules help maintain this continuous column of water that is pulled up the plant:

- **Capillarity** – you have probably observed the way water can rise up a very narrow capillary tube, above the level of water the capillary tube is standing in. The xylem cells also have a very narrow diameter, so capillarity makes a contribution to water being lifted up the xylem in the stem.

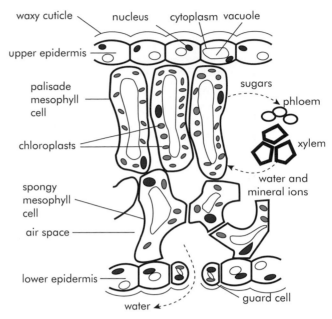

Figure 3.27 Vertical section of a leaf showing internal structure. In this diagram, the arrows emphasise the movement of water, from the xylem into cells, then into air spaces and out through the stoma between the guard cells. Another arrow shows how sugars from palisade cells move into the phloem (and are then transported to other parts of the plant).

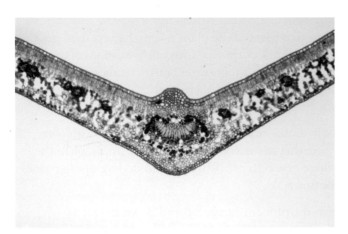

Figure 3.26 Section through a leaf of a plum tree. This photomicrograph passes through the midrib (vein along the centre of a leaf). A network of veins (containing the xylem) in the leaf allows water to reach all cells. Sugars and other substances are transported in the phloem, alongside the xylem. Other cells shown can be identified by comparison with the diagram in Figure 3.27.

- **Cohesion** – individual water molecules are linked to their neighbours by bonds between the water molecules. This means they all pull one another in an unbroken column – something that forms the basis of transpiration pull.

To summarise, we can see that a number of mechanisms contribute to the movement of water through a plant from the soil and out through the stomata of the leaves. This movement comes from a combination of pressure from the root (where water enters by osmosis) pushing it up and pulling from the leaves as a result of transpiration. The whole column of water in the xylem is held together by capillarity and cohesion.

Mineral ions (salts) are also present in the soil water. They dissolve in the water, forming a dilute solution of ions. If the concentration of ions in the soil water is higher than in the root hair cells, then ions can pass through the membrane into the root hair cell by **diffusion** down the concentration gradient.

In many situations, however, a plant needs to move ions into a root cell that already contains a higher concentration than the soil water – that is, against the concentration gradient. To do this they need energy, which is supplied from respiration. This form of transport is known as **active transport**.

Mineral ions (salts) move up the plant, dissolved in the water columns that are pulled up inside the xylem tubes.

Phloem and the transport of sugars and amino acids in the plant

Running closely alongside the xylem is another set of 'tubes', known as the **phloem**. The phloem is made up of several types of cells and these are living cells. The end walls of the cells are present, so the phloem tubes are not hollow (whereas most xylem vessels are hollow).

STUDY TIP

Make a link with Section 6.3 to understand the role of transpiration in the water cycle.

STUDY TIP

Look in Section 1.5 for details on diffusion and active transport.

Phloem transports carbohydrates, in the form of sucrose. It also transports amino acids and some other large organic molecules. Carbohydrates (first as glucose) are synthesised in leaves, in the process of photosynthesis. Some of these carbohydrates are transported away from the leaves, to other parts of the plant.

These carbohydrates may go to parts of the plant where they are used in different ways. They may be stored, converted into other substances or used in respiration. As examples, a potato tuber stores starch, fruits such as grapes and bananas contain sugars, seeds (grains) such as rice, wheat and maize contain starch and many flowers produce nectar, which contains sugars.

In addition to the carbohydrates, **amino acids** are carried in the phloem to parts of the plant where they are used. This may, for example, be in the synthesis of proteins where new cells are being formed in developing tissues (such as young leaves or flowers).

Transport of materials in the phloem can be up and down the plant depending on where the materials are required (unlike xylem, which carries water up only).

■ Factors that affect the rate of transpiration

Look at the internal structures of a leaf in Figure 3.27 and consider them now in relation to loss of water by transpiration. Water vapour in the air spaces just inside the stomata passes through the stomata, by diffusion, into the air outside the leaf. From there it evaporates and water molecules move away from the leaf. Note that relatively little water passes through the waxy cuticle, so most water is lost through the stomata, when they are open.

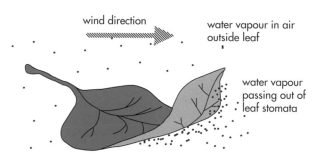

- The greater the wind speed, the more quickly transpired water moves away from the leaf, and the greater the rate of transpiration.
- The higher the temperature, the greater the rate of water evaporation and transpiration from the leaf.
- The lower the air humidity, the greater the diffusion gradient between the leaf and the outside air and, consequently, the rate of diffusion.
- Stomata are closed at night, which reduces transpiration to a minimum.

Figure 3.28 Factors affecting the rate of transpiration from a leaf.

Various factors affect how fast the molecules of water vapour evaporate and move away from the leaf into the air, hence the rate of transpiration:

- **Humidity** – water molecules diffuse down the concentration gradient, from the humid leaf air spaces into the air outside the leaf. If the air outside the leaf is very humid (moist), the rate of transpiration may be very slow, or does not occur at all. If the air outside is very dry (low humidity), evaporation occurs readily and the rate of transpiration is higher.

MATHS TIP

Areas of triangles and rectangles

Area / units² of any triangle = height × ½ base length

Area / units² of any rectangle = length × width

Now go to page 128 to apply this to calculations relating to density of stomata on a leaf.

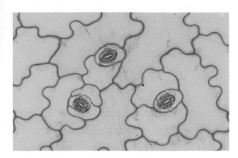

Figure 3.29 Surface view of lower epidermis of leaf of *Kalanchoe* showing stomata and guard cells. The photomicrograph was taken using a light microscope and the stomatal pores are about 20μm long.

Figure 3.30 Pumpkin leaves drooping on a hot day. Leaves droop as a result of wilting and this reduces the loss of water by transpiration. Normally the leaf surface is horizontal or curving upwards.

- **Wind speed** – if it is windy, water molecules are moved away from the leaf and more evaporation can take place, so the rate of transpiration is higher than on a day with no wind.
- **Temperature** – water molecules evaporate faster at higher temperatures so the rate of transpiration is faster on a hot day than on a cool day.
- **Light intensity** – water passes out of the leaf through the stomata, when they are open. Stomata are usually open during the day when it is light but closed at night when it is dark. The mechanism for opening and closing of the stomata is described in Section 3.1.

The structure of the guard cells is such that, when they are full of water, they become turgid and more curved. This allows the stoma between them to open. But when they are flaccid (contain less water), there is no gap between them and the stoma closes. The mechanism is linked to light intensity – so that at night (in the dark) the stomata are usually closed, but they open during the light. This means that very little transpiration occurs at night, when it is dark.

Changes in the rate of transpiration are similar to those you might experience when trying to dry wet clothes outside – they dry faster on a hot, windy day in dry conditions than on a cold, still day in damp conditions. The extra factor for plants is the effect of light intensity on the opening and closing of the stomata.

Sometimes plant leaves wilt if it is very hot or if there is insufficient available water. This actually helps the plant conserve water. When the leaves droop they are no longer held in a position that captures the direct rays of the Sun, so this reduces the direct heating effect of the Sun on the leaves.

■ Investigating how environmental factors affect the rate of transpiration

For a plant growing under natural conditions, loss of water by transpiration is an essential process and helps to maintain the passage of water through a plant. The rate of transpiration can be affected by any of the environmental factors listed above.

You can investigate the effect of the different environmental factors, one at a time, in the laboratory in different ways. Two experiments are shown in Figure 3.31 and Figure 3.32. In each case, we describe the apparatus used to investigate the rate of transpiration as a **potometer**.

Potometer 1 is a simple mass potometer. Any investigation assumes that the loss in mass shown by the balance is due to the loss of water from the leaves.

Potometer 2 uses the movement of the meniscus in the capillary tube as a measure of the uptake of water by the shoot. This method assumes that water lost by transpiration causes more water to be drawn into the stem, hence the movement of the meniscus (or you can watch the movement of an air bubble in the capillary tube).

For both potometers, you can devise ways of providing different environmental conditions and so compare the rates of transpiration in different conditions. Here are some examples:

- Enclose the potometer in a large transparent bag to create more humid conditions close to the leafy shoot.
- Use a hairdryer held close to the potometer to represent windy conditions – you can use different temperatures on the hairdryer or different wind speeds.
- Place the whole potometer in the dark to compare transpiration rates in the dark and light.
- Use a lamp at different distances from the potometer to give different light intensities (but remember that the light may also affect the temperature near to the potometer).

PRACTICAL

Potometer 1 assumes that changes in mass represent water loss from the shoot. The apparatus can be set up in different conditions to investigate the effect of different factors on the rate of transpiration.

MATHS TIP

Percentages

One quantity (n) can be expressed as a percentage of another (N), by setting up the first quantity as a fraction of the second and multiplying by 100:

(n / N) × 100 = the percentage of N represented by n.

Now go to page 133, Qu 5 a) to apply this to calculations relating to percentage change in mass of a leaf in different conditions.

■ Practical activity – using a simple mass potometer

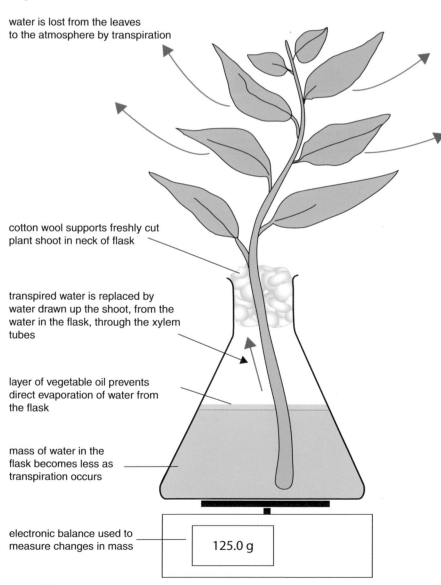

water is lost from the leaves to the atmosphere by transpiration

cotton wool supports freshly cut plant shoot in neck of flask

transpired water is replaced by water drawn up the shoot, from the water in the flask, through the xylem tubes

layer of vegetable oil prevents direct evaporation of water from the flask

mass of water in the flask becomes less as transpiration occurs

electronic balance used to measure changes in mass

125.0 g

Figure 3.31 Potometer 1 – a simple mass potometer.

Potometer 2 can give you accurate readings of the volume of water taken up by the shoot and assumes that this represents water lost by the shoot. The rate of transpiration can then be compared in different conditions. A simpler apparatus can be devised, with a shoot held upright attached to a capillary tube.

Practical activity – using a potometer with capillary tube

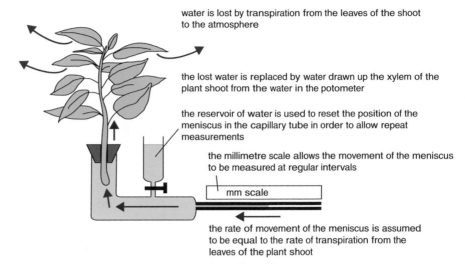

Figure 3.32 Potometer 2 – potometer with capillary tube.

STUDY QUESTION

1 Describe the route taken by a water molecule as it travels through a plant from the soil water around the roots until it reaches the atmosphere outside the leaves. At each stage, state the process or mechanism that is involved in moving the water.

Copy the table and write your answers in the correct sequence. Only the first few lines of the table have been drawn – you must decide how many lines you need.

- Use words from Box 1 to state the route taken.
- Use words from Box 2 to match the process at each stage.

Words may be used once, more than once or not at all.

> **Box 1** (route)
> air space in leaf; atmosphere outside leaf; cell to cell in root; mesophyll cells in leaf; phloem tubes in stem; root hair cells; stomata; waxy cuticle on leaf epidermis; xylem tubes in root and stem; xylem tubes in veins of leaf

> **Box 2** (process)
> active transport; capillarity; cohesion; diffusion; osmosis; root pressure; transpiration pull

Route (box 1)	Process (box 2)
1	
2	
3	

3.5 Transport in humans (1) – blood, structures and functions

Blood is important to us

When you are accidentally cut, you see bright red blood oozing from the wound. Your body contains less than 5 litres of this remarkable fluid and a significant loss could lead to failure of the circulatory system and death.

If you lose lots of blood, in an accident or as part of a surgical procedure, you may need to have the missing blood replaced. This is only possible if you receive a blood transfusion – you are given some blood donated voluntarily by another person. This is because currently there is no artificial substitute for blood (plasma and cells). Originally, a blood donation could only be done directly from donor to patient. Nowadays, the blood can be stored for several weeks until it is required. In the UK and many other countries, blood donors give blood three times a year without receiving any payment. So what cells are found floating in this remarkable liquid and what are the functions of blood in the body?

■ Blood and the blood cells

In humans, materials are transported around the body, between the different tissues and organs, in the bloodstream.

Blood is a complex fluid, composed of **red cells**, **white cells** and **blood platelets**, suspended in a watery fluid called **plasma** (Figures 3.33 and 3.34).

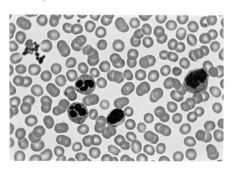

Figure 3.33 Blood cells, seen with a light microscope. This smear is prepared from a drop of blood and illustrates the relative proportions of red and white blood cells. These can be identified from Figure 3.34.

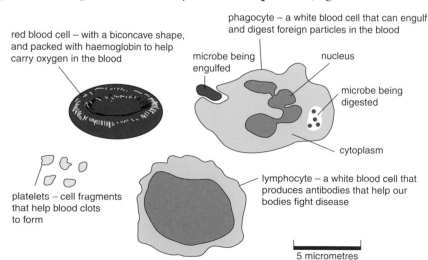

red blood cell – with a biconcave shape, and packed with haemoglobin to help carry oxygen in the blood

phagocyte – a white blood cell that can engulf and digest foreign particles in the blood

microbe being engulfed

nucleus

microbe being digested

cytoplasm

platelets – cell fragments that help blood clots to form

lymphocyte – a white blood cell that produces antibodies that help our bodies fight disease

5 micrometres

Figure 3.34 Different types of blood cell (Note: 1000 micrometres = 1 mm, a micrometre is also written as μm).

MATHS TIP

Ratio

A ratio is a way of comparing the sizes of two or more quantities. It can be expressed as a fraction or numbers:

For example, if there were three blue-eyed children for every brown-eyed child, then the ratio would be:
3 blue : 1 brown,
or ¾ blue : ¼ brown

Now go to page 128, to apply this to a photograph showing different blood cells in a sample of blood.

STUDY TIP

You can make several links to see the importance of the different things transported in the blood: look in Section 2.2 for digested food, in Section 2.3 for respiration, in Section 4.2 for urea and excretion in humans, and in Section 4.5 for hormones.

STUDY TIP

Make a link between the need for iron in the diet and its presence in the haemoglobin molecule (see Section 2.2).

Transport in the plasma

Plasma is the straw-coloured liquid in which our blood cells float. Apart from our blood cells, various materials are carried in the plasma and transported to different parts of the body. The list below summarises what is carried and how:

- **digested food** – dissolved or in suspension in the plasma. These nutrients are carried from where they are absorbed from the small intestine to the cells of the body where they are needed
- **carbon dioxide** – the waste gas from respiration, is transported dissolved in the plasma as hydrogencarbonate ions, from respiring tissues to the lungs
- **urea** – a waste material from the breakdown of proteins. It is produced in the liver and transported to the kidneys for excretion from the body. It is carried dissolved in the plasma
- **hormones** – released into the blood plasma from the endocrine organs of the body. These act as chemical signals to other tissues and organs in the body
- **heat energy** – generated by respiring tissues such as active muscles or the brain, heat energy is carried by warm plasma to cooler body regions, or to the skin where the heat can be lost.

Oxygen transport by red blood cells

Red blood cells get their name from their colour. The colour of the red blood cell is due to the large numbers of **haemoglobin** molecules found in its cytoplasm. Haemoglobin is a **protein** which is able to bind chemically to dissolved **oxygen molecules**. This binding between the protein and the oxygen is easily reversed, so that the haemoglobin readily releases the oxygen molecules (Figure 3.35). Because of this reversible binding reaction between oxygen and haemoglobin, the red blood cells allow the blood to carry far more oxygen than would be possible if the oxygen was simply dissolved in the plasma.

Oxygen diffuses into the lung capillaries and is joined to the haemoglobin molecules in the red blood cells, forming oxyhaemoglobin.

Oxygen diffuses out of the body capillaries, as the oxyhaemoglobin releases it from the red blood cells. It is used for respiration by the body tissues.

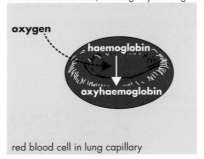

red blood cell in lung capillary

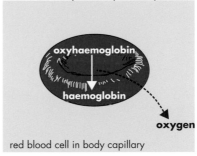

red blood cell in body capillary

Figure 3.35 The diagram shows how haemoglobin binds with oxygen to form oxyhaemoglobin, then releases the oxygen to form haemoglobin again.

The reversible binding reaction between haemoglobin and oxygen allows the red blood cells to pick up dissolved oxygen in the capillaries of the lungs. Here a lot of dissolved oxygen diffuses into the plasma from the alveoli in the lungs. The red blood cells are carried in the blood and can then unload their oxygen in another part of the body, such as in the capillaries of

STUDY TIP

Look at the differences in the structure of arteries, veins and capillaries in Section 3.6 and list the reasons why materials only enter or leave the blood in the capillaries.

STUDY TIP

Refer to Section 1.5 and explain how the shape of a red blood cell would be affected by osmosis if the water content of the blood plasma became either too high or too low.

respiring tissues. From the capillaries, oxygen diffuses out of the plasma into nearby cells, where it can be used for respiration.

A red blood cell usually spends a very short time – probably only about a second – inside a capillary and during that time it has to load or unload its oxygen. The red blood cell has a number of adaptations which ensure that the oxygen diffuses in and out as fast as possible. Here are some of the features that help:

- **biconcave shape** – this shape provides a large surface area in relation to the cell volume, which gives more opportunity for oxygen to diffuse in or out
- **small and numerous** – smaller cells have a larger surface area in relation to their volume, in comparison to bigger cells. This allows a faster rate of diffusion. The blood carries 5 million of these tiny red blood cells per mm^3. Being tiny also helps the red blood cells squeeze through the smallest capillaries in the body
- **no nucleus** – the absence of a nucleus provides extra space inside the red blood cell for millions of extra haemoglobin molecules. This means that the red blood cell can carry extra oxygen, bound to the haemoglobin. However, because there is no genetic information in the cell, it means that repairs cannot be carried out. So the red blood cell only lives for about a month before dying. The dead red blood cells are continually replaced by fresh ones made in marrow of the bones.

Preventing blood loss and the role of platelets

The blood is forced along the blood vessels under pressure, so any damage to the walls of the vessels leads to an immediate loss of blood. When you have a cut through your skin, the walls of the blood vessels are ruptured, and the blood oozes out onto your skin.

The blood contains blood **platelets**, which are small fragments of cells produced by the bone marrow. When blood comes into contact with air, the platelets in the blood explode, releasing a number of chemicals. These chemicals trigger a series of chemical steps that cause the conversion of a protein in the blood (**fibrinogen**), into a meshwork of interlinked **fibrin** molecules. These fibrin molecules form a seal on the broken wall of the blood vessel. This becomes a blood clot and consists of fibrin molecules, trapped red blood cells and platelets (Figure 3.36).

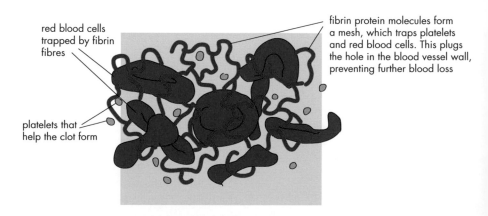

red blood cells trapped by fibrin fibres

fibrin protein molecules form a mesh, which traps platelets and red blood cells. This plugs the hole in the blood vessel wall, preventing further blood loss

platelets that help the clot form

Figure 3.36 Formation of a blood clot.

The clot provides two important functions:

- it prevents further loss of blood from the vessels
- it prevents the entry of pathogens into the bloodstream.

Defence of the body by white blood cells

White blood cells are much less numerous than red blood cells (see Figure 3.33). Their role is to provide the body with protection against the many **infectious diseases** that we encounter during our lives, caused by **pathogens**.

There are a number of different kinds of white blood cell. Each type carries out a different role in defending the body (Figure 3.34). Here are two examples:

- **Phagocytes** are white cells that can directly **ingest pathogens**. They engulf them, drawing them into their cytoplasm. The phagocyte then kills and digests the pathogen with enzymes.
- **Lymphocytes** are white cells that **secrete antibodies** into the plasma. Each lymphocyte can secrete only one specific type of antibody, which has a shape that fits only one type of pathogen cell wall (the antigen). Each of the millions of lymphocytes in circulation has a single antigen that its antibodies 'fit'. This means that a single lymphocyte provides protection only against a single disease.

Antibodies are proteins. They can attach themselves to the surface of a pathogen, at a part known as the **antigen**. This marks the pathogen for destruction. The attachment of an antibody to the pathogen surface triggers a number of chemical processes leading to the inactivation or death of the pathogen.

Antibodies can also make the pathogen cells clump together, which makes it easier for phagocytes to engulf them.

Response to an infectious disease

A series of events happens when a pathogen, such as a bacterium, enters the bloodstream (Figure 3.37a). Initially, the bacterium thrives in the warm, nutrient-rich bloodstream. It grows and divides, and the infected person feels the symptoms of the illness. The increasing numbers of pathogenic bacteria might attack the body tissues and release poisons (toxins) into the blood.

After a couple of days, one of the bacteria may bump into a particular lymphocyte, which happens to recognise the bacterium as a foreign invader. The lymphocyte becomes activated and starts to secrete lots of antibody molecules. The lymphocyte also divides many times to form a clone of lymphocytes. All members of the clone produce lots of the same antibody molecules.

The activated lymphocytes release millions of antibodies into the bloodstream. The antibodies start to attach themselves to the invading bacteria, causing their destruction. Phagocytes engulf and digest the dead and dying bacteria, eventually removing them from the bloodstream. The patient will now have recovered from the disease.

STUDY TIP

Link this with the information about mitosis in Section 5.5 and cloning, described in Section 7.6.

A person who recovers naturally from a disease carries a large number of unused antibodies, specific to that disease, in their blood plasma. They also have a large number of lymphocytes that also specifically recognise the disease. Lymphocytes that recognise a specific pathogen are called **memory cells**. This combination of large numbers of antibodies and **memory cells** means any further attempted invasion by this pathogen will be dealt with immediately. The person has become **immune** to this disease.

If the same pathogen enters the blood on a second occasion, there will be an immediate response as one of the memory cells is likely to meet it quickly. In addition, the response will be even more vigorous, with much larger amounts of antibodies being released into the bloodstream.

STUDY TIP
Look at Section 5.5 and make list of differences between an **inherited disease** and an **infectious disease**.

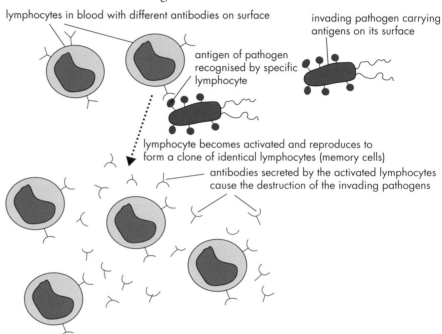

lymphocytes in blood with different antibodies on surface

invading pathogen carrying antigens on its surface

antigen of pathogen recognised by specific lymphocyte

lymphocyte becomes activated and reproduces to form a clone of identical lymphocytes (memory cells)

antibodies secreted by the activated lymphocytes cause the destruction of the invading pathogens

Figure 3.37a) A summary of the immune response mechanism.

Vaccinations

If a person's immune response to a pathogen is inadequate, the infection might cause serious illness or death. Using a vaccination is one way of safely providing protection against an infectious disease. This involves someone receiving an injection of *harmless* pathogen cells in the '**vaccine**'.

The pathogen cells in the vaccine are made harmless by killing them (dead vaccine), or at least by making them unable to grow and divide (attenuated vaccine). Some modern vaccines just contain fragments of the pathogen cell wall, rather than whole cells, but this is enough to provide protection.

The cells and cell wall fragments in the vaccine are recognised by specific lymphocytes in the bloodstream. The lymphocyte becomes activated, triggering an immune response, despite the fact that the vaccine itself is harmless. The result is that lots of antibodies and specific memory cells remain in circulation. This provides immunity to any live harmful pathogens, if they attempt to invade the body.

For some diseases (e.g. tetanus), a single vaccination provides immunity for a few years. For other diseases (e.g. mumps) the immunity lasts a lifetime.

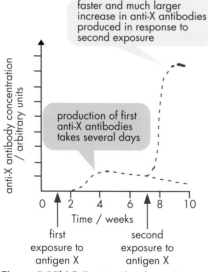

faster and much larger increase in anti-X antibodies produced in response to second exposure

production of first anti-X antibodies takes several days

anti-X antibody concentration / arbitrary units

Time / weeks

first exposure to antigen X

second exposure to antigen X

Figure 3.37b) Following the changes in antibody concentration after one or two exposures to an antigen.

STUDY TIP
Find out which vaccinations are now given to all infants, as a matter of routine, and whether this is the same in all countries.

An international vaccination campaign against smallpox was carried out in many countries. By 1977, this eventually led to the global eradication of smallpox (which is caused by a virus). No cases are known anywhere in the world today, yet in the past, this disease caused the death or disfigurement of millions of people.

In the future it is hoped that vaccination campaigns for other diseases, such as polio, HIV and malaria, will lead to the eradication of these diseases as well. One of the difficulties of achieving this ambition is that these pathogens are always changing their cell surfaces so are not recognised by lymphocytes.

STUDY QUESTIONS

1   The table lists features of various blood components. Copy and complete the table by filling in the missing items in the spaces. (One example has been completed for you.)

Component	Description	Function
Antibodies	proteins dissolved in plasma	stick to foreign materials and help destroy them
Lymphocyte	cell with large single nucleus	
Phagocyte		
	cell fragments with no nucleus	
Fibrinogen		
	cell packed with haemoglobin and lacking a nucleus	

2   Draw a flow chart to show the stages involved in becoming immune to a disease as result of being giving a vaccination. Write comments beside your flow chart to explain what is happening at each stage.

3.6 Transport in humans (2) – heart and blood circulation

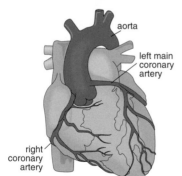

Keeping the heart muscle beating

When you are asleep or at rest, the most active muscle in your body is your heart. This muscular pump pushes blood into your arteries with every beat. The pressure created by these pushes causes your blood to flow throughout your body.

Diseases of the heart are a major problem in the populations of the developed world. When the coronary artery, which feeds blood to the heart muscles, becomes narrowed or blocked, heart muscle cells are starved of the oxygen and glucose they need and they die. A surgical therapy for this problem is coronary bypass surgery – a piece of blood vessel from the leg is sewn onto the heart, allowing blood to bypass the blockage and feed the working heart muscles. The main image shows heart surgery in progress, then the top diagram shows a normal heart and the bottom diagram shows a heart with a bypass.

Some people believe that disease of the coronary vessels, leading to coronary heart disease (CHD), is at least partly the result of a modern lifestyle as it seems more common in more developed countries. What are the factors, including lifestyle features, that make people more likely to suffer from coronary heart disease? Make a list of likely risk factors then consider which may be relevant to you and whether you could change your lifestyle in order to reduce this risk.

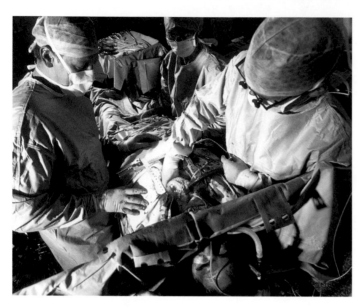

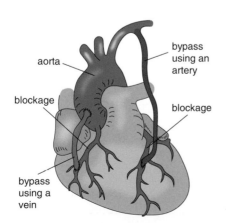

■ The heart

The human heart is about the size of your clenched fist. It is located in the middle of your chest, between your lungs. It beats regularly throughout your life. You can feel its beats by pressing a finger against the inside of your wrist, at the base of your thumb. With each heartbeat, blood is forced into your arteries. This stretches their walls to create the pulse you feel in your wrist.

The pressure generated by the heartbeat forces blood out into a network of **blood vessels**:

- the **arteries**, which carry blood away from the heart towards the various organs in the body
- inside the organs, the **capillaries** allow substances to enter and leave the blood
- the **veins** eventually bring blood from the organs back to the heart.

Structure of the heart

Figure 3.38 shows a vertical section through the heart. You can see that it is made up of four chambers – a left and right atrium (plural = atria), and a left and right ventricle. Note that the wall of the left ventricle is thicker than that of the right ventricle. The arrows show the direction of blood flow through the chambers of the heart. The role of the valves is described below.

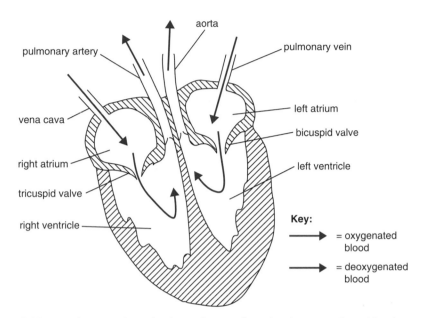

Figure 3.38 Vertical section through a heart showing four chambers together with relevant blood vessels. The arrows show direction of blood flow.

The heart as a pump

The human heart is a **double pump**. The right side sends blood to the lungs (**pulmonary** circuit) and the left side sends blood to the rest of the body (**systemic** circuit). This double pump arrangement ensures that blood is forced quickly through all parts of the **circulatory system**. The sequence of events taking place in the heart is now described. Look at Figure 3.38 to help you follow the stages.

- Blood returning to the heart from the body in the **vena cava** fills the relaxed **right atrium**, and blood returning from the lungs in the **pulmonary vein** fills the relaxed **left atrium**.
- The **heartbeat cycle** of contraction begins when these two atria contract. The right atrium forces blood through the **tricuspid valve** into the **right ventricle**. The left atrium forces blood through the **bicuspid valve** into the **left ventricle**.
- The ventricles now contract. The **right ventricle** forces blood out of the heart, through a **valve**, into the **pulmonary artery**. The pulmonary artery carries **deoxygenated** blood to the lungs. The blood in the **left ventricle** forces blood out of the heart, through a **valve**, into the **aorta**. The aorta carries **oxygenated** blood to the rest of the body.
- The valves in the heart ensure that the blood flows from one place to the next and never goes backwards. The tricuspid and bicuspid valves prevent blood going backwards from the ventricles to the atria. The **semilunar valves** lie near the entrances of the pulmonary artery and aorta. These prevent the blood going backwards from the arteries to the ventricles of the heart. (The semilunar valves are not labelled on the diagram but their position is shown.)

STUDY TIP
Make a link with Section 4.5 for more information about hormones.

STUDY TIP
Look at Section 3.2 and suggest two expected changes to the activity of the lungs during exercise.

STUDY TIP
Look at Section 4.5 and list the other changes to the body that occur when adrenaline levels rise during the fight or flight response.

STUDY TIP
Check in Section 2.3 to make sure you know the word equations for respiration with and without oxygen being available to the muscles.

After atrial and ventricular contractions, the heart relaxes, gradually filling with more blood, before the next heartbeat cycle starts. The heartbeat cycle is measured as the **heart rate**, in beats per minute (bpm). The heartbeat is initiated by a small group of cells known as the **pacemaker**. This lies towards the top of the right atrium. The pacemaker is controlled by two nerves that run from the brain. One nerve causes the pacemaker to increase the heart rate and the other nerve causes it to decrease the heart rate. The heart rate is also influenced by hormones in the blood, such as **adrenaline**.

Exercise, heart rate and supply of oxygen

At rest, a healthy adult human heart might beat 70 beats per minute (heart rate = 70 bpm). This ensures that all the organs of the body are supplied with oxygen, glucose and other substances. During physical exercise, the muscles of the body have to respire more so that they can provide the energy for muscle contraction. More blood is needed to supply up to 20 times more oxygen to the working muscles. The blood also supplies sugars and removes the waste carbon dioxide and heat produced by the muscles.

To achieve this increase in oxygen supply to the muscles, the heart rate increases up to a maximum rate of around 170 bpm (for a healthy teenager). At this increased rate, blood is moved more frequently from muscles to lungs and back again. The volume of blood pushed out of the heart with each beat also increases. These changes in the activity of the heart are coordinated by the **nervous system**. The hormone **adrenaline** also stimulates the heart rate. This hormone is often produced in stressful situations, and helps your body prepare for physical action – part of the 'fight or flight' response.

Once the physical exercise has stopped and the person is resting, the heart rate should fall back to the original resting rate. The time taken for this to happen (the recovery time) is longer for an unfit person than for a physically fit person. A physically fit person will have a lower heart rate and shorter recovery time than an unfit person because the heart muscles (of a fit person) push more blood out of the heart with each beat.

If the exercise is moderate, the supply of oxygen to the muscles might be sufficient to meet all of their needs during the activity. This is called 'aerobic' activity, because the muscle can respire aerobically.

For more intense exercise, the oxygen supply cannot meet the demands being made. For a short period of time, the working muscles can use anaerobic respiration to generate some energy for movement. This produces lactic acid as a waste product and this needs to be removed. Presence of lactic acid in the muscles causes the pain people feel in their muscles when exhausted. Removal of lactic acid requires oxygen and this must be supplied in the blood. This is why, at the end of the exercise, the heart rate may remain higher than usual and only gradually falls back to the resting rate. This supply of extra oxygen after exercise is referred to as 'paying off the oxygen debt'.

Another way the body can further increase the supply of blood to exercising muscles is by diverting some blood from organs, such as the liver, intestines and kidneys, towards the active muscles.

Figure 3.39 Aerobic exercise (left) and anaerobic exercise (right).

■ Practical activity – investigating the effect of exercise on heart rate

Your pulse gives a measure of your heartbeat. You can find your pulse in different parts of your body and a convenient place is in your wrist. If you have a health issue you should discuss the activity with your teacher before doing exercise.

1 Find your wrist pulse, as shown in Figure 3.40.
2 Sit quietly and count the number of pulses in 10 seconds. Multiply this by 6 to find your heart rate in beats per minute (60 seconds).
3 Repeat this three times and calculate a mean value for your heart rate at rest.
4 Do some vigorous exercise, such as running a lap of a playing field or stepping up and down a step once a second continuously for 3 minutes.
5 As soon as you have finished the exercise, find your pulse and count the number of beats in 10 seconds. Wait for 20 seconds, then count your pulse again for 10 seconds.
6 Continue in this way for about 2 minutes. By this time your pulse rate has probably returned to its normal level.
7 For each count, multiply by 6 to give a value for heart rate in beats per minute. Plot a graph of these heart rates against the time after stopping the exercise.

When you have obtained your own results, you can gather together class results and see the range of times for recovery for different people. Then look at the overall patterns and link these to the idea of fitness and recovery time.

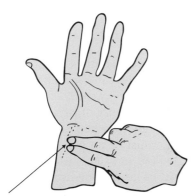

The wrist pulse is found by feeling in the groove in the wrist bones a few centimetres below the base of the thumb. Each pulse is due to the stretching and relaxing of the wall of the radial artery, which takes blood to the hand. Count the number of pulses in 10 seconds and then multiply by 6 to calculate the heart rate in beats per minute.

Figure 3.40 How to find your pulse.

The blood vessels

Blood is propelled around the body inside a closed system of pipes – the blood vessels (see Table 3.2). A single sheet of cells forms the inner lining of all the vessels. A similar sheet of cells forms the thin walls of the smallest ones, the capillaries. Substances such as oxygen diffuse through these thin capillary walls to reach the cells of the body.

The structure of the biggest vessels enables them to carry out their function. **Arteries** carry blood away from the heart under high pressure and velocity, whereas **veins** carry blood back towards the heart under low pressure and velocity. The arteries branch into smaller vessels called **arterioles**.

The veins branch into smaller vessels called **venules**.

The **capillaries** are very small vessels linking the arterioles and the venules. They branch many times and lie in the tissues of the body, among the cells. They have a very small diameter and their walls are thin. It is here that diffusion of substances occurs, between the blood and the fluid that surrounds the cells.

The veins contain regular pocket valves to ensure that blood does not go backwards in them. Sometimes the pressure of body muscles presses on the thin-walled veins and helps to propel the blood in the veins onwards towards the heart.

Key features of the different types of blood vessels are summarised in Table 3.2.

Table 3.2 A summary of key features of blood vessels.

Blood vessel	Function
thick artery wall containing elastic fibres and muscle cells smooth lining promotes fast flow of blood through narrow lumen of artery	**Arteries** carry blood away from the heart, at high speed under high pressure. The pressure fluctuates as the heart pushes and then relaxes. The arteries branch into a network of smaller vessels called **arterioles**. The diameter of the arterioles can vary by the action of muscle fibres in their walls. This can be used to adjust the pressure in the circulatory system, as well as to divert blood from one organ to a different organ. The **arterioles** branch into the capillaries, by which stage the blood is moving slowly and under lower pressure.
thin vein wall containing no muscle fibres flaps of pocket valve that prevent blood flowing backwards large lumen reduces resistance to flow of blood under low pressure	**Veins** carry the blood that leaves the capillaries (inside the tissues and organs) back towards the heart. The blood in the veins begins to move faster than inside the capillaries, but it is under low pressure because there is no pump to keep it moving. The veins have thinner walls than arteries, and lack any muscle or fibres. They also contain pocket valves that prevent the blood being forced backwards in the wrong direction.
single layer of cells form wall of capillary nucleus substances diffuse across capillary wall into and out of plasma red blood cells squeeze through blood vessel in single file	**Capillaries** are found in the tissues of the body. Indeed, every single cell in a tissue is close to a capillary. The smallest capillaries are so narrow that red blood cells have to pass through them slowly in single file. The walls of the capillaries are only one cell thick. This means that diffusion of gases and other substances can take place through the thin capillary wall, between the plasma and the fluid surrounding the cells of the tissues of the body.

STUDY TIP

Examine the plan of circulation
and structure of the left and
right ventricles, and suggest an
explanation for the difference in
thickness of the muscle walls.

STUDY TIP

Link the events in the lungs with
Section 3.2 and how a red blood
cell loads and unloads its oxygen (in
Section 3.5).

Blood circulation

On the simplified plan of the circulation (Figure 3.41), you can see that all
the blood arriving at the **right side** of the heart is **deoxygenated**. This is
because it has given up its oxygen to respiring cells in the muscles
and organs, such as the liver, intestines and kidneys. This blood from the
right-hand side of the heart is then pumped to the lungs in the pulmonary
artery.

In the lung capillaries, the blood gets rid of its carbon dioxide and picks up
fresh oxygen and becomes oxygenated. Then the **oxygenated** blood returns
to the **left side** of the heart in the pulmonary vein. From here the oxygenated
blood is pumped via the aorta to the cells in the rest of the body, where it
provides oxygen for respiration.

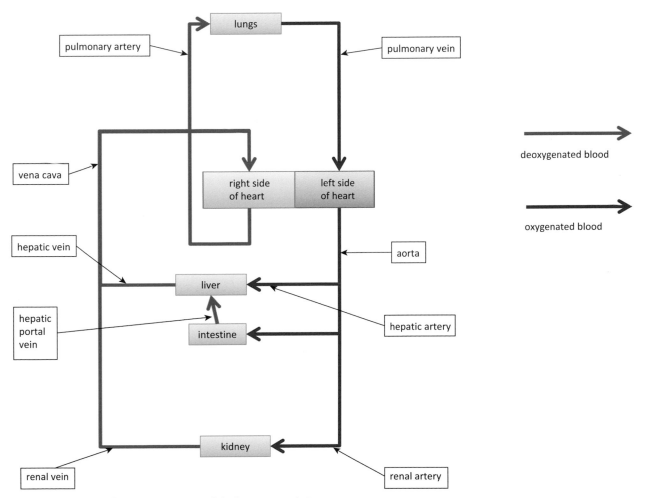

Figure 3.41 A simplified representation of the human circulation system.

In addition to the network of blood vessels throughout the body, there is a
network of **lymphatic vessels**. The fluid that inevitably leaks out of the thin
walled capillaries is collected by the lymphatic vessels and eventually drains
back into the bloodstream.

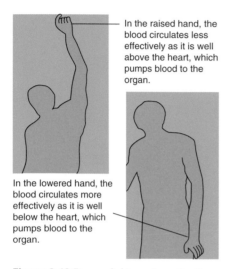

In the raised hand, the blood circulates less effectively as it is well above the heart, which pumps blood to the organ.

In the lowered hand, the blood circulates more effectively as it is well below the heart, which pumps blood to the organ.

Figure 3.42 Finger clicking – investigating circulation and muscle activity.

■ Practical activity – finger-clicking activity

This practical activity demonstrates the effects of arm position on the muscle activity of the forearm muscles. You can link this to how the blood circulation is maintained throughout the body.

1 Raise your arm, as shown in the diagram, and count the number of times you can click your fingers over a period of two minutes.
2 Repeat this finger-clicking activity but with your arm lowered.

The muscles that are used to click the fingers are located in the forearm. When your arm is raised above your head, the blood circulates less effectively as it is well above the heart, which pumps blood to the muscles.

When your hand is lowered (beside your body) the blood circulates more effectively as it is well below the heart, which pumps blood to the muscles. When the arm is lowered, the supply of oxygen to the muscles in the forearm should be better than when the arm is in the raised position, so you would expect to do more finger clicks in the time.

■ Coronary heart disease risks

It is inevitable, as we age, that our circulatory system is affected positively and negatively by various factors — some we can do nothing about and other 'lifestyle' factors about which we can do something. Ultimately, in old age, about half the population of a Western country will die of cancer and half from diseases of the circulatory system. Unfortunately some people will die prematurely of heart disease, particularly coronary heart disease (see introduction on page 118). In this condition, the arteries feeding blood to the heart muscles become narrow or blocked, resulting in a lack of oxygen delivery to the heart muscle cells. As a consequence, these cells may die and the heart may stop beating.

There is evidence that all of the following lifestyle factors help, in the long term, to reduce the chances of premature coronary heart disease:
■ Avoid obesity — ideally keep your body mass index (BMI) between 18 and 25. Carrying the extra weight that comes with obesity puts a strain on the heart. Diabetes is associated with obesity and this can also result in the damage to blood vessels, which makes the job of the heart more difficult.
■ Avoid high blood pressure — keeping it less than 140 / 90 mm Hg, which is when doctors will consider treating it with drugs. This can be done by avoiding excess salt in the diet and by taking regular exercise.
■ Avoid excess cholesterol — excess levels of this substance form part of the plaque that narrows coronary blood vessels by causing their inner walls to swell inwards. They are then more likely to become blocked by a blood clot.
■ Avoid smoking — chemical substances in the smoke actually increase the deposits on the artery walls, raise your blood pressure and make the blood more likely to clot. In addition, the carbon monoxide gas in the smoke reduces the ability of your red blood cells to carry oxygen — making the job of the heart and circulation even more difficult.

STUDY QUESTIONS

1 In some sports, the athlete needs to produce an explosive burst of muscle power in a very short period of time, while other sports require a sustained application of power over a long period of time. For example, an elite 100-metre sprinter finishes the race in under 10 seconds, whereas an elite marathon runner has to run for over 2 hours. These two sports are opposite extremes, one involving almost purely anaerobic respiration by the leg muscles, and the other purely aerobic respiration in the leg muscles.

a) Draw a horizontal line with the word **anaerobic** at one end and **aerobic** at the other end. Then write down the different sports listed in the box below, and arrange them at various points along the line to indicate the extent to which they are aerobic or anaerobic activities.

> 200-metre run; tennis; basketball; hill walking; skiing; weight lifting; judo; football; 1000-metre run; frisbee

b) Choose **three** of the sports listed above and describe what changes you would expect to see in the heart rate of a participant during the sporting activity and immediately after.

2 Study Figure 3.41 and decide which blood vessels have the properties listed.

Property	Blood vessel
Lowest CO_2 concentration	
Lowest oxygen concentration	
Highest absorbed nutrients concentration	
Blood at the highest pressure	
Blood at the lowest pressure	

Summary

I am confident that:

✓ I know which gases are used and produced by photosynthesis and respiration in flowering plants.

✓ I understand why and how the net exchange of carbon dioxide and oxygen differs during the day and at night in flowering plants.

✓ I can describe the structure of a leaf and explain how it is adapted for gas exchange and how the stomata open or close.

✓ I can describe experiments with hydrogencarbonate indicator that can be used to investigate the effect of light on net gas exchange in a leaf.

✓ I can describe (and draw a diagram to show) the structure of the thorax with lungs, including their internal structure.

✓ I know that ventilation of the lungs occurs as a result of pressure changes in the thorax and that the intercostal muscles and diaphragm muscle contract to bring about a change in volume.

✓ I know the importance of the large surface area of the alveoli for gas exchange and can describe and explain other features that allow diffusion between air in the alveoli and the blood in the capillaries.

✓ I can describe how smoking causes damage to the lungs and I am aware of the biological consequences that result.

✓ I can describe an experiment to investigate the effect of exercise on breathing rate in humans.

✓ I understand why diffusion is appropriate for movement of substances into and out of single-celled organisms but not for larger multicellular organisms.

✓ I understand the need for transport systems in larger multicellular organisms and can give some examples to illustrate this.

✓ I know that sucrose and amino acids are transported in the phloem and can describe where the phloem is found in a flowering plant.

✓ I know how water enters a plant through the root hair cells in a flowering plant.

✓ I understand how xylem transports water from the root to the leaves and mineral ions from the roots to other parts of the plant.

✓ I know the meaning of the term transpiration and the importance of transpiration in the movement of water through a plant.

✓ I can describe experiments that can be used to investigate how environmental factors (including humidity, wind speed, temperature and light intensity) affect transpiration and can explain how these factors affect transpiration from a leafy shoot.

✓ I can describe features of red and white blood cells, platelets and the plasma and understand their different roles in the blood, including the transport of materials.

✓ I can describe how red blood cells are adapted for the transport of oxygen.

✓ I can describe the part played by white blood cells in the response of the human body to disease and entry of pathogens.

✓ I understand how vaccination can help to provide protection against infection in the body.

✓ I can describe how platelets contribute to clotting of the blood and the sequence of events that helps prevent loss of blood and entry of microorganisms after a cut.

✓ I can describe (and draw a diagram to show) the structure of the heart and know the route taken by the blood as it passes through the series of chambers and blood vessels.

✓ I understand how the heart pumps blood to the lungs and the rest of the body and understand that heart rate can change in response to exercise and the hormone adrenaline.

✓ I can describe (and draw diagrams to show) features of arteries, veins and capillaries and understand how each type of blood vessel contributes to the circulation of blood in the body and exchange of materials between cells and the blood.

✓ I can describe (and draw a diagram to show) the overall plan of the circulation of blood in the body, including its route through the heart, lungs, liver and kidneys.

✓ I understand how different factors may increase the risk of developing coronary heart disease.

MATHS SKILLS

Orders of magnitude

An order of magnitude is a measure of a quantity, in powers of ten-fold.

For example 10 000 is 3 orders of magnitude greater than 10.

Look at Table 3.1 and state the difference between the carbon dioxide in inhaled and exhaled air, as orders of magnitude.

$= 2$

MATHS SKILLS

Areas of triangles and rectangles

Area (units^2) of any triangle = height $\times$ ½ base length

Area (units^2) of any rectangle = length $\times$ width

Examine Figure 3.29 showing the lower epidermis of a *Kalanchoe* leaf. Assume that the photo measures 7 stomata-lengths $\times$ 11 stomata-lengths, and that one stomata-length = 20 micrometres.

a) Calculate the area of the photograph in micrometres^2.

b) Give this value in mm^2 to the nearest whole mm^2 (remember that there are 1 000 000 micrometre^2 in a millimetre^2).

c) State the density of stomata on the leaf, using appropriate units.

a) area of photo = 30 800 micrometres^2

b) area of photo = 0.03 mm^2

c) Density of stomata = 3 / 0.03 = 100 stomata / mm^2

MATHS SKILLS

Ratios

A ratio is a way of comparing the sizes of two or more quantities. It can be expressed as a fraction or numbers:

For example, if there were three blue-eyed children for every brown-eyed child, then the ratio would be:
3 blue : 1 brown, or ¾ blue : ¼ brown

Examine the photograph of the blood smear in Figure 3.33 on page 112. Count the red cells, macrophages and lymphocytes and give the ratios of each type. Only count whole cells.

173 (+ / - 20) red blood cells : 2 phagocytes : 2 lymphoctyes , or 86 (+ / - 10) red cells : 1 phagocyte : 1 lymphocyte

Example of student response with expert's comments

■ Applying principles

1 A glucometer is an instrument that measures the concentration of glucose in the blood.
The test uses a paper strip that contains an enzyme called glucose oxidase. When a drop of blood is added to the paper strip, the enzyme converts glucose to gluconic acid. This generates an electric current that the glucometer measures and translates into a numerical value.

a) i) In which part of the blood is glucose transported? *(1)*
   ii) There are thousands of different chemical substances in the blood. Suggest why it is only glucose that reacts with the enzyme in the paper strip. *(2)*

b) Give **two** processes in the body that can lead to a change in the glucose level in the blood. State whether the change is an increase or a decrease. In each case, name or describe the blood vessel in which the change might be detected. Write your answers in the table below. *(4)*

Description of process	Increase or decrease in blood glucose level	Blood vessel(s) involved

c) Some people have diabetes and are not able to produce enough insulin. People with diabetes often find it useful to use a glucometer to monitor their blood glucose so that they can take action if necessary.
   i) Where is insulin produced in the body? *(1)*
   ii) How does insulin help reduce the level of glucose in the blood? *(2)*

d) The enzymes used in a glucometer are obtained from cultures of microorganisms. Suggest how the enzymes are produced on a large scale from these microorganisms. *(2)*

(Total = 12 marks)

Student response Total 10/12	Expert comments and tips for success
a) i) In the liquid between the blood cells O	You are expected to know the correct name for the plasma.
ii) The glucose molecules are the <u>only ones</u> <u>that fit</u> ✔ <u>onto the active site</u> ✔ of the enzyme on the strip.	Full marks for this well-expressed answer.
b) Glucose converted to glycogen in the liver with insulin helping – decrease. ✔ Hepatic vein ✔	Full marks.
Respiration in the kidneys — decrease ✔ Renal vein ✔	Respiration takes place in all living cells, so any named organ or group of cells would be correct here. The student has thought about the blood vessels they know and has chosen respiration in the kidneys so that 'renal vein' can be given as the blood vessel involved.
c) i) In the pancreas ✔	Correct.
ii) It causes the liver cells to take up more glucose. ✔	For b) the student said, correctly, that glucose was also converted to glycogen in the liver, with the help of insulin. However, the examiner cannot give marks to answers in another part of the question. To get the second mark, this same answer has to be repeated.
d) An industrial fermenter ✔ could be used. The microorganisms are kept at the right temperature ✔ and provided with nutrient liquid and oxygen. The enzymes are separated from the liquid at the end. (✔)	The student successfully applied knowledge of fermenters to the production of enzymes. The last statement would also have gained a mark but the maximum for part d) had been reached.

■ Understanding structure, function and processes

1 a) i) Draw a labelled diagram of a red blood cell. (2)
   ii) Give **two** ways in which the structure of a white blood cell differs from
        that of a red blood cell. (2)
 b) Red blood cells transport oxygen to the cells of the body.
   i) Explain how the features of a red blood cell help it carry out this function efficiently. (3)

   ii) Name the blood vessel in the human body that contains the highest
        concentration of oxygen. (1)

 c) White blood cells help reduce infection by pathogens in the body. There
    are different types of white blood cell. Choose **one** type of white blood
    cell and describe how it helps reduce the spread of pathogens in the body. (2)

(Total = 10 marks)

Student response Total 8/10	Expert comments and tips for success
a) i) cell membrane haemoglobin drawing ✔ label ✔	Drawing shows circular shape and central hollow (mark 1). Cell membrane labelled (mark 2). Mark scheme does not reward drawing quality, but note neat drawing and labelling, with label line just touching the membrane. Haemoglobin is not a cell structure (it is part of the cytoplasm).
ii) A white blood cell has got a nucleus. ✔ It has no haemoglobin. The lymphocytes are an oval shape ✔ and other white blood cells have an irregular shape.	One mark for nucleus and one for 'oval shape', which is acceptable for the shape. The statement about 'irregular shape' is an alternative way of gaining the same mark. No mark for 'no haemoglobin' because it is a substance in the cytoplasm, not a structure.
b) i) The haemoglobin ✔ can carry oxygen ✔ from the lungs to the muscles. Its shape means that it has a large surface area to volume ratio O so oxygen from the lungs can diffuse in quickly. ✔	Full marks given. No mark for the correct description of the large SA to volume ratio because the feature, the biconcave shape, was not described. However, linking the SA to volume ratio with rapid diffusion of oxygen was considered to be worth the final mark. The shape needs to be described as biconcave (or similar) to gain the shape mark.
ii) The aorta O	The pulmonary vein has the highest oxygen concentration.
c) A white blood cell engulfs bacteria O and destroys them. ✔	The student ignored the instruction to 'choose one type of white blood cell' and did not name the 'phagocyte' as the type that engulfs bacteria. However, 1 mark was given for 'destroy bacteria'. Any appropriate example of a pathogen would be acceptable.

Exam-style questions

1 A student was provided with a carbon dioxide sensor that measures the concentration of carbon dioxide in the surrounding air. This can be linked to a computer so that the readings are recorded directly and plotted as a graph of carbon dioxide concentration against time.

He set up an experiment as shown in the diagram and watched the computer screen on the table beside him.

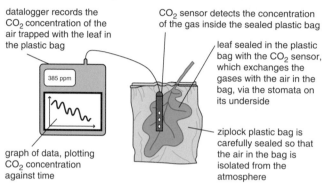

datalogger records the CO_2 concentration of the air trapped with the leaf in the plastic bag

385 ppm

graph of data, plotting CO_2 concentration against time

CO_2 sensor detects the concentration of the gas inside the sealed plastic bag

leaf sealed in the plastic bag with the CO_2 sensor, which exchanges the gases with the air in the bag, via the stomata on its underside

ziplock plastic bag is carefully sealed so that the air in the bag is isolated from the atmosphere

a) The computer graph shows that the level of carbon dioxide inside the bag is falling. Explain why this occurs. [2]

b) The student then opened up the bag and placed two snails inside with the leaf, and closed the bag again.

  i) Draw a graph of his expected results with the leaf and the snails together. [1]

  ii) Explain the results shown by the graph you have drawn. [2]

c) i) Describe how he could modify this experiment to show how light of different intensity affects the carbon dioxide concentration around the leaf. [4]

  ii) Give **one** limitation of using this set up for his experiment with light intensity and suggest how it could be improved. [2]

[Total = 11]

2 The diagram shows part of the human body with the lungs and some internal structures of the lungs.

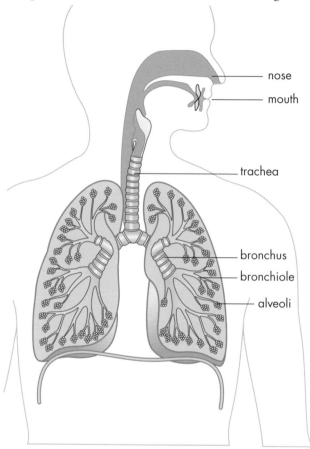

nose
mouth
trachea
bronchus
bronchiole
alveoli

Describe the route taken by a molecule of carbon dioxide in the blood in the pulmonary artery until it is breathed out into the air. In your description, explain the mechanism by which the carbon dioxide molecule moves from the blood and is exhaled from the body. Use information in the diagram to help you. [6]

[Total = 6]

3 **a) i)** Name the part of a flowering plant that transports water up the stem from the roots to the leaves. [1]

  ii) Explain why it is necessary for a flowering plant (such as a bean) to have a transport system, but the protoctist *Chlorella* does not. [3]

b) As part of their gas exchange system, insects have structures called trachea inside them. These trachea are tubes, strengthened with a substance called chitin. They connect directly with the air outside and branch into finer tubes that lie close to the cells throughout the insect's body. Suggest why insects are all relatively small. [2]

[Total = 6]

4 The table lists some exchange surfaces, found in flowering plants or humans. The exchange surfaces allow passage of molecules through them.

Copy the table. For each exchange surface, give the location of the structure, the molecules that pass through it and the direction of movement of the molecules. Write your answers in the table. For some structures there may be more than one answer, shown as 1 and 2. Some boxes have been completed for you.

Structure	Location	Molecules	Direction of movement
villus (villi)	small intestine	named digested food (e.g. glucose, amino acids)	
mesophyll cells (in the light)	leaf	oxygen	
mesophyll cells (in the dark)	leaf	carbon dioxide	
alveolus (alveoli)		1 2	1 2
root hair cells			

[Total = 11]

5 Some students investigated water loss by transpiration from leaves. They collected six leaves of similar size (A, B, C, D, E and F) from a tree.

Each leaf was smeared with vaseline (a grease), as shown in the table, then weighed. The leaves were then strung on a piece of cotton thread.

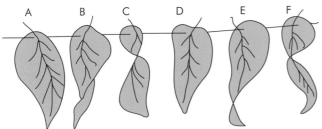

The leaves were left for 6 hours in a laboratory at room temperature, then weighed again. The loss in mass was calculated and expressed as a percentage. The students assumed that the water had been lost by transpiration.

a) The table shows the treatment given to each leaf and the mass at the start and after 6 hours. Calculate the missing values for the percentage loss in mass of the leaves. Two have been done for you. [3]

Leaf	Treatment	Mass / g		Percentage change in mass
		at start	after 6 hours	
A	upper surface smeared with vaseline	1.5	0.6	−60%
B	upper surface smeared with vaseline	2.2	0.8	
C	lower surface smeared with vaseline	2.0	1.9	
D	lower surface smeared with vaseline	1.9	1.7	−9%
E	both surfaces smeared with vaseline	2.7	2.7	
F	both surfaces smeared with vaseline	2.6	2.6	

b) i) Using your knowledge of leaf structure and of transpiration, explain these results. [3]

ii) Suggest why two leaves were used for each treatment. [1]

c) What differences would you expect if the whole experiment had been carried out at night (or in a dark room) rather than during the day? Explain your answer. [3]

[Total = 10]

6 As part of a routine health check, a sample of blood is often taken from a person and the number of red blood cells and white blood cells are counted.

Numbers change under different circumstances, but in a healthy adult, the number of red blood cells is about 5 million per microlitre (μl) of blood. The number of white blood cells is about 5000 cells per μl.

a) i) If you examine a drop of blood under a microscope, which type of blood cell would there be more of? [1]

ii) Draw a labelled diagram of a white blood cell. [3]

iii) Give **two** features of a red blood cell and explain how each feature helps the red blood cell to carry out its function. [4]

iv) Suggest why a check of red blood cell numbers is made before a person gives blood as a blood donor. [1]

b) Occasionally the number of white blood cells rises sharply for a few days. Suggest why this rise in numbers may occur. [2]

c) Some athletes use an illegal process known as blood doping, as a way of boosting the number of red blood cells in the bloodstream. Blood doping involves removing some of their own blood and storing it for several weeks. During this time, the body generates new red blood cells and replaces those that are lost. Just before a big competition, the blood is transfused back into their own body.

Suggest why athletes believe that blood doping could improve their performance in an athletic event. [3]

[Total = 14]

7 The graph shows the heart rate of a running athlete before, during and after a race.

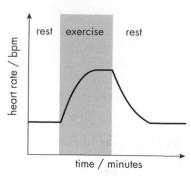

a) Give the word equation that summarises aerobic respiration. [2]

b) i) Describe the changes in the heart rate of the athlete during the race. [2]

ii) Explain the reasons for the changes in heart rate during exercise. [3]

iii) Suggest why the heart rate does not immediately return to the resting rate at the end of the race. [2]

c) Draw a line on the graph to indicate the likely changes in heart rate that you would predict for a fitter athlete taking part in the same race and who finished at the same time as the first athlete. [2]

[Total = 11]

EXTEND AND CHALLENGE

1 International trade in timber has contributed to the introduction of pests into one country from another. Outbreaks of disease can occur when imported timber is carrying spores of fungi or bacteria, which then attack the native trees in a region. Often insect pests are brought in with the timber and these may also carry the spores from one tree to another.

Dutch Elm disease killed a very high number of elm trees in the UK in the 1970s. Now, in the early twenty-first century, another disease known as Bleeding Canker disease is spreading across the UK, killing horse chestnut trees.

Both these diseases involve the growth of microorganisms (fungi and bacteria) within the xylem vessels. A tree can survive for a year or two if only a portion of the xylem vessels is blocked, but once the whole ring of xylem is blocked the tree rapidly dies.

a) i) What is the function of the xylem vessels in a plant?
   ii) How does the structure of xylem vessels make them adapted for their role in the plant?
   iii) If the xylem vessels become blocked, explain why this would lead to the death of the tree.

b) The functional xylem vessels of a tree are located just under the bark. The vessels deeper in the tree provide important support for the tree and it is this part that is used as 'timber'. What feature of xylem makes it so strong?

c) Why do you think imported insect pests are more effective in spreading disease from tree to tree, compared with 'local' insect pests? (Think about food chains and food webs.)

d) Find out about any control measures currently being used to limit the damage being caused by one of the diseases given in the passage.

e) List some uses of timber, then find out more about different types of timber. Summarise your findings in a table. You can start by finding out about 'hardwood' and 'softwood', then include information on:
   • the type of tree each timber comes from
   • where the tree is grown
   • the particular properties (value) of certain trees
   • what the timber is used for.

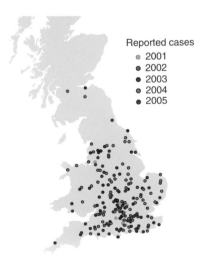

Reported cases
• 2001
• 2002
• 2003
• 2004
• 2005

Map shows geographical location of Horse Chestnut Bleeding Cankers in the UK, reported to Forest Research Disease Diagnosis Advisory Service.

f) Suggest some reasons why there is international trade in timber.

(continued)

2 When fresh blood is spun in a centrifuge tube, the red blood cells (erythrocytes) are precipitated to the bottom of the tube, as shown in the diagram. The red blood cells pack down first, with a thin layer of white blood cells (leucocytes) on top. Above this is the straw-coloured plasma.

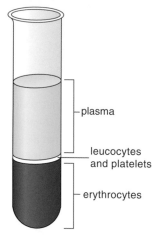

The proportion of the blood composed of red blood cells is called the haematocrit and varies from 40% to 50% in different individuals. People with a haematocrit below 35% are described as suffering from anaemia. People suffering from anaemia often have symptoms including excessive physical tiredness.

One treatment for anaemia involves being given an injection of a hormone (EPO), which stimulates extra red cell production from the bone marrow. This hormone is normally produced when the brain senses lower oxygen levels in the blood.

a) Suggest why anaemia would be expected to give the symptoms described.

b) International athletes often train at high altitude locations in the weeks leading up to an important competition. Others might sleep in a special low oxygen chamber that simulates high altitude. Explain the reasons behind these decisions.

c) Mountain climbers who are intending to climb on the highest peaks usually take several days to walk up to the base of the mountain, rather than simply taking a helicopter flight to a base camp. Explain the reasons behind this decision.

d) If an athlete has an (illegal) injection of EPO, their haematocrit will rise to as much as 60% over the next few days. This puts a lot of strain on their heart. What pressures might cause an athlete to take such a risk?

3 As you get older, the condition of your coronary artery wall deteriorates, as fatty deposits (atheromas) begin to build up in the walls of the vessels. High concentrations of some dietary components, such as saturated fats and cholesterol, speed up the formation of atheromas. The levels of these substances in the bloodstream are influenced partly by the genes you inherit, and also by your diet. Other aspects of your lifestyle, such as taking part in regular physical activity, or smoking, affect the rate of atheroma formation in the coronary arteries.

If an atheroma builds up too much, there is an increased risk that the blood squeezing past the restriction will form a clot, completely blocking the vessel (a coronary thrombosis). The heart muscle cells 'downstream' of the blockage no longer receive blood and begin to die. If a large proportion of the heart muscle is affected, the whole heart may stop pumping (fibrillation) and death occurs.

An early treatment for a person suffering from a fibrillating heart (heart attack) involves an injection of clot-busting drugs. Once they are stabilised, the person may then be given surgery in which a rigid cylinder (stent) is inserted into the narrowed blood vessel in order to widen it permanently. Alternatively, the patient may have heart bypass surgery as described at the start of Section 3.6 (page 118).

Use this book and other sources of information to write an informative illustrated leaflet concerning coronary heart disease (CHD). It should explain to a non-expert:

● the different possible factors leading to heart attacks
● how a blood clot develops in a coronary vessel
● how your personal risk of developing CHD can be reduced
● how stents and by-pass surgery help in the treatment of CHD.

Some hospitals have refused to treat patients for CHD if they refuse to give up smoking as part of their therapy. Do you agree with this decision by the hospitals? Explain the reasons for your answer.

4

Coordination and control

TO THINK ABOUT...

How do we know what is going on around us and how do we respond to our surroundings? How do processes inside the body work together rather than at random? And how do plants respond to changes in their surroundings?

Receiving signals and responding

Scatter a handful of corn among some backyard chickens. Immediately you hear clucking as they burst into activity, pecking this way and that, clustering around the corn. Yet more chickens arrive to join the group and more clucking until the corn is eaten.

A flash of lightning and the loud noise of thunder... a familiar forewarning of a thunderstorm. Both the flash and the crashing noise would probably startle us and alert us to a possible danger.

A group of leaves on a tree – not jumbled in any direction but forming a neat mosaic all facing outwards in a way that captures as much sunlight as possible for the tree.

So what stimulates and controls the sequence of behaviour shown by the chickens? Why does the first chicken alert the others, rather than just eating all the corn that it finds? What coordinates the series of activities described? How do we receive the signals of lightning and thunder and what are the controls inside the body that trigger any responses? What advantage is it for the leaves to have the arrangement shown in the photograph? What are the internal events that make this occur?

This section deals with coordination and control – another essential characteristic that is special to living organisms. While artificial machines can go a long way, none can achieve what living organisms can.

4.1 Excretion in flowering plants

Do plants need to get rid of metabolic waste?

A fundamental difference between flowering plants and animals is that plants synthesise the materials that they require to make up their tissues, whereas animals take in food containing large molecules, which they assimilate in different ways into their body. Plants use simple substances (carbon dioxide, water and certain mineral ions) to synthesise large complex molecules, including carbohydrates, proteins, fats and oils. Animals break these down again into smaller units before building them up again into the particular molecules that make up their tissues.

Excretion is defined as the process that eliminates waste material from metabolic reactions in an organism. But do plants produce waste metabolic materials or do they just synthesise what they need? Are any of the products of reactions in plants so toxic that they need to be eliminated?

Let's look at what goes into and out of a plant when these exchanges take place and whether they fit the definition of excretion.

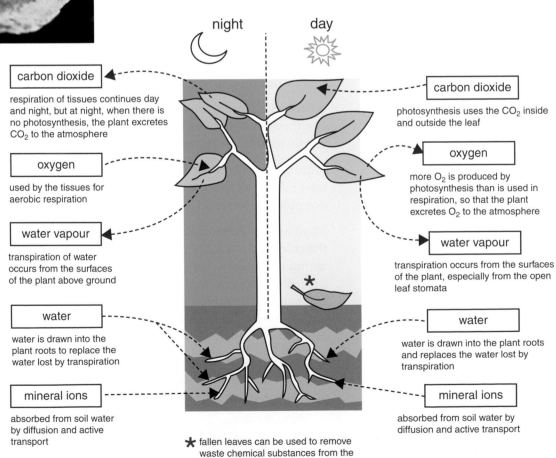

night   day

carbon dioxide

respiration of tissues continues day and night, but at night, when there is no photosynthesis, the plant excretes CO_2 to the atmosphere

oxygen

used by the tissues for aerobic respiration

water vapour

transpiration of water occurs from the surfaces of the plant above ground

water

water is drawn into the plant roots to replace the water lost by transpiration

mineral ions

absorbed from soil water by diffusion and active transport

carbon dioxide

photosynthesis uses the CO_2 inside and outside the leaf

oxygen

more O_2 is produced by photosynthesis than is used in respiration, so that the plant excretes O_2 to the atmosphere

water vapour

transpiration occurs from the surfaces of the plant, especially from the open leaf stomata

water

water is drawn into the plant roots and replaces the water lost by transpiration

mineral ions

absorbed from soil water by diffusion and active transport

✱ fallen leaves can be used to remove waste chemical substances from the plant

Waste materials from metabolic reactions in plants

The diagram on the previous page summarises the processes that lead to excretion in plants. It shows what goes into a plant and what comes out – sometimes, but not always, these materials are involved in metabolic reactions. You can see that the 'waste' excreted by the leaves of a plant depends on whether or not they are exposed to light. The stomata are very important structures for these materials passing into and out of a leaf. Whether the stomata are open or closed also depends on light intensity. In the diagram, movement of the gases shown is by diffusion, so excretion in plants cannot be described as active elimination of materials. We look at each process in a little more detail below.

As with all living things, leaf tissues respire all the time, using oxygen and producing carbon dioxide as a waste product. However, during the day when there is sufficient light, the quantity of carbon dioxide used in photosynthesis far exceeds the quantity produced in respiration. At the same time, the quantity of oxygen generated as a waste product of photosynthesis far exceeds the quantity used in respiration. The net effect during the day is that excess oxygen is excreted through the stomata. At night, respiration uses oxygen and produces carbon dioxide. The net effect during the night is that some carbon dioxide is excreted.

So oxygen and carbon dioxide can both be waste products in the same plant organ. But they can also both be substances involved in metabolic reactions that take place in the leaf cells of the plant. It all depends on the intensity of light.

The water vapour lost by transpiration is mostly a result of the transpiration stream pulling water through the plant. Very little, if any, represents metabolic water from respiration in cells.

The diagram also shows leaves falling from the tree. Some substances are broken down in plant cells and these may produce chemical substances that are not required by the plant and which cannot be converted into another useful compound. So these are waste materials. Such unwanted products can be stored in dying tissues, such as leaves in the autumn. The wonderful colours often seen in leaves during the autumn are mostly the result of the waste chemical substances that they contain (see photographs on page 138).

STUDY QUESTION

1   Use this question as an opportunity to make links to the other parts of the book that give more details of each of the processes shown in the diagram of the tree. This should help you to understand these processes in the context of excretion in a flowering plant.

Work your way around the tree and for each molecule (or ion) shown, find out how it moves in or out of the plant. Use the list below to help you (working around the diagram clockwise from bottom left). Summarise your information in a table. References to the relevant sections are given in brackets.

- mineral ions (salts) into root (Section 3.4)
- water into root (Section 3.4)

- water vapour out of leaf by transpiration (Section 3.4)
- carbon dioxide from respiration out of leaf (Sections 2.3 and 3.1)
- carbon dioxide into a leaf and used in photosynthesis (Section 2.1)
- oxygen from photosynthesis going out of a leaf (Section 2.1)

You can also make links to diffusion (Section 1.5) and how stomata open and close (Sections 3.1 and 3.4).

4.2 Excretion in humans

Different ways of getting rid of waste materials

Red flour beetles never drink. They live, as their name suggests, in dry flour, so they never have access to liquid water. Their food is dry, so the only water they get comes from cellular respiration. Remember, in respiration, glucose is broken down into carbon dioxide and *water*.

Many insects with such a limited supply of water cannot afford to make urine to get rid of (excrete) waste products, such as urea. Instead, they convert urea to insoluble uric acid and store up this waste inside their bodies. By the end of its life, 10% of the body mass of an insect may be made up of this waste uric acid.

Similarly, a growing chick embryo cannot step outside its egg to urinate. So it also converts urea to uric acid and stores it in a special sac (the allantois) in part of the egg. This waste is left behind when the chick eventually hatches. In fact, adult birds also get rid of their waste as uric acid – they excrete it as a paste and the uric acid gives it a white appearance.

So what about humans? What do we produce as waste materials? How do we get rid of waste materials and which organs of the body are involved in excretion?

■ What is excretion?

Excretion is defined as the removal of waste products from the **metabolic reactions** of the body. Metabolic reactions are the many chemical reactions that take place inside the cells of living organisms. Many of these metabolic reactions produce waste products that are not required by the cells (and may even be toxic if allowed to accumulate). **Excretion** should not be confused with **egestion**, which is the expulsion of food that has never been absorbed into the body cells. This material is known as faeces and is egested from the anus.

The metabolic wastes produced by the body include:

- **carbon dioxide** and **water** produced by the aerobic respiration of sugars in cells
- **urea** produced by the liver when it breaks down amino acids (which in turn have come from proteins that are no longer required in the body)
- various other substances, including excess minerals absorbed from the diet (such as sodium) and medical drugs (such as antibiotics).

waste materials must be removed from the body for a variety of reasons. These include:

- the waste material might reach toxic concentrations
- because of the osmotic effects of increased concentration of body fluids
- the loss of space in storing waste products.

The main routes for metabolic wastes to leave the body are:

- **the lungs** – for excretion of carbon dioxide and water
- **the skin** – for excretion of water and excess mineral ions, such as sodium
- **the kidneys** – for excretion of water, urea, mineral ions and some other substances.

STUDY TIP

You can find details of the role of the lungs in excretion in Section 3.2, and the role of the skin in sweating (as part of thermoregulation) is covered in Section 4.3.

The kidneys as excretory organs

One third of the blood pumped out of your heart into your aorta eventually goes to your two kidneys. The kidneys carry out several important functions that keep us healthy. The kidneys remove **waste materials**, such as **urea** from the blood and keep the **water content** of the bloodstream constant (**osmoregulation**).

Each kidney is attached to the dorsal wall of the abdominal cavity. The urine produced is carried in the **ureters** to the **bladder**, and from the bladder to outside the body in the **urethra**. The bladder is a muscular sac, which can expand to store urine produced and can hold it for several hours. Eventually the bladder becomes full and its muscular walls squeeze the contents. When you relax the circle of sphincter muscles at the exit of the bladder, the urine is forced down the urethra and the bladder empties to the outside. Figure 4.1 shows these structures as well as the blood vessels entering the kidney (the **renal artery** from the aorta) and leaving the kidney (the **renal vein** into the vena cava).

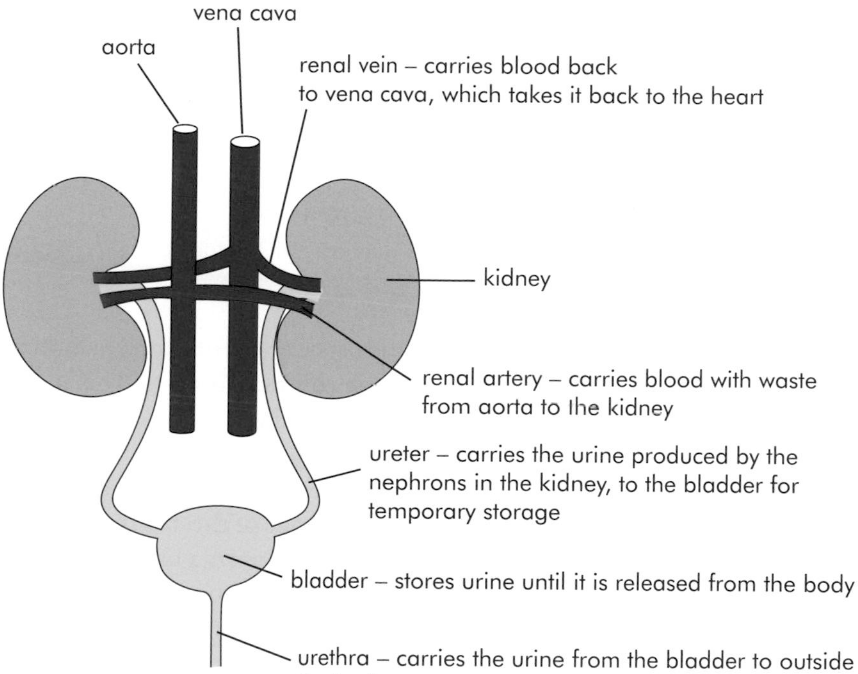

Figure 4.1 Kidneys, bladder and related blood vessels.

STUDY TIP

Make a link to Section 3.6 (circulation of the blood).

Look at Section 5.3 (reproduction in humans) – and notice the relationship of the urethra with the reproductive organs in males and females.

Make sure you get the spelling right for three rather similar words: **ureter** and **urethra** (linked to the kidney) and **uterus** (part of the female reproduction system).

How the kidney removes urea from the blood

Inside each kidney, the renal artery branches into smaller arteries, each of which leads to a knot of capillaries called a **glomerulus**. These glomerular capillaries sit inside a cup-shaped structure, known as the **Bowman's capsule**. This is a double-walled structure that leads into a tube, known as the kidney tubule. The whole structure is called a **nephron** (Figure 4.2).

There are millions of these nephrons, making up the bulk of the structure of the kidney. Leaving a Bowman's capsule, the tubule coils many times (the proximal convoluted tubule), then there is an uncoiled U-shaped portion (the loop of Henlé), which leads into a second coiled portion (the distal convoluted tubule). This passes into a collecting duct, which links with other tubules, then passes to the bladder. Figure 4.2 shows these structures for a single kidney tubule. You can also see how the branches from the renal artery wrap around the kidney tubule, before joining up to form the renal vein.

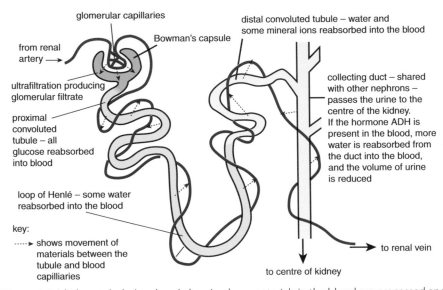

Figure 4.2 A kidney tubule (nephron) showing how materials in the blood are processed and how urine is produced.

The pressure of the blood is high in the renal artery as it enters the **glomerular capillaries**. This high pressure is partly because the diameter of the blood vessels leaving the glomerulus is smaller than the diameter of blood vessels entering the glomerulus. This high pressure forces a lot of water, as well as some smaller molecules (such as urea, mineral ions and glucose), out of the capillaries into the cup-shaped **Bowman's capsule**. Larger components of the blood (such as red blood cells and proteins) are too big to pass out of the capillaries and they remain in the blood. The filtration process that occurs here is described as **ultrafiltration**. Only water and small molecules leave the blood and this forms the **glomerular filtrate** in the Bowman's capsule. This is the first step in the formation of urine.

The filtrate passes through the walls of the Bowman's capsule and into the tubule. As it passes along the tubule, the composition of the filtrate changes because certain substances are taken back into the blood. This process is known as **reabsorption**. The changes as the filtrate passes along the tubule and into the collecting duct are summarised below and also illustrated in Figures 4.2 and 4.3:

- **proximal tubule** – 100% of the glucose in the filtrate is actively transported back into the blood, as well as some mineral ions
- **loop of Henlé** – water is reabsorbed back into the blood
- **distal tubule** – more mineral ions and water are reabsorbed into the blood
- **collecting duct** – is shared with many other nephrons and here, during the final step, even more water is taken back into the blood.

PRACTICAL

Practical activity – investigating the presence of glucose in simulated urine

Several samples of simulated urine are prepared, using water with a little yellow food colouring added. These can be tested for the presence of glucose using Benedict's solution (as described in Section 1.4). 'Healthy' samples are just water and yellow colouring. 'Unhealthy' samples are created by adding 1% glucose solution to the coloured water mixture. The samples model the urine of people suffering from excessively high blood sugar, as happens in diabetes (see insulin and blood sugar control, Section 4.5).

By this stage the urine has become quite concentrated. The collecting ducts lead towards the centre of the kidney. The urine is carried from this region in the centre of the kidney, along the ureter to the bladder for temporary storage. The processed blood returns to the body in the renal vein.

Urine contains water in which waste substances are dissolved. The volume of urine produced varies according to the immediate needs and activities of the body. An important waste product found in urine is urea, produced by the liver from the breakdown of excess amino acids in the diet. You are likely to excrete a total 2 g of urea per day dissolved in your urine. Excess mineral ions, particularly sodium, and a variety of other substances, such as detoxified drugs, are present in the urine. The variation in the colour of your urine from day to day is due to its fluctuating concentration of dissolved substances.

Figure 4.3 gives a summary of the events that occur in a nephron, from the ultrafiltration of blood in the glomerular capillary to the passage of urine to the bladder.

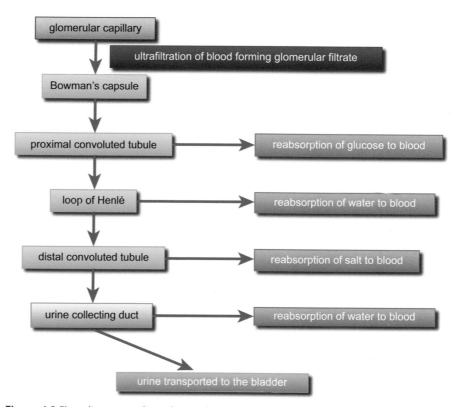

Figure 4.3 Flow diagram to show the production of urine in a nephron.

Controlling body water content (osmoregulation)

When playing an energetic sport, you lose a lot of water, partly by sweating and partly as water vapour in your exhaled air. Even doing moderate exercise, at 35 °C a person may lose around 1.5 litres in an hour. To prevent dehydration, this lost water must be replaced by drinking fluids. That is quite a lot to have to drink! The percentage of water in your body mass is somewhere between 50% and 60% (higher in men than in women). If you cannot replace the lost fluid immediately, or if you drink far more fluid than you need, there is a **homeostatic** system that helps to adjust the water content of your body. This is described as **osmoregulation** – the control of the water and salt concentration in the body fluids.

Table 4.1 summarises typical water intake and loss in an adult in a day.

Table 4.1 Typical water intake and loss in a day for an adult.

Water gains per day / cm³		Water losses per day / cm³	
drinking	1200	urine	1600
water contained in food	1000	sweating and breathing out	800
cellular respiration producing water	300	faeces	100
Total	**2500**	**Total**	**2500**

The kidneys and water homeostasis

Simply breathing involves a continual loss of water from the lungs. The quantity of water lost by other routes (sweating and urine) varies a lot. When we carry out exercise or when the environment is hot, we lose extra water by **sweating** – this is a mechanism for cooling the body and preventing overheating.

This water loss is balanced by a reduction in the volume of urine that is formed. A hormone, known as **ADH** (**antidiuretic hormone**) is involved in this homeostatic mechanism. We get rid of excess water in the urine we excrete. Receptors in the base of the human brain monitor the water content of our blood.

If the water content of the blood falls lower than normal, the brain receptors stimulate the release of ADH from the pituitary gland of the brain. The ADH causes more water to be reabsorbed from the collecting duct in the nephron into the blood. As a result, the kidneys produce a smaller volume of urine. Less water is excreted and the urine is more concentrated. The effect of this is that the water content of the blood is restored to normal.

When we drink more fluid than we need, the water content of our blood rises above normal. As described above, this change is detected by the receptors in the brain, which stimulate a reduction is the release of ADH from the pituitary gland. This results in the kidneys producing a larger volume of urine, so that we excrete more water. This returns the water content of the blood to normal.

STUDY QUESTION

1 Round up of words and vocabulary

In Section 4.2 you meet a number of new terms, some of which are structures in the kidney or urinary system, and others describe processes that take place in the body. You need to be sure you are familiar with all these words and know about the part they play in excretion in humans. A list of some terms is given below.

For each of the words in the list give the meaning of the term or write something about it to show you understand why it is used in this topic. If it is a structure, draw a diagram to show where you would find it and describe its function. Check through one block at a time.

ADH	metabolic reaction
aorta	nephron
bladder	osmoregulation
Bowman's capsule	proximal tubule
distal tubule	renal artery
egestion	renal vein

excretion	ultrafiltration
glomerulus	urea
glomerular filtrate	ureter
homeostasis	urethra
kidney tubule	uterus
loop of Henlé	vena cava

4.3 Coordination and response in living organisms

Tolerating changes in the environment

Some organisms can survive exceptionally harsh conditions. They are called extremophiles and include a number of species of bacteria, such as those that live in the boiling water of hot springs (image on left). The enzymes of these bacteria can operate at temperatures from 30 °C to 94 °C and in a pH range from 6 to 10.

Some multicellular animals and plants are able to survive being completely frozen, and seem unharmed when they thaw out. The little water bear (shown on right) is only about 1 mm long and normally lives in clumps of moss. It is one of the toughest animals on the planet – it can survive being frozen and being dried out. It has even survived trips into outer space (without a spacesuit), courtesy of NASA.

But for the majority of complex animal and plant life forms, the physical limits of life are strict. As humans, we can travel to and survive in some extreme environments, but inside our body the range of tolerance is very narrow. So what are the limits? How do we keep within the range that allows cells to function and what mechanisms are involved in maintaining the right levels, in particular of temperature and water content in the human body?

■ Homeostasis

The term **homeostasis** is used to describe the control of fluctuations in the internal environment in the body of an organism. The **internal environment** includes factors such as water content of body fluids, temperature and pH. This internal environment must be carefully kept within certain narrow limits if an organism is to remain healthy – if the limits are exceeded the organism may die.

The potential range of environments that different species can occupy is largely determined by the ability of their homeostatic systems to keep their internal environment stable. This is true of large organisms such as trees and is also true of the animals that dwell among them. Birds and mammals are the only vertebrate groups that can actively maintain their body temperature when their surroundings become too cold or too hot.

However, other groups of animals can show some control. This is, for example, shown by reptiles, which can adjust their body temperature by seeking sunshine or shade – a behavioural response. Plants are literally rooted to the spot so cannot escape, but some use the transpiration of water from their surfaces, or the drooping of wilted leaves, as a means of providing cooling.

The concept of homeostasis can be applied to plants as well as animals, but in Section 4.3 we limit the discussion to control of temperature in humans. Control of water content in the body is another example of homeostasis and is considered in Section 4.2.

Control of body temperature in humans

For humans, a change in core body temperature of more than 2 °C can be fatal. The body must make a coordinated response to any rise or fall in body temperature. One reason for this limited tolerance is the effect on enzymes and their response to changes in temperature.

If you are ill and someone wants to measure your temperature, they do not just measure the temperature of your skin. Instead they measure the temperature under your tongue or inside your ear. This is because the **skin surface** temperature of the human body can change noticeably from one moment to the next. However the **core body temperature** – that of the blood circulating around the organs inside the body – is kept close to 37 °C.

If you have a fever and your body temperature is abnormally high, you might be given a drug, such as paracetamol, that will help to cool you down, or you might be bathed with cool water. It is important to ensure that the core temperature inside the body does not rise too high outside its limits.

When you take part in a physical sport, such as running or tennis, your active muscles respire more glucose to provide the energy for contracting the muscles. This respiration releases **heat** as a by-product. The heat warms your blood, which then carries the heat away from the muscles. The blood carries the heat from all your internal organs to the body surface. Here heat may be lost to the surroundings. If too little heat is lost, the temperature of your blood begins to rise above normal.

The blood circulating into the base of the brain is constantly monitored by **temperature receptors**. Any rise in temperature stimulates these temperature receptors. The brain then coordinates a cooling response.

The temperature control responses involve sweat glands, blood vessels in the skin and, to some extent, the hairs on the skin:
■ **Sweating** – sweat is a watery fluid, containing some mineral salts (including sodium chloride, which is why sweat tastes salty) and urea. The sweat gland draws water out of the blood capillaries in the skin and sweat passes up the sweat duct to the surface of the skin. The liquid evaporates from the skin surface and to do this (change from liquid to vapour) heat is taken from the skin. This causes cooling. If weather conditions are very humid, sweat does not evaporate easily and you do not benefit from cooling in the same way as you would in a dry atmosphere. Note that sweating continues all the time, but *increases* when body temperature rises above the normal range.

Check in Section 4.2 for details about osmoregulation.

Check in Section 1.4 for information on the effect of temperature on enzyme activity.

STUDY TIP
To help you remember the meaning of homeostasis, 'homeo' = same and 'stasis' = standing still (or same position).

STUDY TIP

Figure 4.4a Sweating is a cooling mechanism.

Figure 4.4b The body takes action to conserve heat in a cold environment.

- **Blood vessels in the skin** – these can contribute to temperature control by changing the volume of blood that flows near the skin surface. When body temperature rises above normal, the diameter of the arterioles (branches from the arteries) increases. This allows more blood to flow through the blood capillaries near the skin surface. Heat carried in the blood can then be lost by radiation. Conversely, when the outside temperature is cold and the body needs to conserve heat, the arterioles constrict in diameter and less blood flows in the capillaries near the skin surface. The term **vasodilation** is used when the arterioles dilate (widen) and the term **vasoconstriction** is used when the arterioles constrict (become narrower).
- **Hairs on the skin** – when the temperature is cold, the hair muscles in the skin contract and the hairs are raised upwards. This allows them to trap a layer of air close to the skin surface and this layer of air helps to insulate the skin surface. The converse occurs when it is warm. While this makes some contribution in humans, the effect of insulation is more obvious in mammals that have fur.

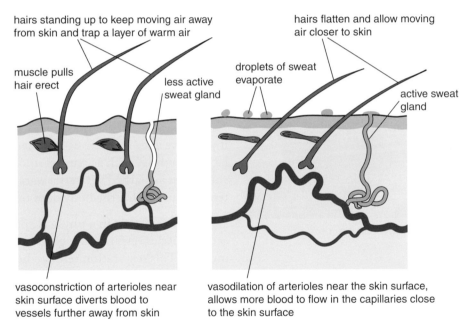

hairs standing up to keep moving air away from skin and trap a layer of warm air

hairs flatten and allow moving air closer to skin

muscle pulls hair erect

less active sweat gland

droplets of sweat evaporate

active sweat gland

vasoconstriction of arterioles near skin surface diverts blood to vessels further away from skin

vasodilation of arterioles near the skin surface, allows more blood to flow in the capillaries close to the skin surface

Figure 4.5 How structures in the skin help with temperature control – when cold (left) and when hot (right).

The sweat glands and skin capillaries are **effector organs**. They act on the messages they receive from the brain and produce an appropriate **response**.

As well as these physiological responses of the body you might alter your **behaviour** – taking off a layer of clothing, having a rest or drinking something cold if you are too hot. Conversely, in cold conditions, people wear more protective clothing to insulate from the cold. Houses may be heated (or cooled) to protect against extremes of heat or cold. These are some of the ways that have enabled people to inhabit regions of the world at temperatures outside what would be comfortable without these behavioural responses.

Table 4.2 summarises responses to falling or rising temperatures in the blood in the body. It includes reference to **shivering**, which is a familiar response in cold situations. The extra activity of shivering muscles generates heat from respiration and this contributes to warming the body.

Now go to page 168 to apply this to some data collected in an investigation on skin temperatures.

MATHS TIP

Mode, median and range

These terms are alternative forms of average taken from sets of data.
The **mode** is the most frequently occurring value in a set of data.
The **median** is the middle value in a set of data ranging from lowest to highest values (so half the data values are below the median value, and half are above it).
The **range** is the value of the difference between the largest and smallest values in a set of data.

Table 4.2 A summary of responses to rising or falling temperatures.

Change in core temperature	Effector organ action	Explanation
Temperature rises above normal	sweat glands secrete more sweat onto skin	sweat evaporates from skin using warmth of body
	arterioles near skin surface dilate, more blood flows through capillaries	warm blood is diverted close to skin surface and heat is lost to the air by radiation
	body hairs flatten	allows cool air to circulate closer to the skin, taking away heat
Temperature falls below normal	sweat glands secrete less sweat	less evaporation of sweat, so less warmth used from the skin
	arterioles near skin surface constrict, less blood flows through capillaries	warm blood is diverted further away from the skin surface and less heat is lost to the air by radiation
	body hairs become erect	this creates an insulating layer of still air near the skin, keeping cool air further away
	muscles shiver	extra respiration in the shivering muscles generates heat to help warm the body

STUDY QUESTIONS

1 Check the meaning of each of the following terms and write definitions for them:
   - homeostasis
   - osmoregulation
   - vasoconstriction
   - vasodilation.
2 How is heat generated in the body and how is *extra* heat generated when you are cold?
3 Why is it important that human body temperature does not fluctuate much from its normal temperature? (Check the effect of temperature on enzyme activity.)
4 Explain how sweating cools the body.
5 List the steps that take place in the arterioles in the skin:
   - when the body is hot
   - when the body is cold.
   Explain how these actions help to control the temperature in the body.
6 Suggest why it is important to wear a warm hat, gloves and warm socks in cold weather.
7 'Hypothermia' may occur when the body temperature falls below 35 °C and can be dangerous if the person does not recover quickly.
   Work out why hypothermia may occur if a person is wearing wet clothes, even though the surrounding temperature is not very low. (Think about conditions that affect the rate of transpiration in plants in Section 3.4.)

4.4 Coordination and response in flowering plants

Figure 4.6 This bean plant responds to the stimulus of touching a support by curling around it.

Sensitivity of plants

Mimosa is a touch-sensitive plant. It grows naturally in tropical regions. When something touches its leaf, the plant instantly folds its leaflets and the whole leaf droops down. This probably protects the leaf by scaring off herbivorous insects that land on it. Trackers in the jungle can spot the route taken by the animal they are hunting by following the folded leaves of mimosa plants. The speed of this plant response is unusual. It happens because special cells at junctions in the leaves suddenly lose their water and collapse. The result is that the leaflets fold and the whole leaf collapses.

Section 4.4 looks at some more familiar responses of flowering plants to their surroundings (responses to the direction of gravity and of light), but scientists are finding out more about other responses of plants. It is now known that plants can send out and detect chemicals in the air so that they can 'sense' when their neighbours are being attacked or damaged.

What else might plants respond to? Why is it an advantage for plants to respond to their surroundings? Make a list of possible things that plants might respond to, then choose one stimulus and think how you would test whether or not a plant can respond to it.

■ Responding to stimuli

Flowering plants are literally rooted to the spot. They cannot run away from danger or chase food to eat. But to survive, they still need to respond to their surroundings. Their leaves need to be in a good position to absorb light for photosynthesis and a plant shoot gradually turns towards the Sun. The tendrils of a pea plant reach out and secure the plant to a nearby support to climb up. A flower develops in response to changing seasons and flower petals open and close according to the time of day. These are just a few examples of ways in which plants respond to stimuli in their environment (note 'stimulus' = singular and 'stimuli' = plural).

Many plant responses are relatively slow, taking minutes or hours, rather than the milliseconds of some human reflexes. The response of the mimosa, described at the start of this section, is unusual in plants.

Tropisms – directional growth responses

The directional responses made by plants involve changes in the growth rates of plant tissues. For example, if cells on one side of the plant shoot elongate more than cells on the other side, the result is that the shoot curves. These growth responses enable a plant organ, such as a root or a shoot, to grow towards or away from the direction of a stimulus.

If the growth is towards the stimulus, we describe the tropism as **positive** and if the growth is away from the stimulus, we describe the tropism as **negative**.

Table 4.3 gives a few examples of tropisms and summarises responses of roots and shoots to direction of light and of gravity. Note that responses to gravity are also described as 'gravitropisms'.

Table 4.3 Examples of tropisms in plants.

Stimulus	Name of response	Examples
Gravity	positive geotropism	seedling roots grow down, towards centre of the Earth
	negative geotropism	seedling shoots grow upwards, away from centre of the Earth
Light	positive phototropism	seedling shoots grow towards a light source
	negative phototropism	seedling roots grow away from a light source

STUDY TIP

You might hear 'plant hormone' used, but 'plant growth regulator' is the correct term. However, you may find it useful to compare the part played by plant growth regulators (in plants) and hormones (for example, in humans).

To understand how a shoot (or root) grows and curves in response to a stimulus, we can look more closely at the regions of a plant shoot where new cells are produced by cell division and where growth occurs. The region of cell division lies just below the shoot tip. Initially, each new cell is tiny but eventually they swell up, elongating to their mature size. The process of cell elongation is affected by **plant growth regulators**. One example of a plant growth regulator is the chemical substance known as **auxin**.

In a plant shoot, auxins are produced in the tip and are transported down from the tip towards the region of cell expansion (Figure 4.7). If the concentration of auxins is equal on all sides of a growing shoot, the expansion of cells on all sides is the same and the shoot continues to grow in a straight line. If, however, there is more auxin on one side, those cells elongate more and the shoot curves.

The stimulus to which a plant responds must be detected by receptor cells and these are usually located in the growing points of a plant, such as the root tips and shoot tips. The receptors then control the distribution of auxin to the regions of growth, behind the tip.

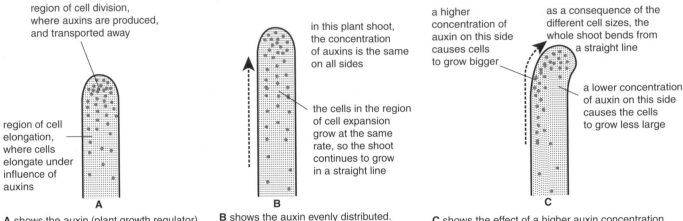

A shows the auxin (plant growth regulator) in the shoot tip and spreading to the region of cell elongation below the tip.

∴.ₐ = auxin molecules

B shows the auxin evenly distributed. Cell expansion is the same for all cells in the region of cell elongation. The result is that the shoot grows straight.

C shows the effect of a higher auxin concentration on one side of the shoot. The result is that the cells in the region of cell elongation grow at a faster rate and the shoot bends.

Figure 4.7 Elongation of cells in the growing region of a shoot under the influence of auxin (a plant growth regulator). The dashed arrows show the direction of growth.

Geotropic responses of roots and stems

Plant roots are not normally exposed to light, as they grow in the darkness of the soil. In the soil, plant roots grow downwards, allowing them to obtain water and mineral ions and helping to anchor the plant in the ground. This is an example of **positive geotropism** (gravitropism) – growing towards the centre of the Earth.

A plant shoot growing out of a germinating seed under the soil grows upwards, eventually reaching the light. This is an example of **negative geotropism** (gravitropism).

As a result of these geotropic responses, the roots and shoots of germinating seeds in the soil always grow in the right direction – so that shoots and their leaves eventually reach the light and roots eventually reach water and mineral ions (Figure 4.8).

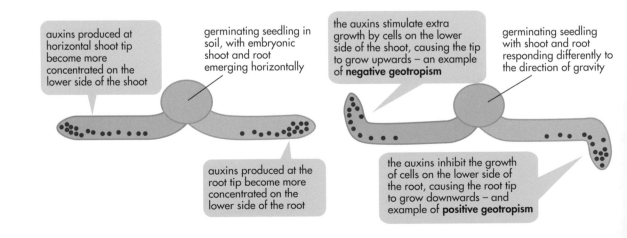

Figure 4.8 The different geotropic responses of roots and shoots is a result of their different sensitivities to the plant growth regulator auxin.

STUDY TIP

Look in Section 2.1 to remind yourself about ways that leaves are adapted to receive light for photosynthesis.

Positive phototropism in stems

An example of phototropism is the positive phototropic response of wheat coleoptiles. (A coleoptile is the first shoot that emerges from certain plants, such as wheat, maize and grasses.)

A summary of an experiment is given in Figure 4.9. This shows that the signal that passes from the tip is a chemical substance. It also shows that the signal that passes from the tip *cannot* be an electrical signal similar to that carried by nerve cells in animals. The positive phototropic response results from the transport of auxin. The stem tip detects the direction of the light. This results in auxin being transported from the lighter side to the darker side of the stem behind the tip. This then leads to uneven growth and the bending of the stem towards the light source.

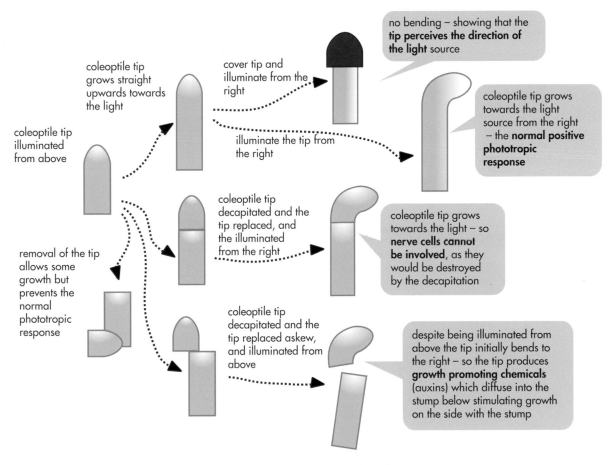

Figure 4.9 Some observations on the phototropic responses of wheat coleoptiles.

STUDY QUESTION

1 a) Flowering plants respond to the direction of light and of gravity. List the responses to each of these stimuli by:
   i) shoots
   ii) roots.
   In each case, describe the response as a tropism and state whether it is a 'positive' or 'negative' response. In each case, suggest an advantage (to the plant) of response.

   b) A response of shoots is to 'bend' towards the direction of light. Answer these questions to help explain how this occurs.
   i)   How does a shoot 'bend' towards the direction of light?
   ii) Where, in a shoot, does this growth response take place?
   iii) What role does auxin play in this response?

   c) Make a table to compare responses of flowering plants and humans. You may have to look in Section 4.5 to help you. In your table, include reference to:
   ■ the types of stimulus
   ■ how or where the stimulus is detected
   ■ how information about the stimulus is transferred to other parts of the plant or human body
   ■ the speed of the response
   ■ the part of the plant or human body that responds to the stimulus, and how it responds.

4.5 Coordination and response in humans

What eyes can see

You are able to read this sentence because rays of light have bounced off the page and into your eyes. Here the rays are focused onto light-sensitive receptors. Nerve impulses containing information about the shape of each letter pass to a special visual region at the back of your brain. Other regions of your brain help you to interpret the letters, to work out the words and their meaning in the written sentence. You did not have to learn to see – this is programmed into your developing nervous system. You do learn to read and to use language to help you understand and communicate with others. Learning allows us to be flexible in dealing with the changing world around us.

Some sort of vision is widespread in animals. Some can see in much lower light levels than we can. Others can see wavelengths of light that we cannot see, such as ultraviolet. Yet others see in much more detail than us. Compare the acute vision of a bird of prey with our own attempts to see a small bird in a tree on the other side of a field. These images show different eye structures in an arboreal frog, a gecko lizard and a Spanish jumping spider. In each animal, the eye plays a key part in detecting signals as part of its response to the surroundings.

The evolution of eyes gives animals a chance to respond to danger *before* they physically meet it. Humans, like other animals, have different senses that give us information about the world around us. What stimuli do we receive and how does the body coordinate the information and make an appropriate response?

■ How we respond to a stimulus

It is amazing how quickly you respond to a painful stimulus. Suppose you place your hand in the hot steam from boiling kettle – without thinking about it and in a fraction of a second, you withdraw your hand from the steam. This is a form of behaviour called a **withdrawal reflex**. Only after you have pulled your hand away do you start to feel the pain. This response does not require a conscious decision on your part, so the reaction time is much reduced. Any delay would mean more damage to the hand from the hot steam. Other responses of the body do not have to happen so urgently.

■ Nervous and hormonal control systems

Two control systems work together to help the human body respond to stimuli: the **nervous** system and the **hormonal** system.

A painful stimulus is just one example of the many stimuli to which the human body must respond during a normal day. Whenever there is a **change** in the external or internal environment of our body, this acts as a **stimulus** and the nervous and hormonal systems **coordinate** a suitable **response** to this change.

Nerve signals are electrical impulses. They are carried by nerve cells (neurones) at speeds of up to 100 metres per second. This allows a rapid response to a stimulus, like the pain-withdrawal one.

Working in harmony with the nervous system is the **endocrine** (**hormone**) system. Hormones are **chemical** substances produced by endocrine glands. These chemical substances provide a signal that triggers a response. Hormones are carried by the **blood** and so circulate around the whole body. This can take a few seconds, so the hormonal control system deals with slower events than the nervous system.

Nerves and the central nervous system

The central nervous system (CNS) is made up of the brain and spinal cord, which lies in a canal running through the bones (vertebrae) of the backbone (spine). From the central nervous system, nerves (neurones) spread out to all the regions of the body. Thus the CNS acts as a central coordinating centre for the impulses that may come in from any part of the body.

A **stimulus** is received by a sensory (receptor) neurone. Some receptors are specialised to receive particular responses. When a receptor is stimulated, it produces electrical impulses. The impulses then pass along a **sensory neurone** to the central nervous system (CNS) – either the spinal cord or the brain. Here the incoming nerve signals are passed to a **relay neurone** and this links to other neurones, depending on the response required. The relay neurone links to a **motor neurone**, which carries the impulse to the **effector**. The effector carries out a **response** and is often a muscle (hence the name 'motor' neurone, suggesting movement, a frequent action following a stimulus). A gland can also be an effector.

The electrical impulses are passed from one neurone to another across special connections called **synapses**. A synapse is a slight gap between one neurone and the next. Relay neurones have hundreds of synaptic connections to other neurones in the CNS. The synaptic connections allow information about the stimulus to be shared between relay neurones. This enables a fast response to be made.

This pathway is summarised in Figure 4.10. Note that a relay neurone may also be called an intermediate neurone – a name that helps you remember its role in the pathway.

Now go to page 168 to apply this to some data collected in an investigation on speed of conduction of nerve impulses along axons with different diameters.

> **MATHS TIP**
>
> **Determining the slope and intercept of a linear graph**
> The **slope (gradient)** of a straight line on a graph shows how the magnitude of the *y*-variable changes in relation to the *x*-variable.
> An **intercept** (where a straight line passes through an axis) on a graph shows what happens when the magnitude of one variable = 0.

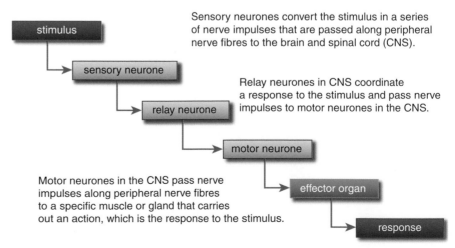

Figure 4.10 Pathway of nerve impulse from stimulus to response.

Neurones are specialised cells, compared with a generalised animal cell (Section 1.3). Neurones have a region called the **cell body**, containing the nucleus, and long thin extensions of cytoplasm from the cell body called **axons**. Some axons are very long, perhaps 1 m or more in length. Electrical impulses are carried rapidly along these axons towards their destination.

At the end of the axons, or on the cell body, there are short branches called **dendrites**. These make connections with other cells at the **synapses**. These features are shown in Figure 4.11. Note the different positions of the cell body in the three neurones – sensory, motor and relay. You can see also that the relay neurone and adjacent parts of the sensory and motor neurones lie in the CNS (central nervous system) – either in the brain or the spinal cord. In the body, the neurones are grouped into bundles of the long axons, making up a nerve fibre.

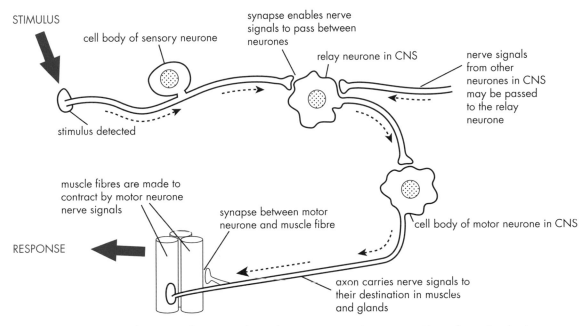

Figure 4.11 Structure and position of sensory, relay and motor neurones in a nerve pathway from stimulus to response. Dashed arrows show direction of nerve impulses.

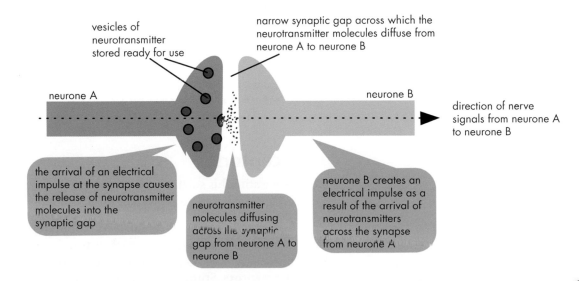

Figure 4.12 How a chemical substance (a **neurotransmitter**) helps to carry the information from one neurone to another.

The reflex arc

The neurones involved in a response to a stimulus are named according to their role in the chain of events.

- **sensory neurones** – receive and detect the stimulus, and send electrical impulses along their axons towards the central nervous system (CNS)
- **relay neurones** – located in the CNS. These receive signals from the axons of the sensory neurones and then coordinate a suitable course of action by sending signals to nearby motor neurones.
- **motor neurones** – also located in the CNS. They receive signals from the relay neurones and then send signals along their axons towards the **effector organ**(s), which **respond** to the stimulus.

Examples of reflex responses include pain-withdrawal, blinking and coughing. These responses help us to avoid serious injury, such as damage to the eye or choking and, therefore, help in our survival. These reflex responses, and the nerve cell circuits that organise them, are specified in our genes. That is why we do not have to learn reflex responses and can carry them out without thinking – any delay caused by thinking could be harmful!

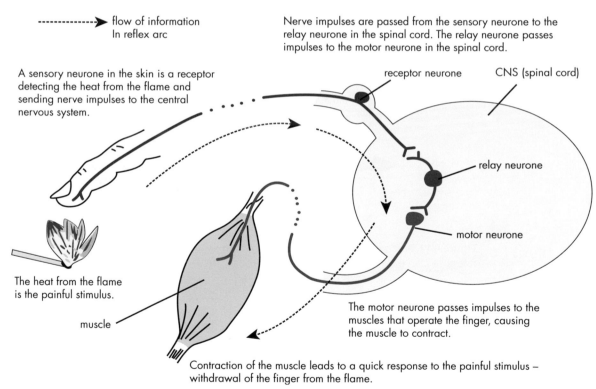

flow of information
In reflex arc

Nerve impulses are passed from the sensory neurone to the relay neurone in the spinal cord. The relay neurone passes impulses to the motor neurone in the spinal cord.

A sensory neurone in the skin is a receptor detecting the heat from the flame and sending nerve impulses to the central nervous system.

receptor neurone   CNS (spinal cord)

relay neurone

motor neurone

The heat from the flame is the painful stimulus.

muscle

The motor neurone passes impulses to the muscles that operate the finger, causing the muscle to contract.

Contraction of the muscle leads to a quick response to the painful stimulus – withdrawal of the finger from the flame.

Figure 4.13 Pathway for a reflex arc (e.g. withdrawal reflex).

■ Receptors

To detect a stimulus such as a ray of light, a sound or a smell outside our body, we use sense organs which contains suitable **receptors**.

The **skin** is our largest sense organ. It contains a variety of receptors that can detect a range of external stimuli, such as **touch**, **pressure**, **heat** and **cold**. The skin can also let you know if there is damage to the skin – something detected by **pain receptors**. Stimuli that cause pain are likely to be dangerous, so the sooner we respond to them the better.

Some receptor organs are specialised in the detection of one particular kind of stimulus. The **eye** is a specialised sense organ that allows us to detect the stimulus of light.

Structure of the eye

Vision is a fantastic human sense. It allows us to detect the light from objects that are near or far away from us. This means that we can see something coming in our direction and take action on the basis of what we see. We are able to see in **colour** and, because we have two eyes, we have **stereo vision**. This allows us to judge accurately the distance an object is from us.

Each eyeball is set into a bony eye socket, covered by eyelids, and this provides protection. Under the upper lid of the eye there is a **tear gland**, from this tears ooze onto the eyeball. The tears contain a natural antiseptic enzyme (lysozyme), which breaks down bacterial cell walls.

Figure 4.14 shows the structures of the eye. Most of the outer wall, the **sclera**, is white and opaque. This tough coat helps keep the eye in shape and provides a place of attachment for the muscles that move the eye. At the very front of the eye, the white sclera changes into the transparent **cornea**. The cornea allows rays of light to pass into the eye and through the hole in the iris, known as the **pupil**. The **iris** is the coloured part of your eye and may be blue, brown or various colours in between. It controls the amount of light that passes through the pupil.

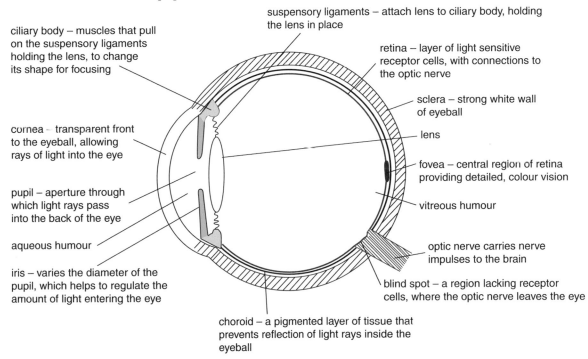

ciliary body – muscles that pull on the suspensory ligaments holding the lens, to change its shape for focusing

suspensory ligaments – attach lens to ciliary body, holding the lens in place

retina – layer of light sensitive receptor cells, with connections to the optic nerve

sclera – strong white wall of eyeball

cornea – transparent front to the eyeball, allowing rays of light into the eye

lens

fovea – central region of retina providing detailed, colour vision

pupil – aperture through which light rays pass into the back of the eye

vitreous humour

aqueous humour

optic nerve carries nerve impulses to the brain

iris – varies the diameter of the pupil, which helps to regulate the amount of light entering the eye

blind spot – a region lacking receptor cells, where the optic nerve leaves the eye

choroid – a pigmented layer of tissue that prevents reflection of light rays inside the eyeball

Figure 4.14 Structures of the eye, seen in vertical section through the eye.

Between the cornea and the lens there is a watery liquid, the **aqueous humour**. The **lens** is transparent and denser than the aqueous humour. It plays an important part in focusing the light rays. Behind the lens, the rest of the eye is filled with a jelly-like liquid, the **vitreous humour**. In this part of the eye, the sclera is lined with a pigmented layer of cells, called the **choroid**. This layer helps to prevent reflection of light rays off the internal walls of the eye. Inside the choroid, the inner layer of the eye is the **retina**. This is made up of receptor cells that receive the stimulus (light) and pass electrical impulses along nerves, to the brain, where the information is converted into something we can 'see'. The retina includes two types of receptor cells. **Rods** detect in black and white (and in low light intensity). **Cones** detect colours (and are stimulated in light intensity higher than that needed for rods).

The neurones from the receptor cells in the retina collect to form the **optic nerve**. This leaves the eye to link into the CNS. There are no receptor cells at the point at which the optic nerve leaves the eye, so this is called the **blind spot**. A region of the retina, known as the **fovea**, contains densely packed cones and is a region where the eye sees particularly good detail and in colour.

Focusing the light in the eye

Light rays from an object pass into the eye through the transparent cornea. The cornea is covered by a thin sheet of protective cells, the **conjunctiva**. The rays of light are bent as they enter the front of the eye. This is part of the focusing process that eventually projects the rays onto the retina at the back of the eye.

The **lens** is another structure that contributes to focusing the rays of light. The lens can change shape and this helps in the focusing process. The lens is suspended by ligaments attached to a ring of muscles called the **ciliary body**. When the lens is thin and flat, it can focus more distant objects onto the retina but when it is fatter (more spherical) it focuses closer objects onto the retina (Figure 4.15). The changes that help adjust the shape of the lens are brought about by the relaxation or contraction of the ring of **ciliary muscles**. The process is described as **accommodation** and the following steps summarise what happens.

Distant object
- ciliary muscles relax
- tension on suspensory ligaments increases (pulled tight)
- lens pulled flatter and thinner

Close object
- ciliary muscles contract (the ring of muscle has a smaller diameter)
- the suspensory ligaments become slack (do not pull on the lens)
- lens springs back to be fatter (more spherical)

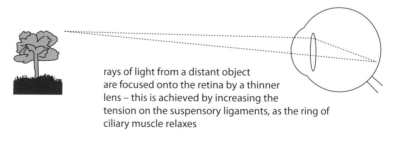

rays of light from a distant object are focused onto the retina by a thinner lens – this is achieved by increasing the tension on the suspensory ligaments, as the ring of ciliary muscle relaxes

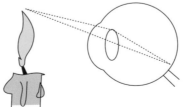

rays of light from a nearby object are focused onto the retina by a fatter lens – this is achieved by relaxing the tension on the suspensory ligaments, as the ring of ciliary muscle contracts

Figure 4.15 How the lens adjusts when focusing on distant and near objects.

Controlling the amount of light that enters the eye

The iris lies between the cornea and the lens, and regulates the total quantity of light entering the eye. The iris is a coloured ring of muscle with a hole in the middle (the pupil) through which light rays must pass before they get to the lens. The diameter of the pupil can vary, depending upon the state of two sets of muscles in the iris – the radial and circular muscles.

Figure 4.16 shows how the action of these muscles can control the diameter of the pupil. If there is a lot of light, the circular muscles contract and the radial muscles relax. The pupil becomes smaller (constricts) and less light enters the eye. When there is too little light, the opposite happens to the muscles so the pupil becomes larger (dilated) and more light enters the eye.

The reflex control of the iris begins with intense light falling on the retina. This causes cells in the brain to send a signal to the iris muscles to act so that the pupil constricts and less light enters the eye. The size and shape of the iris of an animal can indicate whether it is nocturnal (active at night) or a daytime animal. (Look back to the photographs at the start of Section 4.5 and see what you can work out about these animals.)

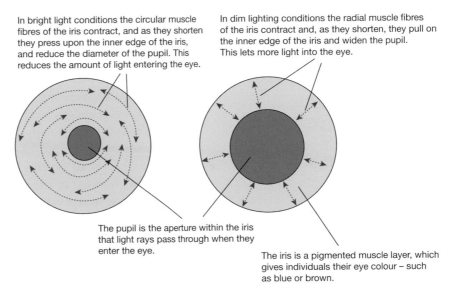

In bright light conditions the circular muscle fibres of the iris contract, and as they shorten they press upon the inner edge of the iris, and reduce the diameter of the pupil. This reduces the amount of light entering the eye.

In dim lighting conditions the radial muscle fibres of the iris contract and, as they shorten, they pull on the inner edge of the iris and widen the pupil. This lets more light into the eye.

The pupil is the aperture within the iris that light rays pass through when they enter the eye.

The iris is a pigmented muscle layer, which gives individuals their eye colour – such as blue or brown.

Figure 4.16 How the muscles of the iris control the diameter of the pupil. The arrows in the coloured part (iris) represent the contraction of muscles.

■ The skin and its role in temperature regulation

Reference is made to the skin on page 158 where it is described as our largest sense organ. Receptors in the skin are sensitive to changes in temperature, as well as to touch, pressure and pain. Structures in the skin also have an important role in temperature regulation in the body – an example of homeostasis. Details of this are given in Section 4.3, as part of the discussion on homeostasis.

■ Hormones and the endocrine system

Reference is made on page 156 to hormones as chemical substances that help control certain activities and processes in the body. They are produced in glands known as **endocrine glands**. The hormones that they secrete pass into the blood and are carried to different parts of the body where they have an effect.

Some examples of reproductive hormones (**testosterone, oestrogen, progesterone, FSH and LH**) are discussed in more detail in Section 5.3. **ADH** (antidiuretic hormone) and its role in osmoregulation is discussed in more detail in Section 4.2. Adrenaline and insulin are considered here.

Adrenaline and how it prepares the body for action

You are almost certainly familiar with the effects of the hormone adrenaline. This hormone is made by the adrenal glands found on the top of each kidney. During stressful moments, when you feel emotions like fear or anger, the brain sends nerve signals to the adrenal glands, which respond by producing adrenaline. This can happen in only a few seconds.

Adrenaline goes to all parts of your body and prepares it for physical action. The changes in the body are often described as the 'fight or flight' response. Adrenaline causes a wide range of changes including the following:

- an increase in heart rate – this ensures an improved supply of oxygenated blood from the lungs to the active muscles of the body
- dilation of the blood vessels inside the muscles, allowing more blood to circulate through them, delivering oxygen and glucose
- triggers the breakdown of stored fat in cells under the skin. These cells release fatty acids and glycerol into the blood, which transports them to the active muscles of the body. The fatty acids are respired to provide the muscles with energy for movement
- triggers breakdown of stored glycogen to glucose in the liver and muscle cells. Glucose released by the liver is transported to the active muscle cells. The glucose is respired to provide the muscles with energy for movement.

STUDY TIP

Look at Section 2.3 which deals with respiration. Link what you find to the changes that happen in the body during the fight or flight response.

Insulin and blood sugar control

Most of us regularly have access to food. Nevertheless, we tend to eat at certain times (morning, midday and evening) and not at other times. In between meals we have other things to do besides eat. This pattern of eating and fasting means that the levels of essential nutrients, such as glucose, in our bloodstream fluctuate. Normally we absorb glucose and other nutrients from the small intestines into the bloodstream during the two hours after eating a meal. The food must be digested before it is absorbed.

STUDY TIP

Make a link to Section 2.2 (Digestion and absorption).

STUDY TIP

Look at Section 1.5. Now draw the expected appearance of a blood cell if the blood glucose level exceeded normal limits.

Link to homeostasis in Section 4.3.

Glucose levels in the blood must be controlled within safe limits. If the concentrations of nutrients such as glucose become too high, they alter the water balance of the blood. Osmoregulation (mainly in the kidneys) is needed to restore the concentration within the required level. On the other hand, if glucose levels fall too low there is a risk that brain cells lack oxygen. If the brain has too little glucose, it shuts down, leading to a coma and eventually death.

The **pancreas** and **liver** work together to smooth out blood glucose levels through the day and night. The pancreas has a role in producing digestive enzymes (see Section 2.2), but this organ also plays an important part in the regulation of blood glucose levels.

After a meal, or eating a sugary sweet, glucose is absorbed into the blood. As the blood glucose level rises above normal, cells in the **pancreas** detect the increase and start to secrete the hormone **insulin** into the blood. Insulin stimulates the muscles and liver to take up glucose from the bloodstream and store it as **glycogen**, a polymer of glucose. This uptake of glucose by liver and muscles reduces the concentration of glucose in the blood. Once the level returns to normal, the pancreas cells stop secreting insulin.

These events are summarised by the sequence:
- after a meal, increased glucose in the blood
- increase in glucose detected by cells in pancreas
- pancreas secretes insulin into the blood
- insulin stimulates conversion of glucose to glycogen (in the liver) and increased uptake of glucose into muscle cells.

If blood glucose levels fall too low, a different hormone is released by the pancreas. This hormone (glucagon) stimulates the liver to break down its stores of glycogen into glucose, and this glucose is released into the bloodstream. The blood glucose level thus returns to normal. (Note the spelling of glucagon and do not confuse it with glycogen.)

For different reasons, some people cannot produce enough insulin, so are unable to regulate their blood glucose level. This is one cause of diabetes. One way of overcoming this is to give daily doses of insulin; production of human insulin using genetically modified bacteria (see Section 7.5) is an important way of producing supplies of suitable insulin.

■ Summary of nervous and hormonal control

Table 4.4 compares coordination and control by the nervous and hormonal (endocrine) systems.

Table 4.4 A comparison of nervous and hormonal control and coordination within the body.

Nervous	Hormonal
signals are electrical impulses	signals carried as chemical substances
carried along nerve cells	carried in the blood
carried along nerve axons to a specific site, so that only a single cell or organ is affected	carried in the blood to the whole body, so it is possible that every cell or organ in the body is affected
carried from receptors in sense organs, along nerves to the CNS, then to the effector	produced by special glands (endocrine glands)
fast response	slow(er) response
effects last short time	may have long-term effects

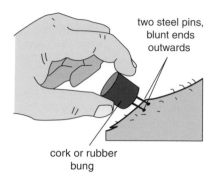

two steel pins, blunt ends outwards

cork or rubber bung

Figure 4.17 Testing the sensitivity to touch of the skin on different parts of hand and arm.

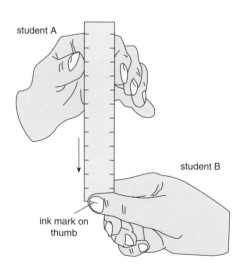

student A

ink mark on thumb

student B

Figure 4.18 Testing reaction time, using the ruler-drop test.

■ Practical activity – testing the sensitivity to touch of the skin on different parts of the hand and arm

Two students (A and B) work in pairs.

1 Stick two steel pins into a cork or rubber bung so that the projecting (blunt) pin heads are at a fixed distance apart (say 1 cm), as shown in Figure 4.17.
2 Student A gently touches the skin of student B a number of times, while student B looks away. Sometimes student A touches with one pin head and sometimes with both pin heads.
3 On each occasion, student B is asked to say how many pin heads have been used. The accuracy of student B's answers gives an indication of the sensitivity of the skin in that region.
4 Compare results on different regions for one student, then similar regions on different students.

Accessible regions of the skin are the finger tips, palm and back of the hand, inside and outside of forearm and elbow.

■ Practical activity – testing reaction time

Two students (A and B) work in pairs. Student A drops the ruler as a surprise and student B tries to catch it as soon as possible.

1 At the start, student A holds the ruler as shown in Figure 4.18. Student B has an ink mark on the thumbnail and this is lined up with a mark on the ruler.
2 Student A drops the ruler, which falls under gravity. Student B catches it as soon as possible and notes the position on the ruler of the thumb. This shows the distance in mm that the ruler has fallen.
3 The reaction time can be calculated in one of two ways:
   ■ from the equation

$$\text{time} = \frac{2y}{g}$$

where y = distance fallen in cm, g = 980 and time is measured in seconds

   ■ reading a value for reaction time from the chart (Figure 4.19).

You can extend the activity and compare the reaction times of the same student after a series of tries or of different students. You can also devise other situations, such as listening to music, chewing or having a drink containing caffeine and test the effect of these on reaction time. *Note that drinks should not be consumed in the laboratory.*

Estimating the speed of transmission of electrical impulses in the body

Measure the distance from the eyes to the top of the head. Under this is an area of the brain that controls movement of voluntary muscles. Also measure the distance from the top of the head to the middle of the forearm, where the muscles controlling the fingers are located. Add these two distances together and this gives a rough idea of the distance over which nerve impulses travel during your reaction to the ruler being dropped. Divide this distance by your average reaction time, to estimate the speed of conduction of information by the neurones.

PRACTICAL

This is another simple practical activity but one which has wider applications with reference to the effects of other factors on the reaction times of different people. You can also devise other ways of recording reaction time, say, with a timer in response to the sound of a bell. It may help you understand the distance within the body that a nerve impulse has to travel, from receptor to effector (stimulus to response).

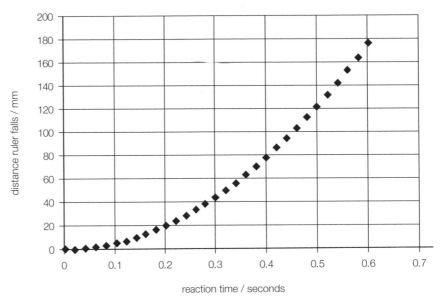

Figure 4.19 Chart for conversion of distance fallen to reaction time.

STUDY QUESTIONS

1 The table refers to some responses shown by the body. Copy and complete the table to show the response, the stimulus, where the stimulus is detected and how the body responds. Some have been completed for you.

Response	Stimulus	Where stimulus is detected	How the body responds (muscle or gland)
blood sugar level reduced	blood sugar level too high		
fight or flight response	physical threat to body	eyes / brain	
	too much light	retina	
eyes focus on nearby object	movement of object		
	painful heat		muscle

2 Copy and complete the table to summarise information about some hormones. For each hormone listed, state where and when it is produced, and describe how the body responds to it. For some hormones, there may be more than one response, so check carefully to make sure you have included all the relevant effects for those you have studied. For some hormones, you may need to look in other parts of the book.

Hormone	Where and when produced	Effect(s) in the body
ADH		
adrenaline		
insulin		
oestrogen		
progesterone		
testosterone		
FSH		
LH		

Summary

I am confident that:

✓ I understand that carbon dioxide is a waste product from respiration but that when the light intensity is great enough for photosynthesis to occur, the carbon dioxide is used by the cells of the plant leaf so that there is no net excretion of carbon dioxide from the stomata of the leaf.

✓ I understand that oxygen may become a waste product from photosynthesis when the light intensity is high enough, even though the cells in the plant leaf continue to use oxygen in respiration. The net effect is that oxygen is excreted from the stomata of the leaf.

✓ I know that the lungs are an excretory organ for the products of respiration.

✓ I know that some excretion occurs through the skin.

✓ I know that some waste products are excreted from the kidney.

✓ I can describe the structures of the urinary system, including the kidneys, ureters, bladder and urethra.

✓ I can describe the structures of a nephron, including the Bowman's capsule and glomerulus, the convoluted tubules, loop of Henlé and the collecting duct.

✓ I understand how ultrafiltration occurs in the Bowman's capsule and how water and glucose are selectively reabsorbed back into the blood as the filtrate passes along the tubule into the collecting duct.

✓ I can explain that urea is a waste product from protein metabolism and that it is excreted in the urine.

✓ I can describe how the hormone ADH is involved in osmoregulation by controlling reabsorption of water in the tubule from the glomerular filtrate so that the composition of the blood remains within appropriate limits.

✓ I know that living organisms respond to changes in their environment, and some of the ways that living organisms respond.

✓ I understand that the term homeostasis describes how the internal environment in a human is kept within certain limits and that examples of homeostasis are the control of body water content and of body temperature.

✓ I know that for a coordinated response, a receptor receives a stimulus and this leads to a response in an effector.

✓ I can describe how sweating can help to cool the body, so that when body temperature rises, more sweating occurs, resulting in increased evaporation of water from the skin surface (hence cooling the body).

✓ I can describe how the arterioles leading to the blood capillaries near the skin surface can alter the blood flow and hence control heat loss from the body.

✓ I understand that the term vasoconstriction means that less blood flows in the skin capillaries near the skin's surface and that vasodilation means that more blood flows near the skin's surface, allowing less or more heat to be lost by radiation.

✓ I can describe geotropic responses in plants, in which roots grow downwards towards the direction of gravity and stems grow upwards away from the direction of gravity.

✓ I can describe the positive phototropic responses in stems in which plant stems grow towards the light and know that plant roots are negatively phototropic.

✓ I understand the role of auxin in the phototropic responses of stems.

✓ I can describe the differences between the hormonal system and nervous system in humans.

✓ I know that the nerves in the sense organs in the body are linked through to the central nervous system (CNS), which consists of the brain and spinal cord.

✓ I understand how a receptor organ receives a stimulus and then sends electrical impulses along the nerves to the CNS and from there to generate rapid responses as appropriate to the stimulus.

✓ I know that neurotransmitters have a role in the passage of nerve impulses across a synapse.

✓ I can describe a reflex arc, illustrated by withdrawal of a finger from a hot object, and understand how the response is made.

✓ I know that a reflex arc consists of a receptor, sensory neurone, relay neurone (which lies in the CNS and links with other neurones) and a motor neurone to the effector (which may be a muscle or a gland).

✓ I can label a diagram to show the structures in the eye and understand how these are involved in receiving a light stimulus.

✓ I can explain how the ciliary body can change the shape of the lens when the eye focuses on a near or a distant object.

✓ I can explain how the iris in the eye adjusts in the response of the eye to changing light intensity.

✓ I can describe the roles and effects of the following hormones:
1. ADH in osmoregulation
2. adrenaline in the 'fight or flight' response
3. insulin in the control of blood sugar level in the body
4. testosterone, progesterone, oestrogen, FSH and LH as reproductive hormones (as described in Section 5.3).

MATHS SKILLS

Mode, median and range

These terms are alternative forms of average taken from sets of data. The **mode** is the most frequently occurring value in a set of data.

The **median** is the middle value in a set of data ranging from lowest to highest values (so half the data the data values are below the median value, and half are above it).

The **range** is the value of the difference between the largest and smallest values in a set of data.

The following readings of skin temperature were taken immediately after some forearm skin had just been licked once to moisten it.

Skin surface temperature measured at different times					
time of reading in minutes	0	1	2	3	4
temperature / °C	27	22	21	24	27

a) What is the mode temperature?
b) What is the median temperature?
c) What is the range for the temperature data?

(a) 27 °C
(b) 24 °C
(c) 6 °C

MATHS SKILLS

Determining the slope and intercept of a linear graph

The **slope (gradient)** of a straight line on a graph shows how the magnitude of the *y*-variable changes in relation to the *x*-variable.

An **intercept** (where a straight line passes through an axis) on a graph shows what happens when the magnitude of one variable = 0.

Measurements of nerve impulse conduction speed in different-sized axons are presented on the graph.
Answer the following questions:

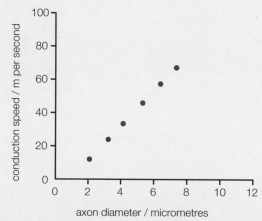

a) Describe the relationship between axon diameter and conduction speed.
b) Estimate the slope of the graph, giving appropriate units.
c) Estimate the size of axon that would have no conductance at all.
d) Estimate the speed of conduction in an axon of diameter 11 micrometres.

(a) directly positive relationship
(b) 10 m per second / micrometre
(c) 1 micrometre
(d) 100 m per second

Example of student response with expert's comments

■ Understanding structure, function and processes

1 a) The diagram shows a kidney nephron and its blood vessels.
Blood is filtered in the capsule and glomerulus part of the tubule.
   i) Name the artery that carries blood from the aorta to part A. *(1)*
   ii) The artery bringing blood to A is wider than the blood vessel leaving the capsule and between the two vessels the blood travels through a network of capillaries. Suggest how this helps the filtration process. *(2)*

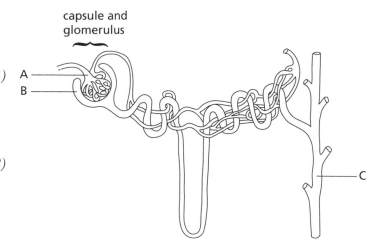
capsule and glomerulus

A

B

C

b) The table shows the concentrations of proteins, urea and glucose in the liquid in parts A, B and C.
For each substance below, describe and explain the changes in concentration measured in the different parts of the nephron.
   i)   Protein *(2)*
   ii) Glucose *(2)*
   iii) Urea *(2)*

c) The data given in the table are for a healthy person on an average day. Describe and explain the changes you would expect in the composition of the liquid in part C in the following cases.
   i) The person ate a meal containing a very large amount of meat. *(2)*
   ii) The person took in the same amount of food and drink on a very cold day. *(2)*

Substance	Concentration / g per litre		
	in part A	in part B	in part C
protein	75.0	0.0	0.0
glucose	1.0	1.0	0.0
urea	3.0	3.0	21.0

(Total = 13 marks)

Student response Total 10/13	Expert comments and tips for success
a) i) Renal artery ✔	
ii) The blood is under high pressure. ✔ The small molecules like urea are filtered through the capillary walls O into the capsule.	The high blood pressure increases the rate of filtration. If the student had mentioned the term 'ultrafiltration' this would also have been worth a mark.
b) i) There is a lot of protein in A but none in B or C. ✔ The protein molecules are too large to pass through into B. ✔	Try to use correct scientific terms, such as 'high concentration' rather than 'a lot of', although examiners give credit where possible.
ii) Glucose concentration is the same in B as in A, but there is none in C. ✔ The glucose is reabsorbed ✔ in the first coiled tubule.	'First coiled tubule' is equivalent to 'proximal convoluted tubule'. Although there are several acceptable names for the parts of the kidney, it is better to use the name in the specification.
iii) The small urea molecules pass from the blood into B. ✔ Later the concentration increases in the collecting duct C. O This is due to water being reabsorbed. ✔	No mark for the second sentence, as no statement that concentrations in A and B are the same. The student had described this correctly for glucose, but needed to repeat it for urea. Each question part is marked separately, so, if relevant, you must repeat phrases to give full answers.
c) i) Meat contains carbohydrate, so the urine would contain more sugar. O	Although meat does contain carbohydrate (glycogen), which will be digested to glucose, none of it will appear in the urine because all glucose in the kidney filtrate is reabsorbed into the blood in the proximal convoluted tubule.
ii) The urine would be more dilute. ✔ In cold weather less water is lost as sweat, so more will be lost in urine ✔ to keep the concentration of the blood constant.	Full marks.

Exam-style questions

1 The table gives the content of glucose, urea and calcium ions in the blood entering the kidney, in the glomerular filtrate and in the urine of a person. Values are given in mg per 100 cm³.

Component / mg per 100 cm³			
Component	Blood	Glomerular filtrate	Urine
glucose	100	100	0
urea	26	26	1 820
calcium ions	4	4	5

a) i) Which of the components provides energy for the body? [1]

ii) Which of the components is a metabolic waste product? [1]

b) Explain why the figures for each component are the same for the blood concentration and glomerular filtrate. [3]

c) i) If the person drank a large volume of water, far more than was needed by the body, predict what would happen to the figures in the urine column. [2]

ii) Describe the processes taking place in the body to support your answer to (c)(i). [3]

[Total = 10]

2 a) Give the meaning of the term **homeostasis** and use examples to explain why it is important in the human body. [4]

b) i) Explain how sweating helps to cool the body. [2]

ii) Suggest why, on a hot day in a dry atmosphere, you feel more comfortable than in a humid atmosphere when the temperature is the same. [3]

c) Mountaineers may experience extreme conditions of cold and wind. Often they wear clothing that uses several light layers rather than thick heavy garments, with a waterproof outer layer.

i) Suggest the advantage of several thin layers rather than thick heavy garments. [2]

ii) Suggest the importance of the waterproof outer layer. [2]

[Total = 13]

3 Auxin is a plant growth regulator and it can affect the growth of wheat coleoptiles (the first shoot of young seedlings of wheat plants).

The diagram shows an experiment into the effects of auxin on the growth of a wheat coleoptile. Lanolin is a grease that sticks to plant surfaces. In the experiment, auxin was mixed with some lanolin and the auxin could then diffuse into the cells of the coleoptile. The seedlings were held at the top of tubes containing water, but only the seedling is shown in the diagram.

A 'blob' of the mixture of lanolin and auxin was stuck to one side of a coleoptile, as shown in the diagram on the left. The seedings were examined after 3 hours.

Possible results of the experiment are shown in diagrams A, B, C and D.

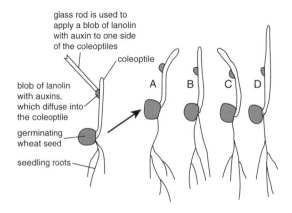

a) i) Which of the results, A, B, C or D, would you expect after 3 hours? [1]

ii) Give reasons to support your answer and explain what is happening with respect to growth in the coleoptile to produce this response. [3]

b) If the lanolin with auxin mixture had been placed in a complete circle around the coleoptile, which result would you expect? Give a reason for your answer. [2]

c) During the experiment, what light conditions should have been used? Explain your answer. [2]

d) Describe a control that should be included in the experiment. [2]

[Total = 10]

4 The diagram shows a section through the human eye. Structures in the eye are labelled with the numbers 1 to 9.

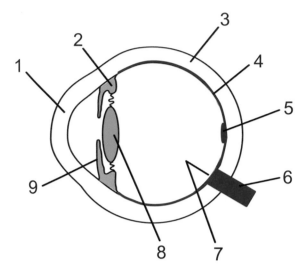

ii) Which number on the diagram shows the part where the light rays focus? [1]

   A 3

   B 5

   C 6

   D 7

iii) Which number on the diagram shows the part that provides the colour of the eye and circular muscles that contract in bright light? [1]

   A 1

   B 2

   C 3

   D 9

a) i) Which number on the diagram shows the transparent layer at the front of the eye? [1]

   A 1

   B 2

   C 8

   D 9

b) Explain how the ciliary muscles in the eye change the shape of the lens so that it can focus on a near object (such as a page of writing) after looking at a bird across a field. [3]

[Total = 8]

EXTEND AND CHALLENGE

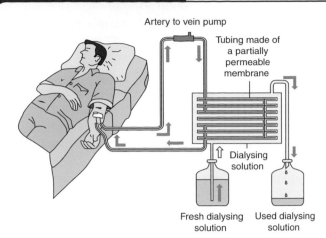

Artery to vein pump

Tubing made of a partially permeable membrane

Dialysing solution

Fresh dialysing solution

Used dialysing solution

1 When total kidney failure occurs, a patient may be offered kidney dialysis two or three times a week. During these dialysis sessions, which last several hours, waste is removed from the blood. However for the person, this is much less efficient and less healthy than having even a single working kidney. So in time, most patients would hope to receive a donated kidney – either from another living person or from a deceased person. There is always a shortage of donor kidneys for the patients who could benefit from them.

Diffusion across a partially permeable membrane is the principle employed in the dialysis machine. Blood from the patient is removed from a vein, and passed through a series of tiny tubes made of selectively permeable material. A (dialysing) solution on the other side of the membrane receives substances from the blood, including waste urea, by diffusion. Some substances, such as glucose, are prevented from leaving the blood, by adding them to the dialysing solution. The treated blood is returned to the patient.

a) What is meant by the term 'partially permeable'?

b) Why is blood removed from the patient's vein, rather than from an artery?

c) Suggest how the composition of the dialysing fluid in the machine should be adjusted to allow only urea to be removed from the blood, and to ensure that no glucose is removed.

d) Explain why lots of tiny tubes are used in the machine, rather than a few larger tubes.

e) Suggest reasons why it is 'less efficient and less healthy' to undertake kidney dialysis rather than having a working kidney.

f) Find out why it is not possible to transplant the kidneys of any deceased person into the body of any patient. What other factors may need to be considered before doing a transplant?

g) The UK government wants to have a system whereby it is presumed that anyone who dies can have their organs harvested for transplants, and that anyone who does not want this to happen must register their disagreement. Do you agree with this suggested change of policy? What problems might it raise?

2 If the leaves of growing plants are starved of light, they begin to lose their green colour, becoming pale yellow. They also grow tall and spindly, as shown in the photograph. This condition is known as **etiolation**.

In natural conditions, seeds usually germinate in the soil or perhaps they lie under leaf litter. While the developing shoots are in the dark they remain yellow, but when they reach the light the green colour develops. The tip of the shoot is often bent like a hook in the soil, but unbends and grows vertically upwards when it reaches the light. Plants that grow in plenty of light tend to be shorter than those growing in dimmer conditions.

In horticulture, traditional methods of growing vegetables and other crops have exploited these growth responses in plants. Rhubarb is often 'forced' in the dark and its stalks are used as a substitute for fruit. Forced rhubarb grows taller than if grown in the light. It is more tender and it is ready to eat earlier in the season. Similar techniques are used on a worldwide basis – examples include blanching of salad crops such as celery, chicory, endive or Chinese chives.

(continued)

a) Name the green pigment that develops and summarise its role in the plant.

b) Suggest how the etiolation response of germinating seedlings is a useful adaptation in the natural environment.

c) What advantages might a plant get from being short rather than tall – providing it is getting enough light?

d) Check the effect of light on the distribution of auxin produced in the tip of a shoot (described on page 151). How can you link this to the different heights of plants as described in the passage and shown in the photograph?

e) Find out about some traditional methods of growing vegetables that use this response of etiolation. Look for information about at least **two** oriental vegetables, and **two** from Europe. When and where are these plants grown and what is the benefit of these practices?

3 Today, an increasing number of adults in the developed world have diabetes. The risk of developing the condition as an adult is more likely in obese people.

Diabetes is a disease that involves the failure of the system that controls glucose levels in the body. A person with diabetes is unable to produce sufficient insulin to keep the blood sugar under control. As a result, their blood glucose reaches dangerous levels. One effect of this is to damage blood vessels. In the long term, if left untreated, damaged blood vessels in the retina can result in the death of cells in the retina, leading to blindness. Poor blood supply to the lower leg and feet sometimes leads to infection and may necessitate amputation of the limb. Diabetes is a serious condition and places huge strains on medical resources of different countries.

Various treatments have been developed to try to enable patients to regain the ability to control their blood glucose. One important treatment involves careful checking of blood glucose and injections of insulin at times when the glucose level would be expected to rise. Another solution could be to transplant healthy insulin-producing pancreas cells into the patient.

a) Explain briefly how insulin helps to control levels of glucose in the blood.

b) If obesity seems to be a major risk factor for the development of diabetes, what preventative measures might be taken to reduce the problem – by individuals and by governments? (First you should find out about the factors that are likely to lead to obesity.)

c) The damage to blood vessels takes place over years and the diabetic person does not directly feel any symptoms from their high blood glucose levels. Why does this make education about the disease difficult?

d) Currently, the method used to check blood for glucose involves pricking a finger to collect a drop of blood. This is then put into a machine that does the measurement. The insulin might need to be injected into the skin. Suggest possible hygiene problems that may arise with this testing and treatment regime.

e) Transplants of a healthy pancreas into a person with diabetes have been performed successfully. Suggest why it is not possible for healthy pancreas cells to be transplanted into every patient.

f) Find out more information about diabetes, say from the website of a diabetes support organisation. Use this information to produce a leaflet for your community explaining how individuals can reduce their risk of diabetes.

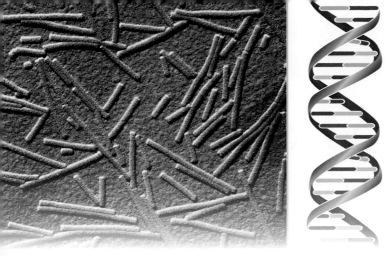

5

Reproduction and inheritance

TO THINK ABOUT ...

DNA is now part of our everyday vocabulary. But when its molecular structure was first described in the 1950s, it signalled a revolution for our understanding of inheritance.

What do you know about DNA? How can this molecule hold all the information that codes for the features of every individual – plant, animal or microorganism? How is the information passed on to maintain continuity from one generation to the next? And what evidence can you describe to show that species change over time?

Reproduction and the continuity of living organisms

A virus injects its nucleic acid into a host cell. In this case, the tobacco mosaic virus in the image contains RNA and this gets forced into a plant cell. That's about all a virus does, but it is enough to program the cell to make more identical viruses. The host suffers and the virus is responsible for causing disease in the plant or animal. This act of reproduction explains why viruses are sometimes included in classification of living things.

In the early 21st century, we all know the story about the discovery of the nucleic acid DNA, and its structure. The now familiar double helix is a way of representing this. We use this to help us understand the basis of inheritance and change over time. So what is in the DNA molecule that is so important in determining the characteristics of the next generation? How is DNA passed on?

How can we explain common features that can be traced through a family pedigree? How do living things maintain their continuity so that we recognise the next generation of individuals as part of the same species?

Fossil evidence gives us information about earlier life forms – dinosaurs, mammoths and others that are now extinct. How do species change over time? How can we explain the disappearance of former species and appearance of new ones?

How did Mendel manage to work out remarkably coherent laws of inheritance without any knowledge of DNA? How did Darwin piece together his theory of evolution by natural selection from his many observations, but without any understanding of Mendel's work on inheritance or our 21st century views on DNA and what it does?

5.1 Reproduction in living organisms

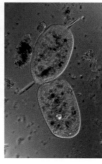

Continuity from one generation to the next

Every individual living organism ultimately dies. But before doing so, if individuals can reproduce, they ensure that their offspring will take their place in the world. Some (or all) of the genes of the parent organism(s) are transmitted to their offspring and onwards to future generations. The images emphasise the next generation of plants, animals and microorganisms.

So how does reproduction allow for continuity as well as change? In human populations, what features are the same from one generation to the next and what differences appear? Imagine a world in which death of humans could be prevented by medical technology – what sorts of social problems would have to be addressed if humans could choose to become immortal?

■ Asexual or sexual reproduction?

Living organisms all share the ability to reproduce and make more like themselves. Reproduction is the process that ensures the species continues from one generation to the next. The information controlling what the offspring look like is contained in the DNA. This essential information divides so that some is passed on to each of the cells that make up the next generation.

For some, particularly for microscopic organisms, reproduction occurs simply by dividing into two. In other organisms, a piece breaks off and is able to grow into a new individual. This can occur in many plants and some animals. These are both methods of **asexual reproduction**.

Most plants and animals reproduce by **sexual reproduction**. This occurs as a result of two special cells joining together. These special cells are the sex cells, known as the **gametes**. One is from the male and one from the female. The offspring receive some DNA from the male parent and some from the female parent and this DNA determines the characteristics of the offspring and ensures there is continuity to the next generation.

Asexual reproduction

For a single-celled (unicellular) organism, such as a bacterium or an amoeba, cell division equals reproduction. When the single (parent) cell divides, both of the cells produced (daughter cells) contain a copy of all of the genes from the parent single cell. Cells or offspring produced by asexual reproduction are the result of a division by mitosis, so that the offspring are genetically identical to the parent. Each daughter cell is a genetic clone of the parent cell. The only possible source of genetic variation between parent and daughter cells is from mutations that might occur, though mutations are rare and usually harmful.

> **STUDY TIP**
> You can find more details about DNA in Section 5.4.

> **STUDY TIP**
> Refer to Section 5.5 and make a link with meiosis and gamete formation.

> **STUDY TIP**
> Check Section 5.5 for information about mitosis.

Asexual reproduction is not restricted to single-celled organisms. Many plants and animals use asexual reproduction as part of their life cycle in addition to sexual reproduction. Asexual reproduction is often faster and more efficient than sexual reproduction. If a parent organism is perfectly adapted to its current environment, then its cloned offspring are also likely to be perfectly adapted.

Humans can only reproduce by sexual reproduction. But in the wider world, many animal and plant species use both asexual and sexual reproduction. Here are some examples:

- Aphids (greenfly) are insects and often become pests on crops because they feed on sap from the phloem. They often reproduce asexually during the spring and early summer when conditions are favourable. This means there is a rapid population boom when there is plenty of food. These asexually produced offspring are genetically identical. Later in the summer they start to reproduce sexually, giving genetically varied offspring.
- A strawberry plant can reproduce sexually, producing flowers that give rise to seeds. The plants can also reproduce asexually by runners – horizontal stems that grow across the surface of the soil and put down roots to form a new plant. The asexually produced plants are genetically identical, whereas those grown from the seeds give genetically varied offspring.
- Most reptiles reproduce sexually, giving rise to offspring as a result of mating between a male and female parent. However, a few are unusual in that they show asexual reproduction and can develop from unfertilised eggs. An example is found in rock lizards in the Caucasus mountains. These colonies of lizards are all females and they show far less variation than is found in similar populations of lizards that reproduce sexually.
- Figure 5.2 shows the edge of a *Bryophyllum* leaf. The tiny little plantlets growing on the edge of the plant leaf eventually drop off onto the ground. If they are lucky enough to find a suitable space and moisture to grow, they develop into a natural clone of the parent plant.

Sexual reproduction

Key features of the reproductive cycle in sexual reproduction are summarised in Figure 5.3(a) for animals and Figure 5.3(b) for flowering plants. This cycle applies to humans and to sexual reproduction in most plants and animals.

The term **diploid** refers to cells that contain the full set of pairs of chromosomes (2n). In humans, this is 23 pairs of chromosomes, giving a total of 46 chromosomes. When gametes are formed by **meiotic cell division**, they each contain half the number of individual chromosomes and are described as **haploid** (n). In humans, this is 23 individual chromosomes.

At **fertilisation**, when the gametes fuse, each contributes a single set of chromosomes to the **zygote**, so that the diploid number is restored. In humans, each gamete contains 23 chromosomes, giving a total of 46 chromosomes in the zygote and all body cells of the embryo and adult.

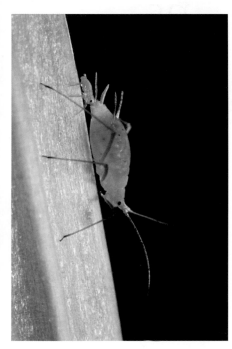

Figure 5.1 Aphids can reproduce both asexually and sexually.

STUDY TIP

Make a link with cloning in Section 7.6.

Figure 5.2 A leaf of Bryophyllum – the tiny plantlets along the edge eventually drop off and grow to form clones of the Bryophyllum plant.

STUDY TIP

Look in Section 5.4 to find out the meaning of the terms **haploid** and **diploid**, what **gametes** are and what happens at **fertilisation**.

177

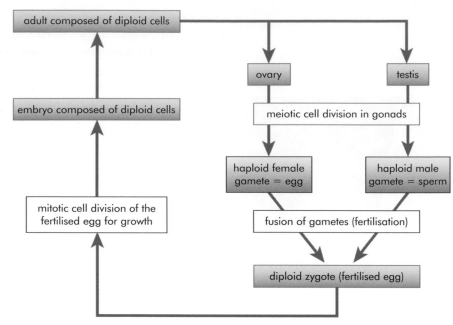

Figure 5.3 (a) Reproductive cycle for sexual reproduction in humans.

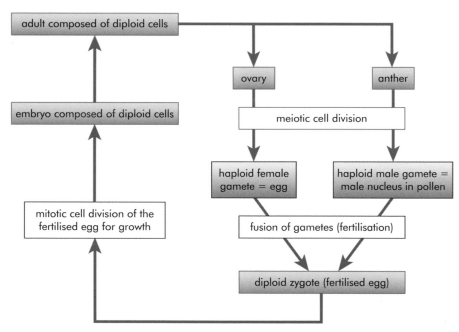

Figure 5.3 (b) Reproductive cycle for sexual reproduction in flowering plants.

Sexual reproduction ensures genetic variation in the offspring in several ways:

- Each of the two parent cells contributes 50% of their DNA to the offspring – so the offspring cannot be identical to either parent.
- The production of sex cells (gametes) by meiotic cell division in the sex organs of each parent results in new combinations of the alleles on the chromosomes that go into the daughter cells.
- Chromosomes are randomly distributed into the gametes during meiotic cell division.

In an uncertain world, it is an advantage to an individual if its offspring are genetically varied. It increases the chances of at least some offspring being well adapted to the next environmental change.

STUDY TIP
You can find more details about natural selection and evolution in Section 5.6.

In the natural world, the environment continuously changes over time and species evolve by natural selection. If none of the individuals in the population are suited to the environment, the species may become extinct. The process of natural selection depends on the genetic variation between individual offspring and this variation is mainly the result of sexual reproduction mechanisms.

Comparing asexual and sexual reproduction

Table 5.1 summarises some differences between asexual and sexual reproduction in flowering plants and mammals.

STUDY TIP
Note that in flowering plants, as for animals, sexual reproduction involves the fusion of two cells (usually haploid, n) to form a zygote (usually diploid, 2n). But in flowering plants, sometimes the two cells (pollen and egg cell) come from the same plant and sometimes the pollen comes from a different plant (but of the same species).

Table 5.1 Comparison of asexual reproduction in flowering plants and sexual reproduction in flowering plants and mammals.

Asexual reproduction (flowering plants)	Sexual reproduction (flowering plants and mammals)
one parent organism	two parent cells involved
produced by mitotic cell division	produced from fusion of two haploid gametes (gametes produced by meiosis)
offspring are genetically identical (clones)	all offspring are genetically unique, different from both parents
only source of genetic variation is mutation	offspring show variation
often produce large numbers of offspring in short time	produces limited numbers of offspring and reproductive cycle may be slow
offspring grow near the parent, competing for the same resources	flowering plant offspring (seeds) can be dormant and are dispersed to new habitats (seed dispersal)

STUDY QUESTIONS

Section 5.1 gives you an introduction to the two sections (5.2 and 5.3) that look at reproduction in more detail. Reproduction in flowering plants is considered in Section 5.2 and Reproduction in humans is considered in Section 5.3. There are also links with Cloning (Section 7.6).

Use this opportunity to make sure you are familiar with all the terms used in Section 5.1 and make appropriate links with relevant topics elsewhere, as indicated above. Here are some questions to guide you and you can follow some of the **Study tips** to help you find the information.

1 What is DNA? Explain how it contains information that controls the characteristics of an organism. Include reference to RNA in your answer.

2 Where are the male and female gametes formed in each of the following?
   a) in flowering plants
   b) in humans

3 Name **four** single-celled organisms.

4 Write down the meaning of the terms **haploid** and **diploid**.

5 Make a table to compare **mitosis** and **meiosis**. In your table, include information about whether the daughter cells are diploid or haploid.

6 a) How does the male gamete reach the female gamete in each of the following?
      i) a flowering plant
      ii) a human
    b) Where does fertilisation occur in each of the following?
      i) a flowering plant
      ii) a human

7 Give **two** advantages of asexual reproduction and **two** advantages of sexual reproduction. Your answers can refer to plants or animals. You should focus on advantages and disadvantages to the species rather than to the individual organism.

5.2 Reproduction in flowering plants

People and flowers

Flowers are very important to people. We offer them as gifts, as a sign of welcome or in memory of people or past events. Flowers become part of our celebrations and often have symbolic significance in different rituals. We enjoy their colours and decorate our houses, streets, parks and gardens with them.

Worldwide there is an enormous diversity of flowers – some large and showy, others small and inconspicuous. The photographs show a tiny selection from the flora in Yunnan province, an area of 39 400 km^2 in southwest China. Yunnan is rich in its diversity with over 15 000 species of flowers, yet this is only a fraction of the worldwide total over 400 000 known plant species.

So how and why has this diversity evolved? What is it that makes flowers so important in the natural world, as well as in our daily lives? The secrets of flowers lie in their role in plant sexual reproduction and the complex relationships that have evolved with the insects that help with pollination. This in turn leads to the production of seeds (and fruits) that provide continuity for the next generation.

Before you go further into the biology of reproduction in flowering plants, have a look around you and make a list of ways you have had anything to do with flowers (or their fruits) in the past week or month.

■ Flowers and their role in sexual reproduction

In sexual reproduction, two cells – the male and female **gametes** – join together to produce offspring. The gametes are also known as the sex cells and they are **haploid**, which means they contain one set of chromosomes. When they join (fuse) together, we say that **fertilisation** has taken place. A **zygote** is formed as a result of fertilisation and this cell is **diploid**. The zygote divides into many cells, and these become the new individual (offspring).

Insect-pollinated flowers

Figure 5.4 shows the structures in a typical insect-pollinated flower.

The list on page 181 gives the way different parts of the flower contribute to sexual reproduction. The descriptions are given from the outside of the flower, working inwards and give the generalised functions of the different parts. Actual flowers may differ from this ideal drawing, particularly if they are wind-pollinated flowers.

STUDY TIP
Check in Section 5.4 to make sure you understand the difference between haploid and diploid.

STUDY TIP
Check in Section 5.1 to remind yourself of how sexual reproduction differs from asexual reproduction.

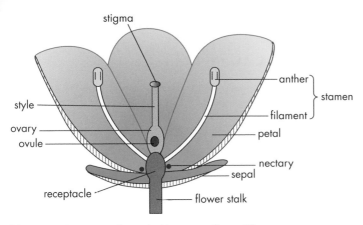

Figure 5.4 Structure of a typical insect-pollinated flower.

Figure 5.5 This geranium flower shows most flower parts typical of an insect-pollinated flower, including nectar guides on the petals.

- **Stem** – supports the flower in a suitable position for pollination.
- **Sepals** – enclose the flower when in bud, often green.
- **Petals** – often brightly coloured to attract insects, which help in pollination. They may provide a 'landing platform' for insects visiting the flower. Some flowers, including wind-pollinated flowers, do not have coloured petals.
- **Stamen** – made up of the anther and filament.
- **Anther** – contains the developing pollen grains and opens when pollen (a dusty yellow powder) is mature. The pollen grains contain the male cells.
- **Filament** – supports the anther.
- **Carpel** – made up of the stigma, style and ovary.
- **Stigma** – receives pollen during pollination.
- **Style** – supports the stigma and pollen tube grows down the style to reach the ovary.
- **Ovary** – contains the ovules.
- **Ovule** – these contain the female egg cells. After fertilisation by the male cell, an ovule develops into a seed.

Most of these features can be seen in the photograph of the geranium flower (Figure 5.5).

Wind-pollinated flowers

Figure 5.6 shows the structures in a typical wind-pollinated flower, such as a grass. The photograph in Figure 5.7 shows a wind-pollinated flower (male flower of maize). This illustrates how the anthers are held on long filaments

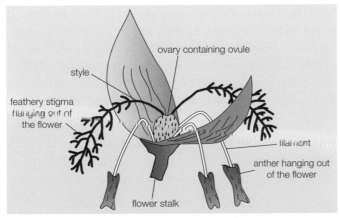

Figure 5.6 Structures in a grass flower.

Figure 5.7 Male flower of a maize plant, showing anthers with long filaments. This is a wind-pollinated flower.

and hang outside the rest of the group of flowers, and the absence of brightly coloured petals.

Table 5.2 Summary of the differences between insect-pollinated and wind-pollinated flowers.

Feature	Insect-pollinated flowers	Wind-pollinated flowers
petals	often large and brightly coloured	small, inconspicuous or absent (replaced by other structures)
nectary and scent	insects feed on nectar (a sugary solution) and are attracted by the scent	none
stamens	enclosed in flower	long filaments so that anthers hang outside flower when ripe
pollen	small with surface features that help it to stick to insects and to the stigma; produced in relatively small quantities	large with smooth surface to help it float through the air; produced in large quantities
stigma	enclosed in flower so that insect brushes past it when entering flower; top surface helps to hold pollen	often feathery and hang outside the flower so increasing chance of catching pollen in the air

Pollination

Pollination is the way that pollen is transferred from the **stamens** to the **stigma**. The pollen does not move by itself, but is carried by different agents – often by **insects**, but also by **wind** and sometimes by water, birds or even mammals (such as bats).

In some cases, the pollen lands on the stigma of the same flower – this is described as **self-pollination**. But in many flowers, there are mechanisms that encourage **cross-pollination**, in which the pollen is transferred to the stigma of another flower. But if pollination is to be successful, this must be another flower of the same species. Compared with self-pollination, cross-pollination produces more **genetic diversity** in the offspring. Many flowers have elaborate mechanisms for encouraging cross-pollination. This increased genetic diversity is one of the advantages of sexual reproduction.

For many flowers, the **pollinating agents** are insects, such as bees (Figures 5.8 and 5.9). Often on a suitably sunny day, bees can be seen flying from one flower to another in a patch of flowers. The bee visits the flower to collect nectar at the base of the petals. As the bee enters the flower, it brushes against the anthers and collects pollen, which sticks to its body. It then visits another flower and as it brushes against the stigma of the second flower, some of the pollen from the first flower is left behind. The bee picks up more pollen and flies on to a third flower to repeat the process, and so on to more flowers. This allows the pollen to be transferred from one flower to another, and usually the bee visits flowers of the same of species in succession.

For wind-pollinated flowers, pollination is more of a random process. When ripe, the anthers open and shed their pollen into the air. The pollen may then be blown in the wind or carried in air currents, and by chance lands on the feathery stigma of a plant of the same species.

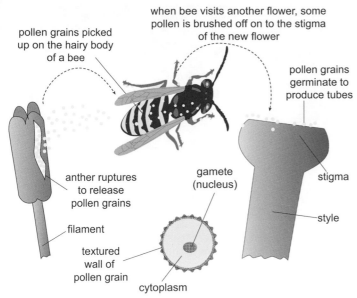

pollen grains picked up on the hairy body of a bee

when bee visits another flower, some pollen is brushed off on to the stigma of the new flower

pollen grains germinate to produce tubes

anther ruptures to release pollen grains

gamete (nucleus)

stigma

style

filament

textured wall of pollen grain

cytoplasm

Figure 5.8 Pollination of a flower by a bee.

Figure 5.9 The bee visiting a hollyhock flower gets covered in pollen and is likely to transfer the pollen to other flowers visited in succession.

Growth of pollen tubes and fertilisation

When a pollen grain lands on the stigma of a flower (of the same species), a pollen tube starts to grow down through the style towards the ovary (Figure 5.10). Digestive enzymes are produced by the tip of the pollen tube, enabling it to grow through the tissues in the style to reach the ovary. You can see pollen tubes growing under the microscope if you dust some pollen grains onto a microscope slide and mount them in a drop of sugar solution (Figure 5.11).

The pollen tube enters the ovary through a small gap known as the micropyle and thus reaches the ovule. There the male nucleus in the pollen fuses (joins) with the female egg cell in the ovule. At this point, **fertilisation** takes place and a **zygote** is formed.

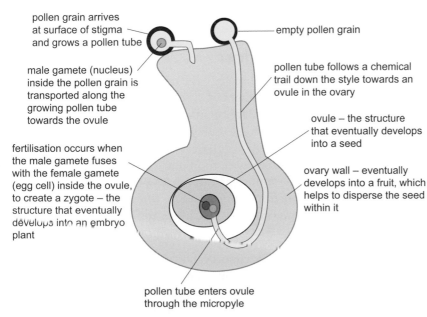

pollen grain arrives at surface of stigma and grows a pollen tube

empty pollen grain

male gamete (nucleus) inside the pollen grain is transported along the growing pollen tube towards the ovule

pollen tube follows a chemical trail down the style towards an ovule in the ovary

ovule – the structure that eventually develops into a seed

fertilisation occurs when the male gamete fuses with the female gamete (egg cell) inside the ovule, to create a zygote – the structure that eventually develops into an embryo plant

ovary wall – eventually develops into a fruit, which helps to disperse the seed within it

pollen tube enters ovule through the micropyle

Figure 5.10 Fertilisation of an ovule by a male pollen nucleus.

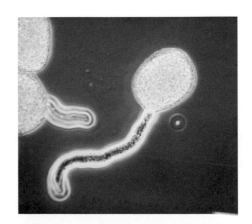

Figure 5.11 Pollen tubes starting to grow, as seen with a light microscope.

Figure 5.12 Stages in development from flower to seeds and fruit in mangetout peas. Pea flower (left), remains of petals and style are visible as the pod develops with seeds inside (centre) and maturing fruit (pod) with seeds (peas) (right). Remains of anthers are visible close to the calyx at the base of the pod.

Figure 5.13 From flower to ripe strawberry.

STUDY TIP

Think about why it is important for seeds to be dispersed away from the parent plant. What would happen if all the seeds fell in the same place? List things that the seeds might be competing for.

From flower to seed and fruit

After fertilisation, the **ovule** (containing the fertilised egg cell – the zygote) develops into the **seed**. The ovule wall becomes the seed coat (**testa**) and parts of the flower surrounding the ovule develop into the **fruit**, which contains the seeds. Often the fruit comes from the ovary wall, but other parts may swell or change in other ways and develop into the fruit (Figures 5.12 and 5.13).

Fruits develop in a variety of forms. Some are sweet and fleshy, familiar as the fruits we eat (apples, grapes, oranges). Others may be hard and dry (acorn, walnut); have a wing that allows them to float in the air (maple, ash); have sticky hooks that get caught onto the coat of passing animals (burrs, goosegrass). There are more variations such as those that allow fruits to float in the water or fruits that 'explode', propelling the seeds some distance from the fruit.

The important thing about fruits is that they provide a mechanism for **dispersal** of the seeds, away from the parent plant.

■ Seeds and germination

The seed contains the zygote (fertilised egg cell) that will develop into a new individual plant. During the stages of development in the ovule, the zygote divides into cells that then develop into the **embryo**. When the seed germinates, the embryo grows into the young **seedling**. Surrounding the embryo are structures known as the **cotyledons**. These contain food reserves that will supply the young seedling with food materials when the seed starts to germinate. (Later the young plant has leaves and can make food by photosynthesis.) Some plants (for example maize, wheat, rice, lily) have **one cotyledon**, whereas others (for example pea, bean, apple, tomato, pumpkin) have **two cotyledons**.

Seeds generally have a period of **dormancy**. During this stage they contain very little water and there is very little internal metabolic activity, but they are still alive. This is an advantage to the plant as the seeds can then be dispersed to other locations and survive quite harsh conditions (such as cold, heat and desiccation), and germinate only when conditions are right.

MATHS TIP

Bar charts and histograms
Bar charts consist of a series of bars of equal thickness, drawn vertically from the *x*-axis of the chart. They are used to represent the **frequency** of items in a series of discrete **categories**. If each category has no logical connection to the others (for example different eye colours), the individual bars are often arranged in ascending or descending order of magnitude. If a series of *x*-axis categories form a logical sequence (for example from small values to large values), then the bars can be placed against each other, to form a **histogram**.

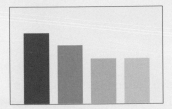

Now go to page 228 to apply this to the results from the investigation with germinating seeds.

PRACTICAL

This is a useful practical investigation to find out the conditions needed for germination of seeds. You can set it up as an experiment to find out what is needed, or compare germination of different seeds in different conditions. Combine results from several sets of tubes and calculate the percentage of seeds that have germinated.

The conditions needed for germination of a seed are as follows:

- water
- oxygen
- a suitable warm temperature.

A few seeds have a requirement for light and, for many seeds, special conditions are required to break the period of dormancy, allowing the processes of germination to start.

■ Practical activity – investigating the conditions needed for germination of seeds

1 Set up 4 test tubes (A, B, C and D) as shown in the diagram.
2 Each tube contains some cotton wool and 10 cress seeds, sprinkled on the cotton wool.
3 Keep the tubes in different conditions, as follows:
   - tube A – dry cotton wool, room temperature, in the light
   - tube B – moist cotton wool, room temperature, in the light
   - tube C – moist cotton wool, room temperature, in the dark
   - tube D – moist cotton wool, cold temperature (in the fridge), in the dark.
   Cap each tube with clingfilm or similar material.
4 Leave the tubes for 3 to 5 days.
5 Observe whether or not the seeds have germinated. Measure the height of the seedlings.
   Note that light may not be essential for germination, but it does have effects on the subsequent development of the germinating seedlings.

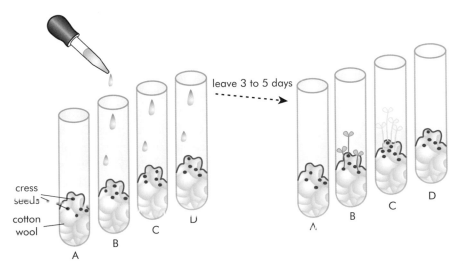

Figure 5.14 Investigating the conditions for germination in seeds.

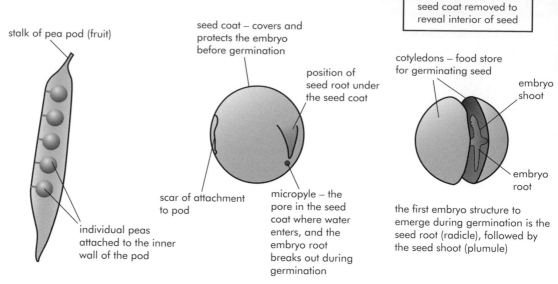

Figure 5.15 A pea fruit (pod) and structures of the pea seeds.

The internal structure of a large seed (with two cotyledons) is shown in Figure 5.15. When conditions are suitable, the seed starts to take in **water** through the small hole known as the **micropyle**. As a result of the water intake, the seed swells until the seed coat (**testa**) bursts. This allows the **radicle** (first root) to emerge, followed later by the **plumule** (first shoot). The radicle and plumule develop from the embryo as a result of cell division and cell enlargement.

The water allows enzymes that were inactive in the dormant seed to become active again. In particular, digestive enzymes act on the stored molecules (such as starch, oil and protein) and convert these into small molecules that the growing embryo can use. For example, starch is broken down into sugars and proteins are broken down into amino acids. Other metabolic reactions begin to take place, including respiration (which releases the energy needed for the growing seedling).

At this stage, the embryo is developing into a young seedling and is entirely dependent on the stored food in the cotyledons for its food supply. As the plumule emerges and leaves begin to develop and open out, photosynthesis can start to occur so that the young seedling is able to make its own food in the form of sugars. Other metabolic reactions follow until the seedling becomes an independent plant and no longer takes food from that stored in the cotyledons. In some seeds, the cotyledons rise above the surface of the soil and become green. These 'seed leaves' also carry out photosynthesis and so contribute food for the early stages of the seedling.

STUDY TIP

Remind yourself of similar reactions taking place in digestion of food in humans (see Section 2.2).

STUDY TIP

Check in Section 2.1 to remind yourself of the equation for photosynthesis.

STUDY TIP

Look in Section 5.5 and compare the outcome of division by mitosis and meiosis. What happens if two cells produced by meiosis fuse?

STUDY TIP

Link natural asexual reproduction and artificial propagation by cuttings with methods for micropropagation and use of tissue culture, described in Section 7.6.

■ Asexual reproduction in plants

If a piece of a plant becomes separated from the parent plant, often it can grow into a whole new plant. This is **asexual** reproduction and the new plants that grow are genetically identical to the parent plant. Cells divide by mitosis to generate new cells in the new plant. Any offspring from the parent plant can be described as a clone, a topic that is discussed in more detail in Section 7.6.

Asexual reproduction can occur naturally and can also be carried out artificially, say by a gardener wishing to increase numbers of a particular plant.

Natural methods of asexual reproduction

One method of natural asexual reproduction is shown in Figure 5.16. The strawberry plant can reproduce naturally by means of **runners**.

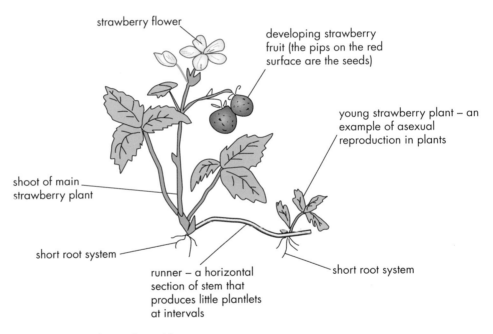

strawberry flower

developing strawberry fruit (the pips on the red surface are the seeds)

young strawberry plant – an example of asexual reproduction in plants

shoot of main strawberry plant

short root system

runner – a horizontal section of stem that produces little plantlets at intervals

short root system

Figure 5.16 Strawberry plant with runners.

The runner is a horizontal stem that grows along the surface of the ground. At certain points, small roots (known as **adventitious roots**) begin to grow into the soil. Soon these adventitious roots take over the function of the main root of the plant, hold the young plant in the soil and absorb water and other materials from the soil. Eventually the first horizontal stem dies and a new plant is established. For this new plant, all cell divisions have been by mitosis.

Artificial methods of asexual reproduction

Sometimes people want to increase the numbers of a certain plant. This can be done by taking **cuttings**, as shown in Figure 5.17. This method is often used by gardeners to increase numbers of a plant, particularly if it has unusual features.

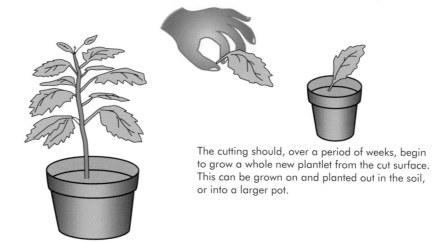

The cut stump of the cutting is sometimes dipped in a mixture of plant growth regulators (in powder) to induce the plant cells to divide and grow. Then the cutting is partly buried in some soil.

The cutting should, over a period of weeks, begin to grow a whole new plantlet from the cut surface. This can be grown on and planted out in the soil, or into a larger pot.

Figure 5.17 Cuttings as means of propagating plants. Cuttings involve the removal of part of a mature plant, such as a leaf or a part of a shoot including a bud. Numerous cuttings can be removed from a large plant. Each cutting is an artificial vegetative clone of the original plant.

A small length of a stem (or a leaf) is cut from the main plant and placed in water or directly into soil. Sometimes plant growth regulators may be added to encourage growth of adventitious roots. As these grow, they take over the functions of a root and the plant becomes independent. It may be necessary to keep the cutting in a suitably humid atmosphere until the new plant is established. The newly established plant can then be planted out into a larger pot or into the soil outside. Many cuttings can be established directly in the soil where they will grow – such as growth of a willow hedge.

As with natural asexual reproduction, all cell divisions have been by mitosis and the new plant is identical to the parent.

PRACTICAL

This practical activity provides a useful demonstration of how artificial methods of propagation can be used and gives an example of asexual reproduction.

■ Practical activity – taking cuttings with busy lizzies

Large shrubby busy lizzie (*Impatiens*) plants have many branches. Branches can be cut off with a number of leaves to be used as cuttings. These cuttings develop successfully in water so can be done without soil.

1 Strip the lowest leaves off each cutting and stand it in a boiling tube containing tap water. Wash your hands afterwards.
2 Place each boiling tube in the light and leave for about a week.

Roots grow from the surface of the cutting in about a week, as shown in Figure 5.18. The steady development of the cuttings is visible through the clear glass walls of the boiling tubes. This can be photographed.

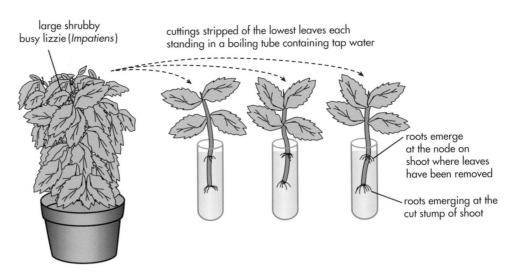

large shrubby
busy lizzie (*Impatiens*)

cuttings stripped of the lowest leaves each
standing in a boiling tube containing tap water

roots emerge
at the node on
shoot where leaves
have been removed

roots emerging at the
cut stump of shoot

Figure 5.18 Cuttings with busy lizzies.

STUDY QUESTIONS

1 a) List **five** flowers you have looked at recently (you can give common names). Say where you saw them and why they caught your attention.

  b) i) List **five** seeds you have eaten recently. Some may have been processed (for example wheat grains to make flour, used in bread or pasta).

   ii) What is the food value to you of each of these seeds and what biological molecules do they contain? In your list, try to include different seeds that provide each of the main food classes.

   iii) Why is the food contained in the seed important to the plant? Name the part in the seed that stores the food.

  c) i) List **five** fruits you have eaten recently. You can include both fruits and vegetables, many of which are fruits in a biological sense (for example tomatoes).

   ii) Did these fruits contain seeds? Note that some commercially produced fruits do not contain seeds (for example 'seedless grapes'). What about bananas – do they contain seeds?

2 a) Now join together the stages listed in Question 1. Then draw a flow diagram to show the progression from flowers to seeds and fruits. This represents the life cycle of a flowering plant.
   ■ Add notes to your diagram to show what is happening and label or list the parts of the plant that are involved at each stage.
   ■ You can set out your flow diagram as a cycle to represent the life cycle of a plant and show how the next generation is started.
   ■ Your flow diagram need not be based on the examples in your lists above but can be based on an 'idealised' flower.
   ■ Remember to show the stage at which fertilisation takes place and state whether each stage is haploid or diploid.

  b) What is the importance to the plant of the fruit stage in the life cycle? Give some examples of fruits that are not eaten by people to help explain this.

  c) Suggest ways that plants which do not produce seeds can be propagated (naturally and artificially).

5.3 Reproduction in humans

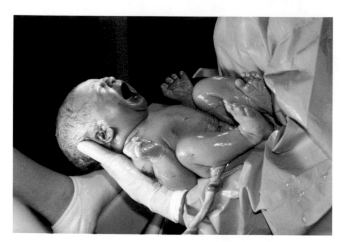

Survival of the human species

As a species, there is no doubt that humans are successful in terms of 'reproductive fitness'. This means that they survive long enough to produce the next generation and pass on their genes to their offspring. As evidence to support this, we have seen that the human population of the world has grown spectacularly and is expected to reach 10 billion by the year 2100. This poses real problems in terms of the ability of the planet to provide the resources needed.

Each individual couple makes their own decision about when to reproduce, but there is a clear pattern – in countries where infant survival rates are good, women choose to have fewer children. Many couples now delay having a baby until later in life, but it is sometimes necessary to use technology to help them. Currently there are more than a million individuals who are 'test-tube' babies. This technology has brought great benefits to individual couples who could not conceive a baby naturally.

However, advances in 'assisted reproductive technology' also raise other questions. Should the limited resources for healthcare be diverted to helping infertile couples when there are already so many people in the world? Is it an abuse of technology for couples to select the sex of their unborn child, or some other characteristic? If you could specify particular features of a future child, what would they be?

■ Humans and sexual reproduction

Humans are mammals. This name tells you that they have mammary glands and a human mother can feed her infant with breast milk. Normally, a human couple produces only one child at a time, rather than say five like a bird (such as a robin) or a million like a fish (such as a cod). Just to replace themselves, a pair of parents needs to rear only two offspring. However, the chance of a newly hatched cod reaching maturity is very much less than that of a human baby.

In sexual reproduction, two cells (the male and female gametes) join together to produce offspring. These gametes are **haploid**, which means they contain one set of chromosomes. When they join (fuse) together, we say that **fertilisation** has taken place. A **zygote** is formed as a result of fertilisation and this cell is **diploid**. The zygote divides into many cells, and these become the new individual (offspring).

Male reproductive system

The male reproductive organs develop during the growth of the embryo. They remain dormant until adolescence, when they become active. At this time the hormone **testosterone** is produced. This causes the body of the male adolescent to develop male secondary sexual characteristics.

The male gametes (sperm cells) are produced inside the **testis**. Each testis contains a large number of microscopic tubules, all linked together. In an adolescent male, new sperm cells are formed by meiotic cell division on the walls of these tubules. From here they pass slowly to a storage area, the **vas deferens**. This process occurs at a lower temperature than core body temperature, which is why the testes are suspended in a sac of skin (**scrotum**) outside the abdominal cavity. This sperm production process continues throughout the life of the adult male and, unless ejaculation occurs, the sperm are steadily shed in the urine.

Figure 5.19 shows the male reproductive organs, how they link in with the urinary system and their position in the body. You can trace the pathway taken by **sperm** from the **testes** where they are produced to the **urethra**, which lies in the **penis**. Sperm are released from the body through the urethra.

> **STUDY TIP**
> Look in Section 5.5 and make sure you understand the differences between mitosis and meiosis.

> **MATHS TIP**
>
> **Numbers in standard form and powers**
>
> Very large and very small numbers are best written in standard form, where the first part is a number between 1 and 10, and the second part is a whole-number power of ten (negative or positive).
>
> For example, 0.00005 becomes 5.0×10^{-5}; and 750.34 becomes 7.5034×10^{2}
>
> Now go to page 228 to apply this to some calculations involving numbers of sperm.

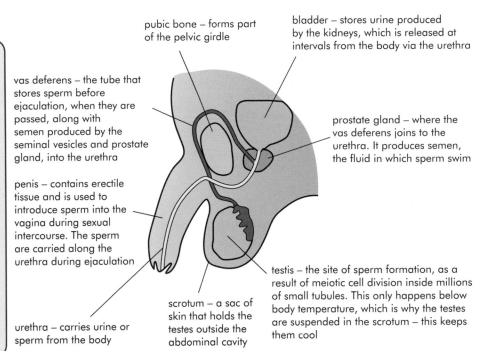

Figure 5.19 Side view of male reproductive organs.

Female reproductive system

The female reproductive organs develop during the growth of the embryo. They remain dormant until adolescence, when they become active. The activation of the ovaries leads to the cyclical production of the hormone **oestrogen** and the cyclical production of the hormone **progesterone**. It also leads to the monthly release of an **ovum** from an ovary (note that an ovum is also known as an **egg cell** – both terms can be used).

This monthly cyclical process is known as the **menstrual cycle**. It carries on until menopause, which generally occurs from the age of about 45 or later. At this time, the menstrual cycle finally stops and a woman can no longer conceive a child. In addition to their direct role in the menstrual cycle, oestrogens cause the body of the female adolescent to develop the female secondary sexual characteristics.

Figure 5.20 shows the female reproductive organs and their position in the body. An **ovum** (**egg cell**) passes from the **ovary**, where it is produced, to the **oviduct** (**fallopian tube**). If fertilisation occurs, the embryo develops in the **uterus**. If fertilisation does not occur, the ovum passes through the uterus and is released from the body through the **vagina**. Note that this is a separate opening and does not link directly with the urethra, as in the male.

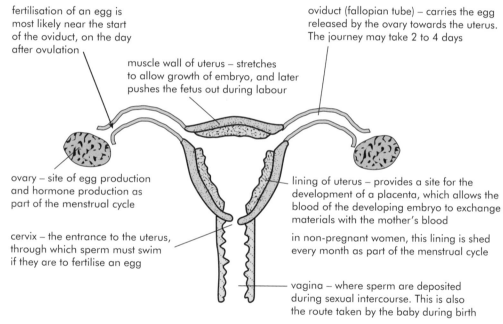

fertilisation of an egg is most likely near the start of the oviduct, on the day after ovulation

oviduct (fallopian tube) – carries the egg released by the ovary towards the uterus. The journey may take 2 to 4 days

muscle wall of uterus – stretches to allow growth of embryo, and later pushes the fetus out during labour

ovary – site of egg production and hormone production as part of the menstrual cycle

lining of uterus – provides a site for the development of a placenta, which allows the blood of the developing embryo to exchange materials with the mother's blood

cervix – the entrance to the uterus, through which sperm must swim if they are to fertilise an egg

in non-pregnant women, this lining is shed every month as part of the menstrual cycle

vagina – where sperm are deposited during sexual intercourse. This is also the route taken by the baby during birth

Figure 5.20 Front view of female reproductive organs.

Menstrual cycle

During her fertile lifetime, a woman produces an egg from one of her ovaries approximately every 28 days. This is part of the menstrual cycle. A number of hormones are involved (Figure 5.21). Some key events in the menstrual cycle and the hormones involved are summarised below:

- **Day 1** of the cycle is taken as the first day the woman sheds the lining of her uterus together with some blood. This is known as **menstruation** and happens with every cycle unless she becomes pregnant.
- The hormone FSH (follicle stimulating hormone) is released from the pituitary gland at the base of the brain. This occurs a few days after the lining is shed. FSH stimulates the growth of a new egg cell in the ovary. The growing egg cell is helped by a cluster of cells around it, called a follicle.
- The follicle cells produce the hormone **oestrogen**. For the first 14 days of the menstrual cycle the level of oestrogen in the bloodstream rises.

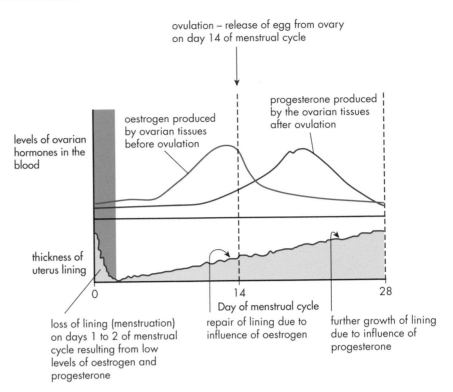

ovulation – release of egg from ovary
on day 14 of menstrual cycle

oestrogen produced
by ovarian tissues
before ovulation

progesterone produced
by the ovarian tissues
after ovulation

levels of ovarian
hormones in the
blood

thickness of
uterus lining

0   14   28
Day of menstrual cycle

loss of lining (menstruation)
on days 1 to 2 of menstrual
cycle resulting from low
levels of oestrogen and
progesterone

repair of lining due to
influence of oestrogen

further growth of lining
due to influence of
progesterone

Figure 5.21 Changes in hormone levels and other events in the menstrual cycle.

- One immediate effect of the rising level of oestrogen, is the formation of a new lining inside the uterus.
- At day 14, the hormone LH (luteinising hormone) is produced by the pituitary gland. LH causes the release of the egg from its follicle in the ovary. This is described as **ovulation**. The released egg is then moved along by tiny hairs into the oviduct.
- LH also causes the follicle cells of the ovary to switch from producing oestrogen, to producing the hormone **progesterone**. One effect of progesterone is to promote the further growth of the uterus lining.
- If fertilisation does not occur, the released ovum (egg) travels along the oviduct over the next 14 days. It is eventually shed with the lining of the uterus at the start of the next cycle. The levels of oestrogen and progesterone fall during the final 2 days of a cycle. This leads round again to the beginning of the cycle, when menstruation occurs.

■ Fertilisation

In the male, during **ejaculation**, millions of sperm cells are forced along the **vas deferens**. Here they become suspended in **semen** – a fluid secreted by several glands, including the **prostate gland**. From the prostate gland, the semen passes into the **urethra**, the tube that leads out of the **penis**. The semen contains sugars that provide the energy required for the movement of the tails of the sperm

During sexual intercourse, semen is ejaculated into the vagina of the female, near to the **cervix**. From here, the sperm cells follow a chemical trail, though a plug of mucus in the cervix, into the **uterus** (womb). They swim across the uterus into the oviducts. Here, if they meet an egg at some point in the oviduct, fertilisation may occur. This is most likely 1 or 2 days after a woman has **ovulated** (released an ovum from an ovary into an oviduct). However, it is *possible* for an ovum to become fertilised at other times as well.

During fertilisation (Figure 5.22), the head of a sperm cell releases enzymes. These digest a path through a protective layer surrounding the egg and the sperm then passes through the egg cell membrane. This is immediately followed by the egg releasing a thick layer of material which prevents the entry of any more sperm cells – only one sperm cell can fertilise the egg. The 23 chromosomes carried by the sperm cell head join the 23 chromosomes carried in the egg cell nucleus to produce a diploid zygote (fertilised egg). The zygote carries 46 chromosomes (23 pairs).

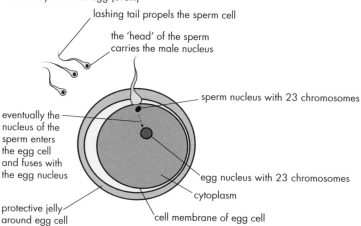

sperm follow a chemical trail, from the cervix, through the uterus and up into the top of the oviduct, where they meet the egg (ovum)

lashing tail propels the sperm cell

the 'head' of the sperm carries the male nucleus

sperm nucleus with 23 chromosomes

eventually the nucleus of the sperm enters the egg cell and fuses with the egg nucleus

egg nucleus with 23 chromosomes

cytoplasm

protective jelly around egg cell

cell membrane of egg cell

Figure 5.22 Fertilisation of an ovum by a sperm.

Figure 5.23 Some sperm cells (image taken with an electron microscope).

Pregnancy

If the egg is fertilised in the oviduct, the woman becomes pregnant. The zygote (fertilised egg) begins to divide to form an embryo. The dividing ball of cells continues down the oviduct for several days, until it reaches the uterus. During its journey towards the uterus, the embryo secretes another hormone (known as HCG). This passes into the woman's bloodstream and stops the menstrual cycle from completing its course. So this stops the monthly shedding of the uterus lining (menstruation). This means that when the embryo reaches the uterus, it is able to attach itself to the thick uterine lining. This is described as **implantation**.

The placenta and amniotic fluid

A **placenta** forms where the embryo is attached (Figure 5.24). The placenta is the organ that allows exchange of materials between the blood of the woman and the blood of the growing embryo. The embryo is surrounded by a layer of fluid, known as the **amniotic fluid**, held in by the **amniotic membrane**. The amniotic fluid acts like a cushion or 'shock absorber' and helps to protect the embryo from bumps to the body of the mother.

For the next 9 months the embryo grows in the uterus. This time is known as **gestation**. Generally the term 'embryo' is used for the first 8 weeks of development, but from about the 9th week it is described as a **fetus**.

During this time of gestation, the developing embryo (then fetus) receives all the oxygen and nutrients it requires via the placenta, from the mother's blood. All the metabolic waste products of the embryo, including urea and carbon dioxide, are passed to the mother's blood via the placenta.

Features of the placenta that help this exchange of materials include the villi. These provide a large surface area for diffusion of materials between the two bloodstreams. Note that the blood of the embryo and that of the mother do not actually come into direct contact or mix, but the capillaries of the embryo's blood system lie close to enlarged spaces (of maternal capillaries) containing the mother's blood.

Requirements of the developing embryo include oxygen, glucose, amino acids, mineral ions and vitamins. These pass from the mother's blood to that of the embryo. The embryo needs to get rid of waste products, including carbon dioxide and urea. These pass from the embryo's blood into that of the mother. For some substances the difference in concentration gradient allows passage of materials to occur by diffusion. For other substances, active transport is involved.

The partially permeable barrier between the mother's blood and that of the embryo (fetus), normally prevents the embryo becoming infected by pathogens, or damaged by substances in the mother's blood. However, some harmful things do cross, such as nicotine, alcohol and certain viruses including rubella (measles).

The placenta also becomes an important source of progesterone as pregnancy progresses, and this prepares the mother's body for the birth of her baby.

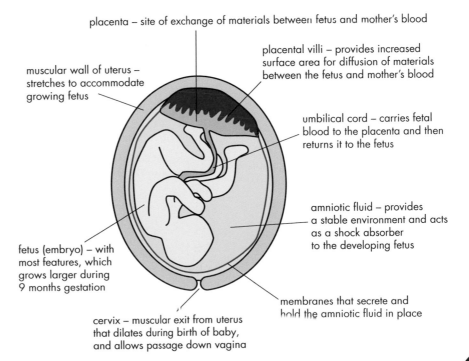

placenta – site of exchange of materials between fetus and mother's blood

placental villi – provides increased surface area for diffusion of materials between the fetus and mother's blood

muscular wall of uterus – stretches to accommodate growing fetus

umbilical cord – carries fetal blood to the placenta and then returns it to the fetus

amniotic fluid – provides a stable environment and acts as a shock absorber to the developing fetus

fetus (embryo) – with most features, which grows larger during 9 months gestation

membranes that secrete and hold the amniotic fluid in place

cervix – muscular exit from uterus that dilates during birth of baby, and allows passage down vagina

Figure 5.24 Developing fetus in uterus.

By 3 months into gestation, the embryo has developed all its key organs in miniature and by this time is known as a fetus. At 9 months, the fetus is born into the world, and becomes known as a baby. It seems that the trigger for birth is a hormone produced by the fetus when it is fully developed.

A woman gives birth to the fetus first, then a little while later she gives birth to the placenta – that vital supporting organ that made the link between the developing embryo (fetus) and the mother.

If a woman breastfeeds her baby, her menstrual cycle does not start again for several months. However, if she does not, the menstrual cycle starts and she is able to become pregnant again.

■ Secondary sexual characteristics in humans

Up until adolescence, apart from the sexual organs, the body forms of males and females are similar in shape and overall appearance. However, the influence of sex hormones produced during puberty leads to permanent obvious differences in body form. These are known as the secondary sexual characteristics. The sex hormones are produced by the testes of males and by the ovaries of females. The following hormones are produced:

- in males – testosterone is produced by the cells between the tubules of the testes
- in females – oestrogen and progesterone are produced by the follicle cells in the ovaries.

The effects of these hormones are summarised in Table 5.3.

Table 5.3 Secondary sexual characteristics in males and females.

Males	Females
pelvis remains narrow – helps with fast running	pelvis enlarges so that hips widen – this allows the passage of a baby through the pelvic girdle during birth
no development of breasts	breasts grow with fat deposits – the breasts provide milk for the newborn baby
greater proportion of body mass becomes muscle rather than fat	greater proportion of body mass is maintained as fat rather than muscle – this provides a supply of nutrients during pregnancy
hair develops on pubic region and under arms, as well as on face (beard)	hair develops on pubic region and under arms
voice box grows larger – gives a lower pitched voice	voice box remains small

STUDY QUESTION

1 The statements in the table refer to hormones involved in the menstrual cycle. Copy the table. For each statement, tick the box(es) in the appropriate column(s) to indicate whether the statement is correct for that hormone.

Statement	oestrogen	progesterone	FSH	LH
stimulates production of follicle cells and growth of an ovum in the ovary				
produced in the follicle cells in the ovary				
produced by the pituitary gland during the first half of the menstrual cycle				
promotes development of new lining inside uterus				
produced by pituitary gland reaching a peak on day 14 of the menstrual cycle				
stimulates growth of uterus lining				
level falls if fertilisation does not occur				
produced in placenta during pregnancy				
prepares mother's body for birth of baby				
causes secondary sexual characteristics				

5.4 Genes and chromosomes

The nature versus nurture debate

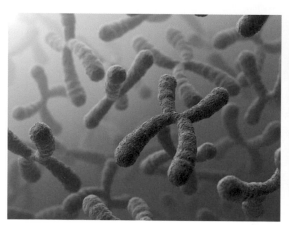 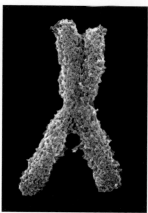 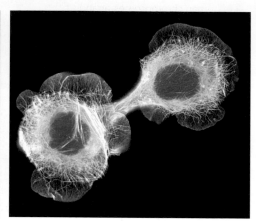

Humans show common features that allow us to recognise that we belong to the same species. But every single human person differs in some way from everyone else.

How are the common features passed on from one generation to the next and what is the biological basis of the differences between people? Is it that each individual is exposed to a different personal environment? Does a suntan get passed on to the next generation, or is that something that just affects one individual and stops there? Or are there genetic reasons for some of the differences between us all?

Scientists now know that some of the differences are due almost entirely to environmental factors, some differences are due entirely to genetic differences, but most are the result of a combination of genes plus the effects of the environment and the interaction between them. Nature could be regarded as the hand of cards you are dealt and nurture as the way you employ the cards you have received. Discuss with other students, which differences between yourselves might result from nature and which from nurture.

Two images show whole chromosomes and these hold the information that is the key to our inheritance. The third image shows a human cell dividing. We need to know how information is contained in the chromosome and how it is passed from one cell to another.

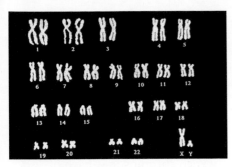

Figure 5.25 The human karyotype – the full set of human chromosomes, arranged to show their shapes and relative sizes. This one is a male (see the XY pair).

STUDY TIP

Make a link to the structure of cells in Section 1.3.

▪ Chromosomes

Every one of the billions of cells that make up your body contains a library of genetic information called the genome. This information is stored inside the **nucleus** of each cell, in the **chromosomes**, as a code within molecules of **DNA** (**deoxyribonucleic acid**).

Chromosome number and homologous pairs

There are **46 chromosomes** inside the nucleus of each human body cell. Each chromosome is a molecule of DNA. Each chromosome has a partner, which looks very similar in its shape and structure, so our body cells contain **23 pairs** of chromosomes (Figure 5.25). The chromosomes in a pair are described as **homologous** chromosomes. One member of each pair came from each of your parents — so one set of 23 chromosomes came from one parent; the other set came from your other parent, giving you a total of 46. Gametes (in humans) each contain one set of 23 chromosomes.

The chromosome number has been worked out for many species. Here are just a few examples to let you see how this number varies: onion = 16 chromosomes; rice = 24; chimpanzee = 48; donkey = 62; chicken = 78.

The chromosome number for the extinct woolly mammoth has been worked out from a frozen carcass and this is found to be 58. Each of these examples gives the total number of chromosomes, so each of these species has half the number of homologous pairs. In the examples given, you can see that onion has eight pairs of chromosomes and a donkey has 31 pairs.

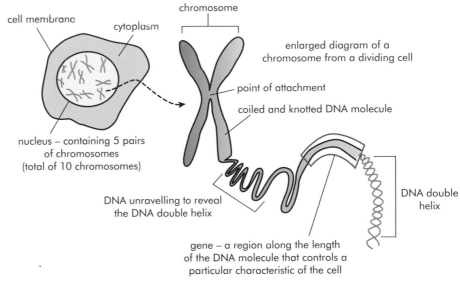

Figure 5.26 Looking more closely at chromosomes – from the cell nucleus to the DNA. Each stage in the sequence shows an enlarged view of the stage before.

A single set of chromosomes (one of each homologous pair) is known as the **haploid** number. We use the symbol 'n' — often written as (n) — to show the haploid number. Two sets of chromosomes (including both members of the homologous pairs) are described as being **diploid** (2n).

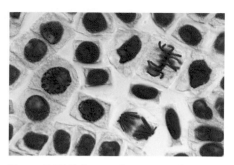

Figure 5.27 Cells near the root tip of a hyacinth, seen with a light microscope. In cells that are not dividing, chromosomes cannot be distinguished in the nucleus, but in cells that are dividing the dark strands of the chromosomes can be clearly seen. In one cell (lower part of photo), the chromosomes are separating and each group will go into one of the two daughter cells produced by mitosis.

Using a light microscope, chromosomes are visible only in **dividing cells** and you can see them more clearly when stained with a dye that sticks to DNA. The reason why chromosomes become visible in dividing cells is because the DNA becomes coiled forming shorter and fatter lengths of DNA. After cell division, the chromosomes uncoil again, becoming less obvious.

Genes and alleles

The genes are arranged along the length of each chromosome. Each of the 23 types of human chromosomes carry thousands of different genes. There are about 22 000 different genes that make up the complete genetic information for a human — or the human **genome**.

A gene is a length of DNA at a specific place on a chromosome, that codes for a particular **protein**. Each protein made in a cell gives a particular characteristic (or inheritable trait) to the individual carrying it.

As described above, chromosomes occur in homologous pairs. This means the position of genes along each member of a homologous pair are identical, and the cells always contains two copies of each gene — one copy on each member of the pair of chromosomes.

Genes often have more than one form, each of which instructs the cell to do slightly different things. These alternative forms of a gene are known as **alleles**. For example, in humans, the gene for eye colour may code for brown eyes or for blue eyes. In pea seeds, a gene for seed shape may code for round peas or for wrinkled peas. Different alleles are the result of subtle changes to the DNA code within a gene, as described below.

So, to look at these two examples again, in an individual one chromosome (of the homologous pair) may carry the allele for brown eyes and the other chromosome of the pair may have the allele for blue eyes. Similarly, the allele for round seeds may occur on one chromosome and for wrinkled on the other chromosome of the pair.

An individual may have two alleles that are the same (for a particular character), such as having the blue eye allele on both chromosomes, or the two alleles may be different. A single individual can only carry two alleles of a gene, but there may be other possible alleles of a gene in the population. For example, in the human population, the blood group gene has three alleles — called A, B and O, yet each individual human carries a maximum of two alleles.

Many of our genes are also shared with most other species of life on Earth — we share 98% of our genome with chimpanzees and 50% of our genes with bananas! Bacterial cells carry a thousand times fewer genes than animals and plants, and all their genes are fitted onto a single circular chromosome.

The structure of DNA

A DNA molecule is made of two strands of **polynucleotides** spiralling around one another to form a double helix shape. A nucleotide is made up of a sugar, phosphate and a base. If you imagine the DNA double helix as a twisted ladder, the sides of the ladder consist of an alternating series of sugars (deoxyribose) and phosphates. The rungs of the ladder are composed of pairs of bases. One base of each pair forms half a rung (Figure 5.28).

STUDY TIP

To help you remember the word, poly = many, hence polynucleotides.

STUDY TIP

Think of a simple way that you can remember the letters for these pairs of bases.

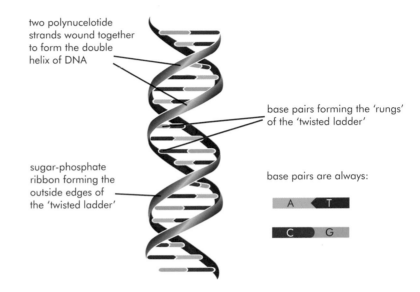

two polynucelotide strands wound together to form the double helix of DNA

base pairs forming the 'rungs' of the 'twisted ladder'

sugar-phosphate ribbon forming the outside edges of the 'twisted ladder'

base pairs are always:

A — T

C — G

Figure 5.28 Structure of the DNA double helix – note the base pairs.

Each base projects from one side of the ladder, towards the middle of the double helix. Here it matches up with a complementary base, projecting from the opposite side of the ladder. There are four bases, usually known by their initial letters, as follows: adenine (A), thymine (T), guanine (G) and cytosine (C).

The bases can only pair up according to the following rules:
   A can only form a pair with a T
   G can only form a pair with a C.

The double helix structure of DNA was proposed in Cambridge by James Watson and Francis Crick in 1953. Later they were awarded a Nobel Prize for this work. They had solved the riddle of the structure of this important molecule and they also realised how it could be accurately copied (replicated) as part of cell division. This discovery opened the way for enormous advances in understanding how DNA works and in the development of techniques of gene technology that have become an important part of scientific work in the 21st century, with applications in many fields.

STUDY TIP

You can find more details about gene technology and genetic modification (GM) in Section 7.5.

STUDY TIP

Look in Section 5.5 for details about cell division and mitosis. Make a link here between DNA and 'mistakes' that lead to mutations – see Section 5.6.

When a DNA double helix is split lengthwise into two halves, the missing half of each ladder is rebuilt. Because A always pairs with T, and G always pairs with C, the sequence of bases in the two new complete molecules of DNA is identical. This is crucial when a cell divides by mitosis as it ensures that each new daughter cell gets a full set of chromosomes (and therefore, the same genes) from the original parent cell.

■ Practical activity – extracting DNA from strawberries

1 Place one large strawberry (or two smaller ones), with stalk removed, into a small, strong plastic bag. Add two or three drops of detergent and a level teaspoon of table salt.
2 Knot the mouth of the bag and then massage the fruit, reducing it to a liquid mush with no lumps.
3 Cut the end of the bag to release the liquid into a funnel lined with muslin.
4 Pour ice cold ethanol gently onto the pink filtrate, forming a clear upper layer in the tube. (Ethanol is hazardous, so follow safety advice from your teacher.)
5 Strands of white DNA precipitate at the junction between the two layers and eventually float off into the upper layer.

Note that other material can be used, depending on the time of year and what is available.

When using ethanol, ensure there are no naked flames or other sources of ignition, and that the laboratory is well ventilated.

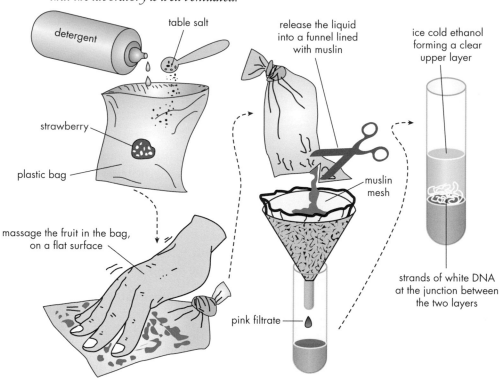

Figure 5.29 Extracting DNA from strawberries

■ The genetic code and protein synthesis

A protein consists of a chain of a hundred or so amino acids. There are 20 different amino acids that can be used in a chain, and each type might be used once, repeatedly or not at all. A gene directing the synthesis of a protein ('coding' for a protein), must specify the exact type of amino acid used for each member of the chain. This information is encoded within the gene DNA as a sequence of chemical bases. The four DNA bases (the 'rungs' of the ladder shown in Figure 5.28) are usually referred to by a single letter code: Adenine (A), Thymine (T), Guanine (G), Cytosine (C).

STUDY TIP

DNA and **RNA** – **differences and similarities**
DNA base pairs lie on the "inside" of the DNA **double helix** and this protects them from random chemical changes, reducing the chance of mutations. Chromosomes last a lifetime and copies are passed on when a cell divides
RNA molecules are **single stranded** and last for a few minutes in the cytoplasm of the cell, before being broken down. Also remember that in RNA, **U** replaces **T** (found in DNA).

STUDY TIP

Transcription and **translation**
Transcribing means making a copy of something, which is what happens when the DNA of a gene is transcribed into mRNA (a nucleic acid copy of another nucleic acid)
Translating means converting something into another form. This is what happens when the mRNA 'message' is **translated** into a protein (a chain of amino acids).

MATHS TIP

Numbers in standard form and powers

Very large and very small numbers are best written in standard form, where the first part is a number between 1 and 10, and the second part is a whole-number power of ten (negative or positive).
For example 0.00005 becomes 5.0×10^{-5}; and 750.34 becomes 7.5034×10^2

Now go to page 228 to apply this to some calculations involving DNA.

The sequence of these bases along the DNA of a gene determines the exact order of the amino acids in the protein. Three adjacent bases (a **codon**) code for a particular amino acid. For example AGA is a **codon**, and TGC is another. In the same way that different letters of the alphabet can be rearranged to make different words, the many different possible arrangements of the four DNA bases provide enough codons to allow the gene to specify all amino acids in a protein. So, for instance, to code for a protein 100 amino acids long, 100 codons composed of a total of 300 bases are required.

Protein synthesis involves several steps (Figure 5.30). The protein is synthesised in the cytoplasm of the cell, but the gene directing its synthesis is held in the nucleus of the cell in a chromosome. When a gene is activated ('switched on'), a matching copy of the DNA base sequence within the gene is **transcribed** (re-written) into another kind of nucleic acid molecule called **mRNA** (messenger ribonucleic acid).

RNA is composed of single-stranded chain of **RIBO**nucleic acid (as opposed to the double chain of **DEOXYRIBO**nucleic acid in DNA) and contains the bases A, C, and G like DNA. However, in RNA the base Uracil (U) occurs instead of the base T (found in DNA). The base sequence of the mRNA provides codons specifying the same amino acid sequence specified by the DNA of the gene.

The mRNA molecule is mobile and flexible. It moves out of the nucleus, into the cytoplasm, where it becomes attached to organelles called ribosomes. The ribosome carries out the synthesis of the protein (a step called **translation**). The ribosome moves steadily along the mRNA molecule like a train on the railway tracks, two codons at a time. Amino acids are delivered to the ribosome, attached to transfer RNA (**tRNA**) molecules.

Each tRNA molecule has a group of three bases called an **anticodon**. For example UCU is an anticodon, and AUC is another. Each different amino acid is attached to a specific tRNA. Each specific tRNA anticodon pairs up only with a specific mRNA codon. As a consequence, different amino acids are delivered to the ribosome by different tRNA molecules, in a precise order, because of the mRNA codon-tRNA anticodon pairing rules. For example, the codon UAU can only pair up with the anticodon AUA.

On the ribosome, the amino acids become chemically linked together with strong bonds. Thus the sequence of amino acids originally encoded in the DNA of the nucleus ends up being synthesised in the cytoplasm of the cell. This chain of amino acids can then fold into a protein. The series of steps described above shows how DNA directs the production of proteins and, in turn, the characteristics that you inherit. The exact pattern of different proteins made by an individual cell determines what the cell does and what it becomes.

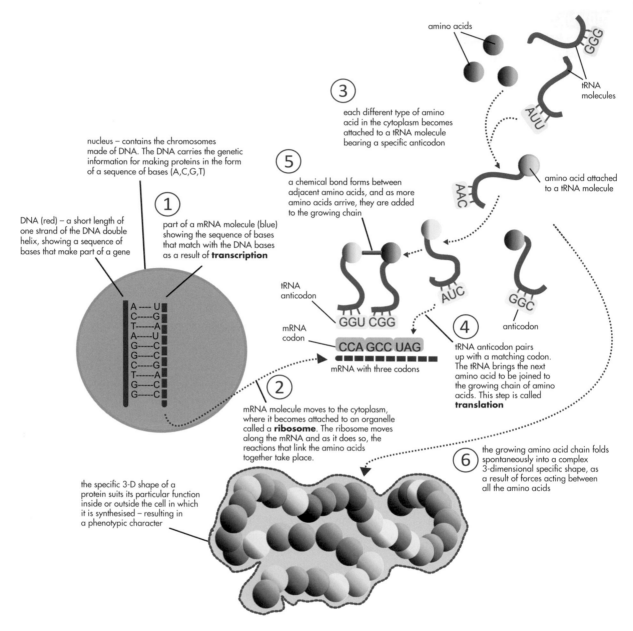

Figure 5.30 Protein synthesis - the diagram illustrates how genetic information in the bases of a DNA molecule undergoes transcription to mRNA followed by translation to form a chain of amino acids as part of a protein (steps numbered 1 to 6).

MATHS TIP

Probability

Probability is a measure of the likelihood that something will happen, ranging on a scale from impossible (= 0) to certain (= 1). On this scale, an even chance of something happening = 0.5.

This can also be expressed as a fraction or as a percentage (so 0.5 = 50% = 1/2). Remember it is still possible for unlikely things to happen!

Now go to page 229 to apply this to some calculations involving bases and mutations in DNA.

STUDY QUESTIONS

1 It is important that you understand and remember the definitions for various terms that are used in Section 5.4. A list of terms is given in the box:

allele; chromosome; diploid; DNA (deoxyribonucleic acid); gene; haploid; homologous chromosomes

a) First cover up the table (below) and write your own definitions for each of the terms in the list.
b) Then copy the table, look at the definitions or descriptions in it and match the correct term to each description.
c) Cover up the table again and write the definitions again, to check that you remember them correctly.
d) Write a sentence using each term or give an example for each term to show that you do understand them.

Definition or description	Term
Thread-like structures in the nucleus, made up of DNA, that carry the genetic information for the organism.	
A molecule with a double helix shape, consisting of two strands of nucleotides held together by pairs of bases. It carries the genetic information in coded form for all the characters in the organism.	
A pair of chromosomes that are the same shape and same size, and carry genes for the same characters in the individual. One of each pair came from each parent.	
A cell containing one set of chromosomes (n).	
A cell containing two sets of chromosomes (2n).	
A region along the length of the DNA of a chromosome that contains the information to control a certain characteristic.	
An alternative form of a gene – occurs at the same position (locus) on the chromosome and controls the same character but in different ways.	

2 a) List the four bases that occur in a DNA molecule. Use their names as well as the letters. Then draw a diagram to show which bases form pairs together.
  b) DNA is described as a 'double helix'. Draw a simple diagram of a DNA molecule to show both strands and the helix shape. On the diagram, label the molecules that form the backbone and show where the pairs of bases fit into the structure.

3) Here are some questions to help you with the stages of protein synthesis and the different words used.
  a) In a DNA molecule, how many bases code for one amino acid? Write down two examples. (Use letters for the bases rather than their names.)
  b) Name the four bases found on an RNA molecule. Use their names as well as the letters.
  c) For the two examples you gave in a) for DNA, write down the corresponding bases that would match them and form a codon on an mRNA molecule.
  d) Now write down the bases that would be found on a tRNA molecule, to correspond with the examples you wrote down in c) and form an anticodon.
  e) Which stage above represents transcription? (Refer to the letters for the part of the question and give a description of the the term in words.)
  f) Where, in the cytoplasm, does translation occur? Describe in words what happens.
  g) Explain how a protein that is synthesised relates to the information contained in DNA molecules in the nucleus of a cell.

5.5 Patterns of inheritance

DNA and the information it contains

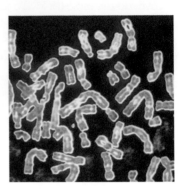

Recent technological advances have allowed scientists to unravel the DNA sequences of thousands of individuals. Much of your individual DNA is identical to that of all other humans, but the tiny variations that do exist are responsible for many of the differences between you and other people.

Scientists can now map these DNA differences in great detail. This has allowed forensic scientists to use a genetic fingerprint to identify somebody with great certainty. DNA fingerprinting is often used to help catch criminals who have left DNA at the scene of a crime.

You can also send a saliva sample (containing cheek cells) to a commercial laboratory and, for a fee, they will examine some of your personal DNA sequences. As part of this service they can tell you about some of the diseases to which you are more (or less) susceptible, compared with the average human. They can also tell you about your geographical family origins.

Would you like to use such a commercial service? What disadvantages can you imagine with having such information? Should everyone have their DNA sequence stored on a database? How is the DNA that is special to every plant and animal species passed from one generation to the next? How are the characteristics shown in the family group (image above) passed from one generation to the next to show both similarities and differences?

■ Genes and inheritance

Each of your 46 **chromosomes** carries hundreds of individual genes. In Section 5.4, we described how **genes** are located along the length of the chromosomes in the nucleus, and that each chromosome is made up of DNA (deoxyribonucleic acid). The gene carries the information that tells the cell how to produce a certain character. Each chromosome has a partner, similar in shape and size, and carrying an equivalent range of genes. These pairs of chromosomes are called **homologous chromosomes** so a human has 23 pairs of chromosomes.

STUDY TIP

Make a link here to Section 5.4 (Genes and chromosomes).

On each of the chromosomes in the homologous pair, there are alternative forms of each of the genes. These alternative forms are known as **alleles**.

For example, there is an allele that gives information for brown eye colour and an alternative allele that gives information for blue eye colour. If both the brown and blue alleles are present together in a person, the eye colour is brown. If both alleles are for blue eye colour, the person has blue eyes.

Using this example, the genes carried by the individual are described as the **genotype** and the 'appearance' of the character (i.e. the eye colour) is described as the **phenotype**. Sometimes you cannot actually see the character, but the term phenotype describes the character within the individual.

We use symbols to represent the alleles, and in this case we can use **B** for the brown allele and **b** for the blue allele. To summarise this eye colour example:

phenotype (appearance) = brown eyes   genotype (alleles) can be BB or Bb
phenotype (appearance) = blue eyes   genotype (alleles) only bb.

This example also shows that the allele for brown eye colour always shows up in the phenotype, but that for blue eye colour does not show up if the brown allele is also present. We say that the brown allele is **dominant** and the blue allele is **recessive**. Deliberately, we wrote the dominant allele with a capital letter and the recessive allele with a lower case (small) letter. This convention is often followed in genetics.

Figure 5.31 uses a diagram to summarise the application of these terms. It shows four genes and their positions on a pair of homologous chromosomes. You can see two more terms used on this diagram. If both alleles for the gene are the same, we say the individual is **homozygous** for that gene. If the two alleles are different, the individual is **heterozygous** for that gene.

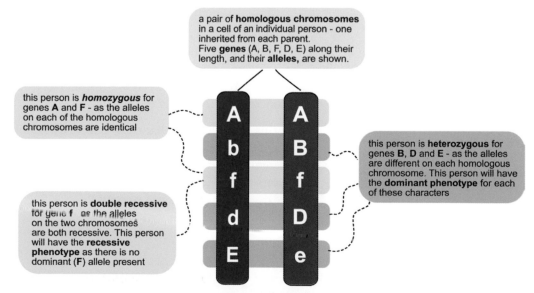

a pair of **homologous chromosomes** in a cell of an individual person - one inherited from each parent.
Five **genes** (A, B, F, D, E) along their length, and their **alleles,** are shown.

this person is *homozygous* for genes **A** and **F** - as the alleles on each of the homologous chromosomes are identical

this person is **heterozygous** for genes **B, D** and **E** - as the alleles are different on each homologous chromosome. This person will have the **dominant phenotype** for each of these characters

this person is **double recessive** for gene **f** as the alleles on the two chromosomes are both recessive. This person will have the **recessive phenotype** as there is no dominant (**F**) allele present

Figure 5.31 A pair of **homologous** chromosomes, illustrating a range of terms.

To complete this list of terms, sometimes both alleles contribute to the phenotype in a way that is neither dominant nor recessive. We describe this situation as **codominance** and we look at examples later in the Section 5.5. The term **incomplete dominance** is also used.

Monohybrid inheritance

The first successful scientific study on inheritance was carried out in an Austrian monastery garden, mainly between 1856 and 1863. Gregor Mendel was a monk but was also interested in plant breeding and patterns of inheritance. He used pea plants and chose to study the inheritance of characters such as flower colour, seed shape and plant height. He was fortunate in his choice because in pea plants these phenotypic features are controlled by **single genes** with two alternative alleles. Mendel had access to 'pure-breeding' plants and he was able to control which plant bred with another by covering the flowers and transferring pollen from flower to flower by hand.

When Mendel carried out his experiments he did not know about cell structure and certainly knew nothing about DNA, but we can now interpret what he did using modern terminology. The term **monohybrid inheritance** refers to the inheritance of a single character, such as seed shape or eye colour, and this was the basis of some of Mendel's experiments.

Inheritance of height in pea plants

Mendel discovered that if he crossed pure breeding tall pea plants with pure breeding dwarf ones, all their offspring were tall. This first generation of offspring is called the F_1 generation. However, when he crossed these F_1 tall offspring with each other to produce the next generation (F_2) offspring, he found that there were some tall and some dwarf. The ratio of tall to dwarf in this second generation was close to 3 tall : 1 dwarf.

We can use a genetic diagram (Figure 5.32) to explain what happened. We follow the convention described above and use the letter **T** for the dominant tall allele, and **t** for the recessive dwarf allele. Remember that a parent has two alleles for this character but passes only one of a pair of alleles into a gamete cell. There is an equal chance (a 50% probability) of it being one or the other. In this cross, the parents were 'pure breeding', so one parent is homozygous for tall (**TT**) and the other parent is homozygous for dwarf (**tt**).

In this example, note that the recessive dwarf allele does not show up in the phenotype of the heterozygous F_1 plants as the dominant tall allele is also present. Yet in the F_2 generation, 25% of the offspring are dwarf. These dwarf plants are homozygous for the recessive dwarf allele ('double-recessives') and the dominant allele is not present.

STUDY TIP

Look in Section 5.5 for more detail about meiosis and the formation of gametes.

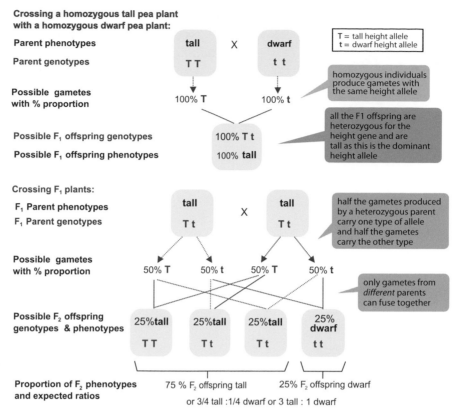

Figure 5.32 Genetic diagram showing the inheritance of the character for height (tall or dwarf) in pea plants.

The percentages beside the phenotypes show the expected percentage of offspring with that genotype. You can express this as a ratio or a fraction. Note that a 3:1 ratio means that three quarters (75%) of the offspring show the phenotype for the dominant allele and one quarter (25%) of the offspring show the phenotype for the recessive allele.

Sometimes a genetic cross is set out as a Punnet square. You can use both systems, but check carefully that you show the genotype and phenotype of the parents, their possible gametes, then the offspring (genotypes and phenotypes). Always use these headings down the side of the page to guide you.

parent (Tt) gametes / parent (Tt) gametes	T	t
T	TT	Tt
t	Tt	tt

Figure 5.33 A Punnett square is another way of showing a genetic cross. In this example, the cross is between two parents heterozygous for tall (each with the genotype Tt). Genotypes for the gametes are on the outside and possible genotypes for the offspring are shown in the boxes.

STUDY TIP

If you are choosing your own letters to use in a genetic cross, make sure you choose letters that look different in capitals and lower case, otherwise your 'working' may get confused.

Codominance

Sometimes, the phenotype of an individual is affected by both the alleles working in combination, rather than only one allele determining the phenotype because it dominates the other allele.

One example is found in snapdragon flowers. If pure (homozygous) white-flowered plants are cross bred with pure (homozygous) red-flowered plants, the offspring have pink flowers. When the pink F_1 plants are cross bred with each other, the F_2 offspring are white, pink and red.

Note that in this genetic diagram (Figure 5.34), different letters are used for the alleles and capital letters are used for both.

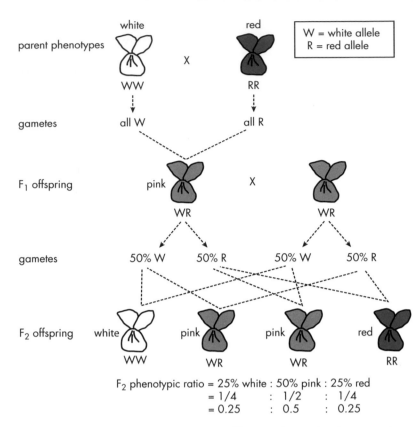

Table 5.4 Blood groups in humans – multiple alleles and codominance.

Genotype	Phenotype
AA	Group A
AO	Group A
BB	Group B
BO	Group B
OO	Group O
AB	Group AB

F_2 phenotypic ratio = 25% white : 50% pink : 25% red
= 1/4 : 1/2 : 1/4
= 0.25 : 0.5 : 0.25

Figure 5.34 Genetic diagram showing inheritance of flower colour and codominance in snapdragon flowers .

Another example of codominance is found in human blood groups. This also illustrates a situation in which there are more than two alleles, though only two would be present at one time in an individual – one on each chromosome inherited from each parent. There are three alleles: A, B and O. A is dominant to O; B is dominant to O; A and B are codominant. Table 5.4 shows how different combinations of these alleles (the genotypes) result in the four possible blood groups (the phenotypes).

Polygenic inheritance

Many of the phenotypic differences we notice between individual humans, such as hair and skin colour, are the result of **polygenic inheritance**. In polygenic inheritance, several different genes all contribute something towards the phenotypic feature. Human height is an example where more than a hundred genes all contribute a tiny amount, affecting approximately 20% of our actual height. That leaves 80% of our height accounted for by environmental factors, such as good nutrition and exercise.

Family pedigrees

The pattern of inheritance of a genetic trait can be followed through a family pedigree diagram, like that shown in Figure 5.35. This involves constructing a family tree and adding notes to individual entries to indicate whether they show a trait or not. This can be used to help work out the genotype of the individuals in the family tree, and therefore the chances of a new offspring being affected by a trait or being a carrier.

MATHS TIP

Bar charts and histograms

Bar charts consist of a series of bars of equal thickness, drawn vertically from the *x*-axis of the chart. They are used to represent the **frequency** of items in a series of discrete **categories**.

If each category has no logical connection to the others (for example different eye colours), the individual bars are often arranged in ascending or descending order of magnitude. If a series of categories form a logical sequence (e.g. from small values to large values), then the bars can be placed against each other, to form a **histogram.**

Now go to page 229 to apply this to some data about blood group frequencies.

When drawing a family tree, the convention is to use a square to show males and a circle for females. Individuals who are unaffected by the trait are shown with no shading but if the individual is affected by the trait shading is used in the square or circle. Usually a key to these symbols is given beside a family pedigree. Roman numerals (I, II, III etc) down the left-hand side of the page show you the successive generations.

The example shown in Figure 5.35 looks at the pattern of inheritance of six-fingered hands, known as polydactyly. The condition is the result of a dominant mutant allele. This means it is expressed in the phenotype in both heterozygous and homozygous individuals. If we use the letter **D** to represent the six-fingered allele and **d** for the five-fingered allele, a person with **Dd** has six fingers. When interpreting a family pedigree, you may not be told what the genotypes are but you have to recognise clues in one or more of the people and then work through the others to unravel the pattern.

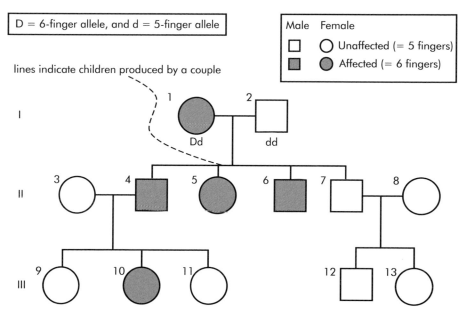

Figure 5.35 Using a family pedigree to show inheritance patterns.

Working through family records and creating such family pedigree diagrams does help scientists to predict whether an allele is dominant or recessive to another possible allele, and to calculate the probability of a couple having an affected child. Some couples may wish to consider inheritance of certain characters when planning a family. Nowadays, it is also becoming more common to sequence the DNA of different members of a family, in order to see if particular alleles are associated in some way with diseases for which the basis is not purely genetic, such as diabetes.

Inheritance of sex in humans

The first question many people ask about a new born baby is: 'Is it a boy or a girl?'. The sex of a baby is determined at the moment of conception (fertilisation), when the father's sperm cell fuses with the mother's egg cell.

STUDY TIP

Check in Section 5.4 for details about karyotypes.

The sex of a human is determined by one of the pairs of chromosomes, known as the sex chromosomes. This pair can be picked out in a karyotype (a picture showing the pairs of chromosomes – see Figure 5.25). Sex chromosomes are given the letters X and Y. The Y chromosome can be recognised because it is shorter in length than the X chromosome. Two X chromosomes produce a girl (XX), and one X and one Y chromosome produce a boy (XY).

We can use a genetic diagram to show how the sex chromosomes are inherited, or passed from one generation to the next (Figure 5.36). The diagram is similar to the genetic diagrams shown for monohybrid inheritance, but remember that the X and Y represent whole chromosomes, not single alleles.

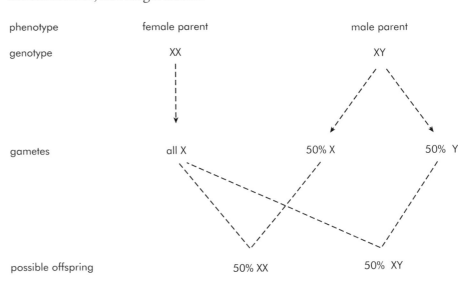

Figure 5.36 Inheritance of sex chromosomes in humans

STUDY TIP

Note that all eggs carry the X chromosome but for the sperm, 50% carry X and 50% carry Y.

MATHS TIP

Probability

Probability is a measure of the likelihood that something will happen, ranging on a scale from impossible (= 0) to certain (= 1). On this scale, an even chance of something happening = 0.5.

This can also be expressed as a fraction or as a percentage (so 0.5 = 50% = 1/2). Remember it is still possible for unlikely things to happen!

Now go to page 230 to apply this to a question on whether a baby will be a boy or a girl.

During the formation of gametes at meiosis, it is chance as to which of the chromosomes in each of the homologous chromosome pairs goes into a gamete. In humans, a female is XX, so this means that all her eggs must carry an X chromosome, as she only has this type of sex chromosome in her body cells. However, for a male (XY), approximately half (50%) of the sperm cells carry an X chromosome and half (50%) carry a Y chromosome. Thus there is approximately a 50% chance that each new baby will be a boy or a girl (summarised in the genetic diagram in Figure 5.36).

Every time a couple conceives a baby, there is a 50% chance of the baby being a boy and a 50% chance of the baby being a girl. This probability does not mean that with two children, one will be a boy and the other a girl. The same chance applies to each baby that is born. Taken on a larger scale than a single family, the patterns become clear though overall, the world's population does have slightly more women than men.

Studying patterns of inheritance requires an understanding of the nature of chance and probability. Just because something is *possible*, it does not mean it is *inevitable*. You may come from a family in which there are lots of boys, or lots of girls, rather than the predicted 1 girl : 1 boy ratio.

■ Cell division

It is best to think of cell division as **nuclear division** as the important thing is for the chromosomes in the nucleus to divide and to do this in a way that reliably passes on the genetic information into the 'daughter' cells. There are two types of cell division as follows:

- In the 'body' cells of the organism – cells divide by **mitosis** and produce daughter cells that are genetically identical to the parent cell. Each daughter cell has the same number of chromosomes as the parent.
- In the 'sex organs' of the organism – cells divide by **meiosis** and produce gametes that have half the number of chromosomes. Each daughter cell may be genetically different from the parent cell.

Only certain cells are able to divide in an organism. In plants, cells that have the ability to divide include cells near the root tip (allowing fresh roots to grow) and cells in the centre of a bud (allowing a new shoot to grow out). In humans, new cells can grow just beneath the top layer of the skin and stem cells in the bone marrow produce fresh blood cells. Many cell types in both plants and animals are unable to divide again after they have been formed and become specialised.

Mitosis

Suppose you are looking at a cell with two pairs of chromosomes – i.e. four chromosomes. When the cell is ready to divide, the DNA in each of these chromosomes doubles inside the nucleus, then divides along its length. These lengthwise 'halves' become the new chromosomes that go into the daughter cells. Because they divided along their length, each new chromosome is an exact copy of the DNA in the original chromosome. The process is summarised in Figure 5.37.

Mitotic cell division occurs in growth and whenever 'repair' is needed. So, once an egg cell has been fertilised by the male sperm (in animals) or pollen nucleus (in plants) all the divisions from the fertilised egg to the embryo, and growth of the embryo into the mature plant or animal, are by mitosis. In this way, identical information is passed on to each daughter cell in the DNA.

Meiosis

We now consider a similar cell with two pairs of chromosomes, but this time it is going to produce sex cells (gametes). In this case, during meiotic division, one chromosome from each pair of homologous chromosomes separates and passes into one of the daughter cells. The daughter cells have only half the number of chromosomes, compared with the parent cell, and along the length of the chromosomes they may have different alleles. This means that the daughter cells are likely to be genetically different from the parent cell. Meiosis produces four daughter cells (Figure 5.38).

Each daughter cell has one member of each pair of chromosomes and they are described as **haploid** (n). At fertilisation, one haploid gamete fuses with another haploid gamete and the **diploid** number (2n) is restored.

STUDY TIP

You can find out more about different types of cells in Section 1.3.

STUDY TIP

This links with Sections 5.1 and 5.2 (Reproduction in living organisms and Reproduction in flowering plants), and you can find more on cloning in Section 7.6.

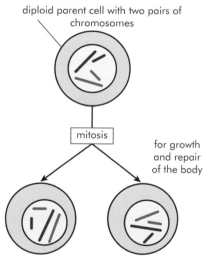

diploid parent cell with two pairs of chromosomes

mitosis

for growth and repair of the body

mitotic cell division produces two daughter cells

Figure 5.37 Cell division by mitosis. Each daughter cell is genetically identical to the parent. Mitosis occurs when cells are produced for growth and repair.

STUDY TIP

Look in Section 5.1 to remind yourself about the meaning of **haploid** and **diploid**, then check in Sections 5.2 and 5.3 to find out the names of the gametes in plants and humans, and where they are formed in each case.

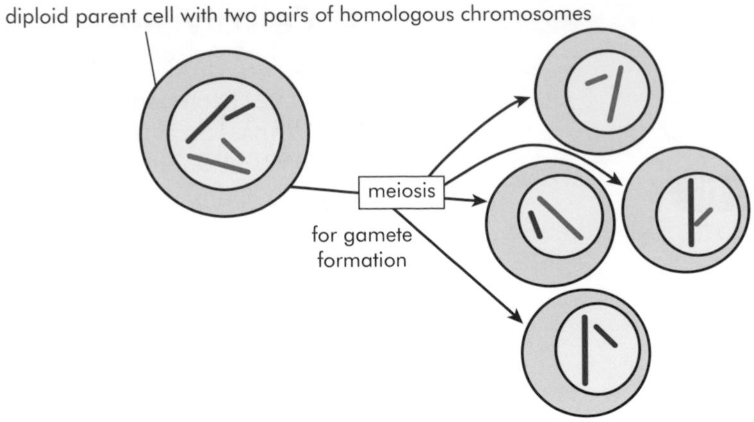

Figure 5.38 Cell division by meiosis. Each daughter cell contains half the number of chromosomes compared with the parent and each is genetically different. Meiosis takes place before the formation of gametes.

When meiosis occurs, the distribution of chromosomes into the daughter cells is a random event. We cannot predict which of the parent chromosomes and which of the parent alleles go into which daughter cell. One consequence of this is that offspring have a different combination of alleles (genetic information) compared with their parents. This leads to genetic variation in the offspring and is an important basis of the phenotypic variation we see across any population of plants or animals. You can compare the genetic variation that results from sexual reproduction with the lack of variation seen as a result of asexual reproduction.

To summarise, Table 5.5 shows the outcome of division by mitosis and meiosis for a diploid organism.

Table 5.5 Comparing division by mitosis and meiosis.

Mitosis	Meiosis
daughter cells genetically identical	daughter cells genetically different, leading to variation
daughter cells have same number of chromosomes as parent cell	daughter cells have half the number of chromosomes compared with the parent cell
daughter cells are usually diploid (2n), but haploid (n) if the parent cell is haploid	daughter cells are haploid (n), or have half the number of chromosomes in the parent cell
cell division produces cells in growth and repair	cell division leads to production of sex cells (gametes)

PRACTICAL

As a modelling exercise, this activity should help you understand the enormous variation that can result from sexual reproduction. Your exercise is just small scale, but brings together the ideas of chance and potential genotypes in the 'offspring'.

Practical activity – modelling genetic variation

This activity models the variation that could result from sexual reproduction involving a genome of 10 genes plus the sex chromosomes. The model helps you to understand the enormous variation that is generated by chance in two parents passing alleles on to their child. Note that in humans, 23 pairs of chromosomes, carrying a total of 25 000 genes are involved in the process of making a baby.

In this model, each gene lies on its own chromosome, most are codominant, and each gene inherited independently of the others. Each 'parent' randomly chooses 10 alleles and a sex chromosome and these contribute to the genome of the 'baby' they make with their partner. The partners then use the table to translate the genome of their baby into a facial portrait.

1 Work in pairs – you both need a photocopy of Table 5.6.
2 Each student personally decides, for each of the eleven 'genes', what they think their personal genotype is. These decisions can be random or deliberate. But for the sex chromosomes, one student should choose the XX genotype (mother) and the other the XY (father).
3 Then individually toss a coin to decide, for each pair of letters in your personal genome, whether you will pass the left allele or the right allele (in the column in the table) to your 'baby'. Mark your chosen letters on the photocopy. This process mimics the production of gametes in meiosis, when there is a 50% chance that one or other of a pair of alleles ends up in a gamete.
4 Now share your chosen alleles with your partner in order to produce a genome for the 'baby'.
5 Construct a portrait of the face of your baby, using the features that would result from the genome you have produced.

Use this table of 10 genes and the phenotypic results of the alleles of those genes present to determine an individual 'baby' face.

Table 5.6 List of features, along with letter codes for alleles, and the phenotypes resulting from allele combinations.

Inheritable feature (letters code for alleles of gene)	Phenotype of baby's face (resulting from individual allele combinations in genotypes)		
Shape of face (A or B)	Round AA	Oval AB	Square BB
Shape of eyebrows (D or d)	Fine DD	Fine Dd	Bushy dd
Form of hair (E or F)	Straight EE	Slightly curled EF	Very curly FF
Facial hair line (G or g)	Straight fringe GG	Straight fringe Gg	Widows peak gg
Colour of hair (J or K)	Black JJ	Blond JK	Brown KK
Size of eyes (L or M)	Small LL	Medium LM	Large MM
Vision (N or P)	Short sight (specs needed) NN	Normal sight NP	Long sight (specs needed) PP
Size of lips (Q or R)	Thin QQ	Thick QR	Medium RR
Freckles on cheeks (S or T)	Abundant SS	Occasional ST	Absent TT
Dimple in chin (V or W)	Deep dimple VV	Slight dimple VW	Absent WW
Sex chromosomes (X or Y)	Female XX	Male XY	

STUDY QUESTIONS

1 In Section 5.4, there is a study question that looks at definitions of terms in genetics. Here are some more definitions. It is important that you understand and remember the definitions for various terms that are used in Section 5.5 together with those in Section 5.4. A list of terms is given in the box:

codominant; dominant; genotype; heterozygous; meiosis; phenotype; recessive; XX; mitosis; XY; homozygous

a) First cover up the table and write your own definitions for each of the words in the list.

b) Then copy the table, look at the definitions or descriptions in it and match the correct term to each description.

c) Cover up the table again and write the definitions again, to check that you remember them correctly.

d) Write a sentence using each term or give an example of each term to show that you do understand them.

Definition or description	Term
The genetic make-up of an individual with respect to the alleles it carries for a particular character.	
The 'appearance' of an individual with respect to the characters it shows – not necessarily visible but the expression of the alleles it carries (its genotype).	
The character for this allele always shows up in the phenotype, even though the alternative allele may be present in the genotype.	
The character for this allele shows up in the phenotype only if the other allele is the same. It does not show in the phenotype if the other allele is different.	
If the two alleles of the genotype are different both make a contribution to the phenotype.	
The alleles at one particular locus (position on the chromosome) are identical on both chromosomes of the homologous pair.	
The alleles at one particular locus (position on the chromosome) are different on the two chromosomes of the homologous pair.	
The two sex chromosomes that are found in female body cells.	
The two sex chromosomes that are found in male body cells.	
Nuclear division that produces two cells that are genetically identical with the same number of chromosomes as the parent cell.	
Nuclear division that produces four cells that are genetically different and with half the number of chromosomes as the parent cell.	

2 Make a copy of the family pedigree on page 211. It refers to the inheritance of six-fingered hands (polydactyly). Use this pedigree to answer the questions and see how you can tackle questions involving inheritance of characters from information in family pedigrees.

a) i) You are told that person 1 is heterozygous for this character. Write in the genotypes of the children in the next generation (generation II).

ii) Explain why it is not be possible for person 1 to be homozygous.

b) i) Write in the genotypes of other individuals in generation III.

ii) What is the probability of person 7 and person 8 having a child with polydactyly? Show how you worked out your answer.

5.6 Variation, change and evolution

Darwin and the Galapagos Islands

The Galapagos Islands are a small group of islands off the pacific coast of South America. They are famous as one of the places that Charles Darwin visited on his travels on HMS Beagle between 1831 and 1836. The many observations made on the voyage were very important to Darwin as he began to develop his theory of evolution by natural selection.

In 1976, scientists on one of the islands were able to record natural selection leading to evolutionary change as they watched. That year, a climatic event called El Nino caused a severe drought on the island of Daphne Major, shown in the top image. This led to the plants on that island producing many large seeds rather than lots of smaller seeds. Birds on the island, called ground finches (middle image), eat seeds but they suffered a population crash because most of them could not eat the large seeds. Just a few of the finches, with larger than usual beaks, were able to eat the seeds and they survived. In the next 2 years, numbers in the finch population increased again, but now most of the finches had large beaks and the small beaks were relatively rare. The scientists had seen a change in a character in the birds of the population in response to the consequences of the drought. They had witnessed an evolutionary change.

So how and why do populations of organisms change over long periods of time? Is it 'survival of the fittest'? What does that phrase mean in terms of the theory of evolution by natural selection?

■ Variation in a population

Look at any group of people in the street. They are all different from each other, yet we recognise them all as being humans – members of the same species. Even if the group happened to include some identical twins, they are still likely to be a little different from each other. Then do the same with a group of plants growing on a hillside. Even if they are of the same species, you can see differences between them – maybe in the size of the leaves or in details of the flower appearance. In both cases, you are noticing the **variation** between members of the same species.

■ Practical activity – variation in a group of people

This activity looks at two sorts of variation in fingers. You can do this activity with a group of people, say with members of your class.

1 Make measurements of finger length, as shown in diagram A in Figure 5.39.
2 Record the relative lengths of the ring finger and forefinger, as shown in diagram B in Figure 5.39. For this you can use 3 categories, depending on whether the forefinger is longer, shorter or the same as the ring finger.

Collect up results for the group and look at the variation in the measurements and observations you have made. You can then find ways to organise the results in a table and display them in a suitable graph.

Measurements taken for A are likely to show continuous variation whereas observations for B give an example of discontinuous variation. You need to use different graphs to display the data.

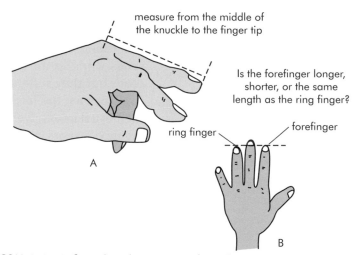

Figure 5.39 Variation in finger length in a group of people.

■ Phenotype = genotype + environment

The **phenotype** ('appearance') of any individual is the result of the interaction between its **genotype** (their genes) and the **environment**. This is discussed in Sections 5.4 and 5.5. In those sections the focus is firstly on the genes that contain the information and secondly on how this information is passed from one generation to the next. At any one time, the actual phenotype of an individual depends on the result of the interaction between the genotype and the environment.

Now we look more closely at variation within a population of individuals of the same species. As an example, suppose you have a set of young plants with identical genotype and you grow them in different conditions – applying different mineral ions in the form of fertiliser, or giving them different lighting or watering. As they grow and mature, the plants are likely to show considerable variation – maybe in height, leaf colour or time of producing flowers or fruits. You can apply similar principles to human populations and understand that, for example, the level of nutrition (underfeeding or excess of food) can lead to differences in the height of individuals and whether they are thin or overweight. You can probably find examples that show how the younger (contemporary) generation is taller than their parents (Figure 5.40). The genotype contains the information and offers the potential for a certain phenotypic character, but the environment often determines whether or not that potential has been allowed to develop.

Figure 5.40 Different generations of the same family

STUDY TIP
Make a link with meiosis and inheritance in Section 5.5.

STUDY TIP
You can find more details about DNA in Section 5.4.

STUDY TIP
Look at Section 1.4 and suggest why changing the shape of a protein such as an enzyme may change the way it carries out its function.

STUDY TIP
Look at Section 5.4 and remind yourself of the way in which the **codons** in the DNA of a gene are **transcribed** and **translated** into a protein, as part of **protein synthesis.**

■ Mutations and the origin of variation

In a population of sexually reproducing organisms the genotypes of all individuals are different. This is the result of the random distribution of chromosomes to gametes during meiosis. It is then chance as to which gametes (from the male parent and female parent) meet at fertilisation and this determines the combination of alleles that come together in the offspring. Each parent provides only 50% of the alleles for the offspring, so each generation is a bit different from the parent generation. Each individual offspring is given a unique genotype – a selection of the alleles passed on from their parents. This shows us why there is variation between individuals in a population.

The origin of all genetic changes is **mutation**. This occurs when the information coded in the DNA of a gene is altered in some way. One or more bases coding for amino acids may be altered – a single base might be might be exchanged for a different one, such as adenine (A) being swapped for say guanine (G). If a codon is changed in such a way, then the amino acid specified by that codon might be different. A change in just one amino acid, among the hundred or so that make up a protein, might cause a major change to its shape and function. This might mean that the mutant protein works less well or not at all. For this reason, some mutations are harmful while others have no effect on the protein and phenotype. In exceptional cases the mutant protein might actually work better than before. This is one way in which alternative versions of genes (alleles) arise, along with alternative phenotypes.

Most of these random changes to a gene cause a change in the protein the DNA codes for. A change to the shape of a protein might mean that it works less well or not at all. This means that most (but not all) mutations are harmful.

Figure 5.41 At the dentist, a nurse stands behind a screen as protection against frequent exposure to X-rays. The patient's body may be protected by a special jacket.

STUDY TIP

You can remind yourself about the need for chlorophyll for photosynthesis in Section 2.1.

Mutations occur naturally and spontaneously. But certain environmental factors can increase the rate at which mutations occur. Such factors are described as **mutagens** (generate or 'cause' a mutation) and include the following:

- ionising radiation – such as gamma rays, X-rays and ultraviolet light
- chemical mutagens – such as the chemicals in cigarette smoke.

Because of their effect on DNA, resulting in mutations, mutagens are considered to be dangerous to living organisms. You can understand why, for example, use of X-rays in hospitals and at the dentist is carefully controlled so that the exposure of people to them is kept to a minimum (Figure 5.41).

Some mutations are so harmful that they are 'lethal'. The individual carrying the mutation is unlikely to survive and this may go back to the zygote or embryo. This means that the mutation is lost from the population, though the same mutation may occur again in another individual at another time. An example is a mutation that results in lack of chlorophyll in plants. This occurs as a recessive allele but if the plant is a double recessive for this condition it cannot survive as it is unable to carry out photosynthesis. In human populations, some mutations are so severe that the fetus is unlikely to survive and dies before birth. In other cases, the baby may be born with certain genetic disorders. With increased understanding of genetic disorders, it is sometimes possible to provide treatment that improves the lives of people with the genetic condition.

If a mutation affects the genes that control cell division, then the cells might begin to divide continuously. This can result in the formation of a tumour. This is the basis of cancer and many mutagens, such as cigarette smoke, increase the risk of cancer.

On rare occasions, a mutation might lead to a protein that does its job better, or in an equivalent manner, compared to the usual version. This is probably how different versions of genes initially arise. If a useful or neutral (harmless) mutation occurs in the gametes of an individual, this might be passed on to their offspring during sexual reproduction. Sexual reproduction ensures that all alleles are shared out among the families of interbreeding individuals in a population, producing unique new combinations of genes in each new generation of offspring. These combinations then spread further through the population. It is this inherited variation that is so important in the changes in populations that can lead to evolution. In individuals produced by asexual reproduction (clones), the only possible genotypic differences between individuals result from new mutations (which are usually harmful).

■ How natural selection leads to evolution

The term **evolution** describes the observation that successive generations of living organisms seem to change over the course of time. Fossil records show us that there were plants and animals in the distant past that are no longer around today. From frozen areas of North America and Siberia we have remains of woolly mammoths that allow us to reconstruct the appearance of an animal that no longer exists today.

Similarly, we have fossils of dinosaurs and sequences of fossils that show us how the modern horse developed in a series of stages from an earlier, much smaller, almost dog-like animal. For the horse, the series of changes, for example in limb structure, height and teeth, can be linked into a sequence of changes over a timescale worked out from the rocks in which the fossils are found. Early horse-like animals date from over 50 million years ago.

Our reconstruction of early human ancestors comes largely from fossil records, going back to perhaps 2 million years ago. Other evidence can be used to show much more recent changes in populations of plants and animals, sometimes over the period of a few decades or less in the contemporary world.

In 1859, Charles Darwin published his book 'On the Origin of Species'. At the time of publication, the book caused some controversy, but it has stood the test of time as the best scientific theory to explain the mechanism by which modern day forms of plants and animals have evolved from common ancestors and to account for the enormous diversity of species in existence. The name Darwin remains closely associated with the theory of evolution even though other people were thinking along similar lines at the time when his book was published. However, Darwin was unable to explain how characteristics were passed from one generation to the next. Our modern understanding of inheritance started with Mendel but wasn't 'rediscovered' until 1900, so Darwin was unaware of this.

The following statements, some close to the wording used by Darwin in his book, set out the steps that lead to the theory of evolution by natural selection.

Observation 1 – Two parent organisms produce more offspring in their lifetime than are required to replace the previous generation.
Observation 2 – Numbers in plant and animal populations remain remarkably constant – they do not continually increase.
Conclusion 1 – Observations 1 and 2 indicate that there is a 'struggle for survival' or that some selection must occur – either for mates or for some environmental factor.

Observation 3 – There is natural variation in a population of sexually reproducing organisms. This variation is genetically determined.
Observation 4 – Some individuals in the population inherit an advantageous set of genes, giving them a phenotype that provides an advantage in a particular environment. These individuals are more likely to survive as they are better adapted to their environment and so pass on their genes to the next generation.
Conclusion 2 – Observations 3 and 4 show how the individuals that are better adapted are more likely to survive and breed and so pass on their genes to the next generation. This is often described as 'survival of the fittest' – and here 'fittest' indicates the individuals that survive long enough to produce the next generation.

In this way, we can see how, over time, the 'pool' of genes in a population can change. There is some selection of those individuals with a combination of genes that are better adapted to the conditions in which the population is living. The outcome, over several generations, may be seen as changes in the phenotypes within the population. These changes are described as evolution. If environmental factors had been different, the outcome of the selection may have gone in a different direction, producing a population with different phenotypes.

STUDY TIP

Make sure you are clear about the terms **evolution** and **natural selection**. Remember that natural selection is the **mechanism** by which evolution is thought to occur, as proposed by Charles Darwin. Evolution is the **process** of change in genotypes and phenotypes in populations, hence in species, that happens over successive generations.

Figure 5.42 Three different-coloured forms of the peppered moth.

Figure 5.43 Camouflage: visibility of different colour forms on the bark of birch trees.

■ Natural selection and evolution in action

In examples given above, we refer to fossil records and this may suggest that evolution is something that happened only in the past. Here we give two examples that relate to more recent situations and which illustrate the way that environmental factors present selection pressures that in turn result in changes (evolution) in the population.

The peppered moth

A well-studied example of the response of a species to changing selection pressures is seen in the peppered moth (*Biston betularia*). This moth is widespread in the UK and is found in two forms – a dark form and a light-coloured form (Figure 5.42). During the day, the moths usually rest on the bark of trees. However, in some parts of the UK, as a result of smoke pollution, the bark of the trees became blackened. This meant that the light-coloured moth variety could be seen more easily by the insectivorous birds that feed on the moths, whereas the dark form was less visible (Figure 5.43). As a result, in the polluted regions, the light variety was more often eaten and became rarer than the dark form. In the unpolluted regions, the light variety remained common (Figure 5.44).

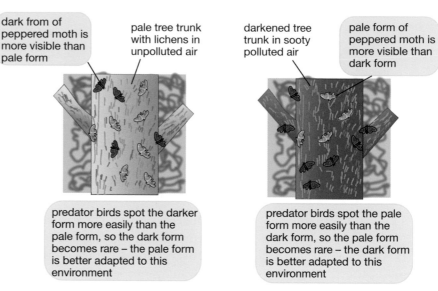

Figure 5.44 Illustration of natural selection in peppered moths resulting in changes to the moth populations in polluted and unpolluted regions.

When smoke pollution stopped and the trunks of trees were no longer darkened, the situation was reversed.

■ Practical activity – illustrating how selection works

This activity provides a model that illustrates selection in action. It can represent the situation for peppered moths on different backgrounds (such as tree trunks) and their visibility (or not) by predators. In this activity, green and red sticks (for example; matches or toothpicks) are used (to be the 'prey'), scattered randomly on an area of 'grassland'. Students take it in turns to be the 'predator' and the teacher acts as timekeeper for the 10-second intervals (a whistle is useful).

1　Mark out an area of 'grassland', 10 m × 10 m, using string or tape. The teacher then randomly scatters 100 red and 100 green sticks across this area.
2　Students work in pairs. Student A is the predator and student B is the scorer. The roles can later be reversed.
3　Student A is timed for 10 seconds and asked to search the grass area for sticks. At the end of the 'search' for 10 seconds, student A gives the sticks to student B.
4　Student B counts the sticks and records the number of each colour of the sticks that were 'captured'. These sticks are retained and not returned to the grassland.
5　Repeat the processes in steps 3 and 4 for several more 10-second periods until student A cannot find any more sticks.
6　Plot a graph to show the total number of each colour of stick captured during each 10-second search. Combine the results for the whole class on the same graph.

Capture rates for each colour of stick during each 10 second time period are likely to be affected by various factors. These include:
■ the ease with which the stick can be seen against the grass
■ the density of the population of sticks remaining on the grass.
Suggest ways in which this model predator-prey system could be improved to make it more like the real world.

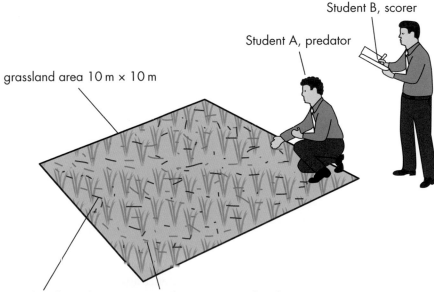

Student B, scorer

Student A, predator

grassland area 10 m × 10 m

100 red sticks and 100 green sticks are scattered in the area.
Figure 5.45 Illustrating how selection works with a predator–prey model.

Resistance of bacteria to antibiotics

A recent example that highlights problems caused by the speed at which natural selection can occur is found with antibiotic resistance. Antibiotics are chemicals (extracted from some fungi) that are able to kill the bacteria that cause an infection. Antibiotics kill bacteria *inside* your body and do not harm your own body cells. They were first developed during the 1940s and transformed the treatment of bacterial infections. Before the 1940s whole hospital wards were devoted to the care of infected people suffering from septicaemia (a bacterial infection of the blood), many of whom would eventually die from it. Use of antibiotics has helped to overcome many bacterial infections.

Bacteria breed very quickly, sometimes in less than 10 minutes from one generation to the next. With such fast reproduction rates, many generations of bacteria can grow in a day. A bacterial population shows variation and might contain just a few mutants, which happen to be immune (resistant) to an antibiotic. Such mutants have been recovered from Siberian permafrost from 35 000 years ago, long before the era of modern medicine. This character would have no advantage if the bacteria are growing in a situation with no antibiotics – in an antibiotic-free habitat.

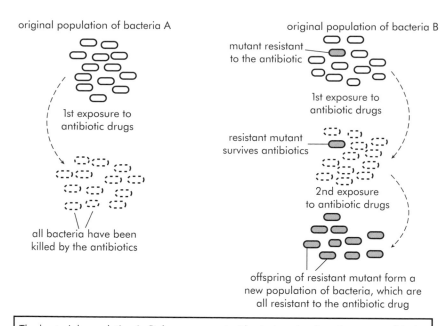

The bacterial population in B shows one mutant bacterium that is resistant to antibiotics. This one survives and produces offspring that are identical to the parent bacterium, so they are also resistant to the antibiotics and survive. This illustrates how the resistant form has been selected and can produce a colony that is also resistant.

Figure 5.46 Illustrating how a population can change and become resistant to an antibiotic. Population A has no resistant bacteria whereas population B shows one bacterium resistant to antibiotics.

But modern medicine has now introduced antibiotics to the habitats of humans. Places like hospitals have huge quantities of such drugs in the patients, on the wards and even down the drains. Many of the bacteria have no defence against the drug, so they are quickly killed. However, if there are just a few resistant bacteria, they are likely to survive. These resistant bacteria now have less competition. So their numbers are likely to increase and not be killed by the antibiotic and could grow to form a new population of bacteria, all of which are resistant to the antibiotic.

The antibiotic has provided a selection pressure, and the species of bacteria has evolved in response to the pressure (Figure 5.46). A similar situation arises with the widespread use of antibiotics in agriculture for livestock animals – both to prevent infection and to treat diseased animals.

The problem of resistance is why most doctors give antibiotics to patients only if they are sure the patient is suffering from a bacterial infection. Antibiotics cannot cure diseases caused by viruses, so there is no point, for example, using them for someone suffering from influenza. The truth is, the more antibiotics there are in the environment, the greater the chance that resistance will evolve in bacteria. It is important that sensible regulation is implemented globally, otherwise the effectiveness of antibiotics as a cure against bacterial infection will be reduced.

STUDY QUESTION

1 The pictures show a woolly mammoth and an elephant. The mammoth is extinct but elephants exist today. The size and appearance of the woolly mammoth has been reconstructed from preserved frozen remains of their bodies and skeletons, and from fossils.

Use information in these pictures to help you explain to a non-scientist how the elephant may have evolved from an ancestor similar to a mammoth. Answer the questions below to help guide you in your answer.

a) i) List **three** features you can see in the pictures that are similar in the mammoth and the elephant.
   ii) List **three** differences you can see between the mammoth and the elephant.
b) i) Summarise the steps that Darwin might have used to explain how populations (or species) change over time.
   ii) Suggest factors that might have been important in the selection of forms that were less like a woolly mammoth and more like an elephant.
c) Suggest reasons that might have led to the extinction of mammoths in the last few thousand years.

Summary

I am confident that:

✓ I can describe the differences between sexual and asexual reproduction, and understand that sexual reproduction gives more opportunities for variation but that asexual reproduction may be an advantage in some situations.

✓ I know which cells are the gametes in flowering plants and in humans.

✓ I know that fertilisation involves fusion of a male and a female gamete to produce a zygote.

✓ I know that a zygote divides by mitotic cell division and develops into an embryo.

✓ I can describe (and draw a diagram to show) the structure of a typical insect-pollinated flower and of a typical wind-pollinated flower.

✓ I understand how each of these flowers is adapted for pollination and how pollination occurs.

✓ I can describe the events from the time the pollen lands on the stigma until fertilisation occurs and understand that this leads to formation of a seed and fruit.

✓ I know how to investigate the conditions required for the germination of seeds and how the germinating seed obtains its food requirements from the reserves stored in the seed until it is independent and can carry out photosynthesis.

✓ I can describe examples of natural and artificial asexual reproduction in plants.

✓ I can describe (and draw a diagram to show) the structure of the human reproductive systems of a male and of a female and understand the functions of the different structures.

✓ I understand how different structures in the human male and female reproductive systems are adapted for their functions.

✓ I can describe key events in the menstrual cycle, and understand the roles of the hormones oestrogen and progesterone and of FSH and LH in this cycle.

✓ I can describe the how the developing embryo obtains oxygen and food through the placenta, and gets rid of waste carbon dioxide and urea, and understand the importance of the placenta for exchange of materials between the embryo and the mother.

✓ I know that the amniotic fluid is important for protection of the embryo.

✓ I can describe secondary sexual characteristics in males and females, and understand that the hormones oestrogen and testosterone are involved in their development.

✓ I know that the chromosomes are located in the nucleus and that the genes are located on the chromosomes.

✓ I understand that a gene is a length of DNA and that a gene contains information that determines a character in the individual.

✓ I can describe the outline structure of a DNA molecule as a coiled double helix and know that it has four bases – adenine (A) pairs with thymine (T) and cytosine (C) pairs with guanine (G).

✓ I know that an RNA molecule is a single strand, containing the bases adenine (A), uracil (U), cytosine (C) and guanine (G).

✓ I can describe how mRNA and tRNA are involved in transfer of information from DNA to assemble amino acids to form a peptide that becomes part of a protein.

✓ I understand the terms transcription and translation as stages in protein synthesis.

✓ I understand that three bases in a DNA molecule code for one amino acid and that a mistake in the DNA sequence may lead to synthesis of a different protein.

✓ I know that an allele is an alternative form of a gene, located at the same position on a chromosome, and that the different alleles for a gene control the same character but in different ways.

✓ I can give a definition of each of the following terms and know how to use them when talking about or describing inheritance of characters: dominant, recessive, codominance, phenotype, genotype, homozygous, heterozygous, diploid, haploid.

✓ I know that human body cells have 46 chromosomes (23 pairs of chromosomes).

✓ I can do genetic diagrams to show how a single character is inherited and predict phenotype ratios from these diagrams.

✓ I can interpret inheritance of characters through family pedigrees that give information about the phenotypes of different members of the family.

✓ I know that the sex chromosomes of a female (human) are XX and males are XY, and can show how the sex chromosomes are inherited.

✓ I can describe mitosis as the division of a cell to produce daughter cells that are identical (to each other and to the parent cell) in chromosome number and genotype.

✓ I can describe meiosis as the division of a cell to produce gametes that have half the number of chromosomes and different combinations of alleles.

✓ I understand how, as a result of fertilisation, it is chance as to which alleles are passed into the offspring and that this leads to variation.

✓ I understand how the phenotype of an individual is the result of its genotype and interaction with the environment in which it lives.

✓ I am aware that variation within a species is a consequence of variation between the genotypes and phenotypes of individuals, and understand how this variation can arise.

✓ I know that mutations occur, which result in changes in the genetic material and that these changes can be inherited but that mutations are rare and often harmful, though some can be beneficial.

✓ I can name some mutagens that increase the chance of a mutation occurring.

✓ I can describe the steps that could lead to changes in species over time, how some individuals in a population may be selected because they are more successful in their environment, and so are more likely to breed and pass on their characteristics.

✓ I can use this understanding (above) to explain the process of evolution by natural selection.

✓ I can describe how populations of bacteria become resistant to antibiotics and use this as an example to illustrate evolution by natural selection.

MATHS SKILLS

Bar charts and histograms

Bar charts consist of a series of bars of equal thickness, drawn vertically from the *x*-axis of the chart. They are used to represent the frequency of items in a series of discrete categories.

If each category has no logical connection to the others (for example; different eye colours), the individual bars are often arranged in ascending or descending order of magnitude.

If a series of categories forms a logical sequence (for example; from small values to large values), then the bars can be placed against each other, to form a histogram.

Decide whether a bar chart or a histogram is the appropriate graphical format for displaying the results of the seed germination practical.

Bar chart with bars group in each category

MATHS SKILLS

Numbers in standard form and powers

Very large and very small numbers are best written in standard form, where the first part is a number between 1 and 10, and the second part is a whole-number power of ten (negative or positive).

For example, 0.000 05 becomes 5.0×10^{-5}; and 750.34 becomes 7.5034×10^2

An adult human male can produce 20 million sperm per cm^3 of semen.

    a)   Write this figure out in standard form.

    b)   Of these 75% will be mobile. Calculate how many sperm this is. Write your answer in standard form.

    c)   A single ejaculation contains 369 million sperm. Calculate the volume of semen in cm^3 this involves.

c) $18.45 \ cm^3$

b) 1.5×10^7 sperm

a) 2×10^7 sperm per cm^3

MATHS SKILLS

Numbers in standard form and powers

Very large and very small numbers are best written in standard form, where the first part is a number between 1 and 10, and the second part is a whole-number power of ten (negative or positive).

For example 0.00005 becomes 5.0×10^{-5}; and 750.34 becomes 7.5034×10^2

    (a)   The DNA inside the nucleus of a single human cell has a total of 6×10^9 base pairs. The distance between each pair is 0.34×10^{-9} metres. Calculate the total length of DNA within the single cell nucleus.

    (b)   There are approximately 1×10^{12} cells in the human body. Calculate the total length of DNA found in the whole human body, using kilometres as units.

(b) 2.04×10^9 kilometres

(a)   2.04 metres

MATHS SKILLS

Probability

Probability is a measure of the likelihood that something will happen, ranging on a scale from impossible (= 0) to certain (= 1). On this scale, an even chance of something happening = 0.5.

This can also be expressed as a fraction or as a percentage (so 0.5 = 50% = ½). Remember it is still possible for unlikely things to happen!

a)   If a single base is selected from a DNA molecule entirely at random, what is the probability of it being adenine (A)?

b)   The chance of a mutation to a DNA base pair during cell division is estimated as 1×10^{-9} per base pair. With 6×10^9 bases in a single cell, how many mutations can be expected per cell division?

b) 6 new mutations per cell division.

a) ¼ or 0.25 or 25%

MATHS SKILLS

Bar charts and histograms

Bar charts consist of a series of bars of equal thickness, drawn vertically from the x-axis of the chart. They are used to represent the frequency of items in a series of discrete categories
If each category has no logical connection to the others (e.g different eye colours), the individual bars are often arranged in ascending or descending order of magnitude.
If a series of categories form a logical sequence (e.g. from small values to large values), then the bars can be placed against each other, to form a histogram.

The table gives blood group frequencies in UK and China

Group	% frequency	
	UK	China
AB	3.0	5.0
A	35.0	27.8
B	8.0	18.9
O	37.0	47.7

Present this information in a suitable graphical manner.

Bar-chart with bars for each group in each country

MATHS SKILLS

Probability

Probability is a measure of the likelihood that something will happen, ranging on a scale from impossible (= 0) to certain (= 1). On this scale, an even chance of something happening = 0.5.

This can also be expressed as a fraction or as a percentage (so 0.5 = 50% = 1/2). Remember it is still possible for unlikely things to happen!

A couple already have two girls, and the mother is planning to have another baby in the belief that it must be a boy this time. Describe what you would say to the mother about her chances of having a boy next time.

The chance of a boy is still the same as before = 50%

MATHS SKILLS

Arithmetic mean (average); mode and median

The arithmetic mean of a set of data, is used to describe the overall result, and allows different sets of data to be compared.

To find the arithmetic mean, add all the values in a set together, and divide by the number of individual values.

Mode and median terms are alternative types of average, taken from sets of data.
The mode is the most frequently occurring value in a set of data.
The median is the middle value in a set of data ranging from lowest to highest values (so half the data the data values are below the median value, and half are above it).

The following table shows the middle finger lengths of eight school pupils of different ages.

Find **(a)** the mean, **(b)** the mode and **(c)** the median of the data set, to one decimal place.

Middle finger lengths of eight individual pupils / mm							
65	58	70	55	67	72	69	55

(a) Arithmetic mean = 63.9
(b) Mode = 55
(c) Median = 65

Example of student response with expert's comments

■ Using and interpreting data

1 White tigers are rare in the wild, but carefully recorded breeding programmes in zoos have led to an increase in their number.

In one zoo, a white tiger cub was born to two tigers with orange fur. In tigers, the allele for orange fur (**F**) is dominant to the allele for white fur (**f**). Tigers with white fur are homozygous for the recessive allele.

a) Explain what is meant by the term **allele**. (2)

b) For the tigers described above, use a genetic diagram to show:
   - the genotype of each parent
   - the gametes they produced
   - the possible genotypes of all the offspring
   - the possible phenotypes of all the offspring. (4)

c) State the probability that the next tiger cub born to these parents would also be white. (1)

d) Suggest reasons why white tigers are rarely found in the wild. (2)

(Total = 9 marks)

Student response Total 7/9	Expert comments and tips for success
a) Alleles are <u>alternative forms of a gene</u> ✔ and are found at the <u>same position on a chromosome</u>. ✔ For example, T and t are alternative forms of the gene for height in pea plants.	Full marks. If you learn the definitions it helps you to give the correct answer. The example is also worth a mark but the maximum for the question has been reached.
b) *parents O* *female gametes* F f ✔*gametes* F: FF, Ff *male gametes* f: Ff, ff white ✔*offspring genotypes* phenotypes: 3 orange: 1 white <u>phenotypes (not linked to genotypes)</u> O	The student used a Punnett square to work out the possible offspring genotypes. This is recommended for more complicated crosses. • A common error is to forget to put in the parents (as here) or to forget to label the gametes or the possible genotypes of the offspring. • The student should have listed the possible offspring genotypes underneath the diagram. The phenotypes of the possible offspring can then be written in below the genotypes. • The student has not indicated the phenotypes (fur colours) for each of the three genotypes. She has only given the phenotype (white) for **ff**.
c) 1 in 4 ✔	Fusion of gametes at fertilisation occurs randomly, and only one of the four possible outcomes gives a cub with genotype **ff**. The probability that the next cub will be white is therefore 1 in 4. In part b), the student recorded the ratio of possible phenotypes as 1 white : 3 orange. This ratio is not a probability, but you can use the ratio to work out that the probability is 1 in 4.
d) Recessive alleles <u>only show in the homozygous recessive</u> ✔ white tiger. ✔ This is only produced from a mating between two tigers who are heterozygous. There are so few tigers that might meet and mate that the chance of this happening is very small.	The marks are given for recognising that the allele for white fur is recessive, so only shows in the phenotype if there is no dominant allele present ('homozygous recessive white tiger').

■ Extended writing

1 Describe how transfer of pollen grains onto the stigma of a flower can lead to fertilisation and fruit formation. *(6)*

Student response
When a pollen grain germinates on the stigma, the pollen tube starts to grow. It produces enzymes to digest its way through the style. Inside the ovary, the male nucleus fuses with the female nucleus in the ovule. The ovule can grow into a seed and the ovary, which contains the seeds, grows into a fruit.

Mark scheme
The mark scheme shows how marks are awarded.
(1) (pollen) tube grows / pollen grain germinates
(2) (grows down) style
(3) digestion / enzymes (involved)
(4) ovary (appropriate ref linked to pollen tube growth / ovules)
(5) (pollen tube / male gamete) enters ovule / eq
(6) through micropyle
(7) (male) nucleus / (pollen grain) nucleus / male gamete
(8) fertilisation / fuse / join + female gamete / nucleus / ovum / egg or ref to zygote formation
(9) ovule becomes seed
(10) ovule wall becomes seed coat / testa
(11) ovary becomes fruit / correct reference to other parts of the flower forming part of the fruit
Total: 6

Student response Total 6/6	Expert comments and tips for success
When a <u>pollen grain germinates</u> ✔ on the stigma, the pollen tube starts to grow.	Mark (1) awarded.
<u>It produces enzymes</u> ✔ to digest its way …	Mark (3) awarded.
… <u>through the style</u>. ✔	Mark (2) is awarded for the route of the pollen tube, not just for the word 'style'.
Inside the ovary … ✔	Mark (4) is not awarded for the word 'ovary' on its own, but for linking the ovary with the ovule and fertilisation. Aim to put the role of the various parts into a logical order, as very often the marks are awarded only if the sequence of events is correct.
… the male nucleus … ✔	Mark (7) awarded for describing the part played by the male nucleus.
… fuses with the female nucleus … ✔	Mark (8) awarded.
… in the ovule. (✔)	Mark (5) could be awarded (but maximum for the question, 6 marks, already reached).
The ovule can grow into a seed … (✔)	Mark (9) could be awarded (but maximum for the question, 6 marks, already reached).
… and the ovary, which contains the seeds, grows into a fruit. (✔)	Mark (11) could be awarded (but maximum for the question, 6 marks, already reached).

See General Advice for Extended Writing questions on page 79.

Exam-style questions

1 The table lists some statements about insect-pollinated and wind-pollinated flowers. Copy and complete the table by ticking in the boxes if the statement is correct for that type of flower. One line has been done for you.

Statement	Insect-pollinated flower	Wind-pollinated flower
has brightly coloured flowers	✓	
large feathery stigmas hang outside the flower		
produces pollen in the anthers		
stigma lies inside the flower and has sticky surface that helps capture pollen		
may have lines on petals that act as guides towards the nectar		
stamens hang outside the flower when ripe		
male cell in pollen joins with female egg cell in ovule		
produces large quantities of pollen with large pollen grains		

[Total = 7]

2 The headline below came from a newspaper.

> **New pesticide link to sudden decline in bee population**

a) i) Explain why growers use pesticides on their crops. [2]

ii) Suggest why there could be a link between the pesticide and numbers of bees. [1]

b) i) Give reasons why a 'sudden decline in bee population' might cause problems for fruit growers. [2]

ii) Describe the events that take place in a flower after it has been visited by a bee, up to the formation of a zygote. [3]

c) Some growers use biological control with their crops. Give **two** advantages and **one** disadvantage of using biological control rather than pesticides. [3]

[Total = 11]

3 In humans, a fertilised egg implants in the wall of the uterus in a woman. A placenta develops and this allows exchange of materials between the developing embryo (fetus) and the mother.

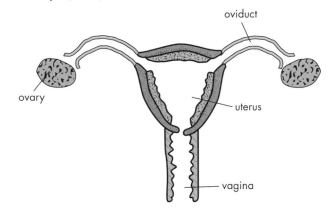

a) i) In the menstrual cycle, day 1 is when the uterus lining starts to be shed. On which day(s) of the menstrual cycle does ovulation occur? [1]

ii) Copy the diagram. On the diagram, draw a labelled arrow to show where fertilisation is most likely to occur. [1]

b) Which of the following hormones is responsible for stimulating ovulation?

A FSH

B LH

C oestrogen

D progesterone

c) i) Explain how the structure of the placenta is adapted to make it suitable for exchange of materials. [2]

ii) Give **one** substance that passes from the mother to the fetus and **two** substances that pass from the fetus to the mother. Copy the table below and write your answers in it. One has been done for you. [3]

From mother to fetus	From fetus to mother
oxygen	

d) Give **two** functions of the hormone oestrogen. [2]

[Total = 9]

4 The diagram shows part of a DNA (deoxyribonucleic acid) molecule.

a) i) Copy the diagram and use a labelled arrow to show the position of the paired bases on the DNA molecule. [1]

ii) One side of the DNA double helix has a sequence of bases as shown below. Under each letter, write the bases that would be found opposite, on the other strand of this DNA molecule. [2]

C A G T C A

b) Protein synthesis occurs in the cell, in several steps. Information contained in a gene determines which protein is synthesised. The steps involve nucleic acids (deoxyribonucleic acid and ribonucleic acid) and amino acids.

Which of the following correctly describes mRNA?

A a double-stranded molecule; bases include cytosine (C) and uracil (U); moves from the nucleus to the cytoplasm

B a single-stranded molecule; bases include cytosine (C) and thymine (T); occurs only in the cytoplasm

C a single-stranded molecule; bases include cytosine (C) and uracil (U); moves from the nucleus to the cytoplasm

D a double-stranded molecule; bases include cytosine (C) and thymine (T); occurs only in the nucleus [1]

[Total = 4]

5 In protein synthesis, information contained in the bases in DNA is 'transcribed' into RNA bases then 'translated' into amino acids in a chain that becomes part of a protein.

a) The table shows just a few of the 64 possible mRNA codons and their corresponding amino acids. Note that one of the amino acids has two possible codons.

mRNA codon	amino acid coded for
UAU	tyrosine
UCU	serine
UCC	serine
CCC	proline
ACG	threonine
CAA	glutamine

i) Write out the sequence of **mRNA** codons that codes for the following sequence of amino acids. Use capital letters in your answer.

Glutamine - Glutamine - Proline - Tyrosine [2]

ii) Write out the sequence of **DNA** bases that corresponds to the mRNA sequence you have written for (i). Use capital letters in your answer. [1]

b) i) A protein unit of haemoglobin (the human red blood cell pigment) is composed of a chain of 141 amino acids. Calculate the minimum number of DNA bases required to code for its structure in the haemoglobin gene. Show your working. [2]

ii) Suggest why a change to just one base within the haemoglobin gene (a mutation), might have severe consequences for the person affected by this mutation. [2]

[Total = 7]

6 White-skinned 'albino' giraffes are very occasionally seen in a herd of wild giraffes. These animals have inherited an allele that means they are unable to make melanin (a dark pigment) in their body cells.

Two giraffes with normal skin colour can produce an albino giraffe but two albino giraffes can only produce albino offspring.

a) Use the symbols **A** or **a** to give the genotype of the albino and explain why the allele for albinism must be recessive. [3]

b) Use a genetic diagram to show how two giraffes with normal skin colour can produce an albino giraffe. Include genotypes of parents, the gametes formed, the offspring genotypes and offspring phenotypes. [4]

c) Albino giraffes usually attract predators and are often killed before they grow up and breed themselves. Suggest why albino giraffes do occasionally occur in giraffe populations. [1]

[Total = 8]

7 Rats have been pests living around human habitation for thousands of years. Rats eat stored food and transmit disease.

Warfarin is a chemical that prevents clotting of the blood. Warfarin has been used as a pesticide to control rats. At first warfarin was effective and, if rats ate food poisoned with warfarin, they died. A number of years later it was noticed that many rat populations had become resistant to the pesticide and the rats did not die if they ate food poisoned with warfarin.

a) Suggest why rats die if they eat food poisoned with warfarin. [2]

b) Explain how populations of rats may have become resistant to the effect of warfarin. [5]

c) Warfarin has now been replaced by another chemical pesticide. Sketch a graph to predict what might happen to population numbers of rats in a limited area when the new pesticide is used regularly near human habitation. [3]

[Total = 10]

EXTEND AND CHALLENGE

1 Plants can help solve crimes

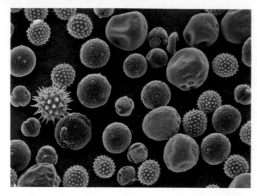

Pollen grains have a very hard and durable outer layer. This can remain unchanged for a very long time, sometimes for thousands of years. This outer layer can have many different patterns, characteristic for a particular species and pollen grains have many different shapes. It is possible to identify the plant species from the characteristic features of the pollen.

Scientists can use evidence from pollen grains in forensic analysis of a crime scene. They have been able to link pollen from particular plant species at a crime scene with pollen found on the clothing of a suspect.

Scientists have also used information from pollen grains preserved in peat deposits to trace the history of plants growing in an area over a long period of time. In some cases this has provided information about changes in the climate.

Find out about one example of the use of pollen in each of these situations:

a) forensic analysis of a crime scene

b) tracing history of vegetation in an area and linking this with changes in local climate.

Write up your findings as a story or as a presentation that could be understood by someone who is not a science specialist. Make sure you explain how the evidence is used and that the science is correct.

2 Down's syndrome and screening for genetic disorders

All the body cells of a healthy human should contain 46 chromosomes. This is because a set of 23 chromosomes is inherited from the gametes (egg and sperm cell) of each parent. Sometimes an error occurs during meiosis at the time that gametes are formed, with the result that a gamete might contain an extra chromosome or have one fewer than the normal number. This is called a chromosome mutation. In these situations, often the zygote produced after fertilisation dies.

Down's syndrome is a condition that occurs when an egg carrying an extra chromosome (24 chromosomes) is fertilised by a normal sperm cell (with 23 chromosomes). The result is that instead of pairs of homologous chromosomes in the zygote, one 'pair' has three chromosomes. Even though the embryo is affected, a baby is born. A Down's syndrome baby may have a range of problems, including learning and physical difficulties – ranging from cases that are severe to babies in which the symptoms are much less apparent.

It is now known that the older a woman is when she becomes pregnant, the greater the risk of her having a baby with Down's syndrome. Tests that should detect a fetus with Down's syndrome are often offered to older women. In some cases, women who are shown to have a fetus with Down's syndrome, may decide to terminate the pregnancy.

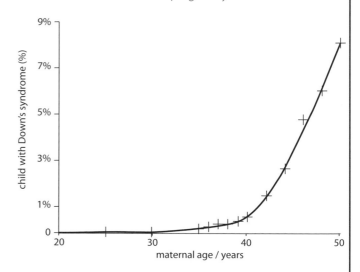

a) Suggest **two** environmental factors that might cause chromosomal mutations.

(continued)

b) Look at the graph and suggest an appropriate age after which a woman should be offered a test for Down's syndrome on her unborn child.

c) Some people find it difficult to make the decision about testing for Down's syndrome and terminating a pregnancy. For example, this might be because the severity of the disorder varies considerably and there is a slight risk in undertaking the tests. Make a list of reasons for and against the decision to test and to terminate a pregnancy. Write this into a paragraph that sets forward different viewpoints.

d) As nations become wealthier, there is a corresponding increase in the age at which women choose to have their first child. Suggest what social factors might be linked to this change.

e) An increasing number of prenatal tests are being developed and used to detect possible genetic disorders in unborn children. Find out about at least **two** of these tests under these headings:
- what is being tested for
- whether the diagnosis indicates risk or whether it detects the presence of a disorder
- how far knowledge of the disorder can help the parents prepare for the birth and care of the child
- how far knowledge of the disorder can help indicate medical treatment that may be available to improve or cure the symptoms.

f) Suggest what difficulties are faced by prospective parents in the future as more prenatal genetic tests are developed.

3 There is considerable publicity about over-use of antibiotics and resistance to antibiotics in bacterial populations.

a) Explain to a non-scientist how this resistance develops in bacterial populations and discuss some examples of the consequences of this resistance.

b) The graph shows the changing rates of prescription of antibiotics by family doctors during the year in the UK.

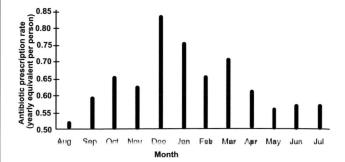

i) Describe the general pattern you see in the results over time.

ii) Suggest why the y-axis of this graph might give an exaggerated impression of the seasonal changes in prescription rate variation.

iii) Nowadays doctors are advised to prescribe fewer courses of antibiotics, even if some patients always expect to be prescribed a pill. If the doctor diagnoses that a patient is suffering from a virus rather than a bacterial infection, the doctor avoids prescribing an antibiotic. What explanation could the doctor give to a disappointed patient when this happens?

iv) Suggest why rates of antibiotic resistance are much higher in countries where antibiotics can be bought directly by patients, without a prescription.

6 Ecology and the environment

TO THINK ABOUT . . .

Ecology is the study of the relationships between living organisms and their environment. Other parts of this book focus on a single organism (plant, animal or microorganism) and how it works inside. Now the focus is on groups of organisms and their interactions with each other and with their surroundings. Try to make a list of ways that plants, animals and microorganisms depend on each other and how the different groups interact with their surroundings.

Living organisms and their ecosystems on planet Earth

At the start of this book, we consider the fact that planet Earth is the only one upon which we know that life exists. We also look at what sets living things apart from non-living ones.

Seen from space, there are patterns in the distribution of living organisms across the face of the planet. Broad bands of colour running across the continents indicate major ecosystems, such as rainforest, tundra or grasslands. Each ecosystem is an interacting community of plants and animals, all interdependent upon one another.

Some species of plants and animals are on the brink of extinction. Could their loss destroy a whole ecosystem? Scientists studying the changing climate of the planet are trying to determine how far the changes are the result of natural cycles and processes, or whether they are the direct result of human activities.

But what holds an ecosystem together? If the risk of ecosystem malfunction is as great as some scientists think, do we, as humans, need to make changes to our current way of life? Can we avoid a catastrophic change in the global climate and can we help conserve the planet as it is for future generations? What could you change about your lifestyle to avoid harming ecosystems and to allow the continuation of the fragile interrelationships that exist within ecosystems?

6.1 The organism in the environment

A focus on organisms in an ecosystem

Look carefully at the photographs. First you are likely to notice the coloured flowers – bluebells in a wood in Eastern England and primulas in high mountains in Western China. In each photo, the flower is a single species, surrounded by many others: plants, animals and microorganisms. The bluebells and primulas are each part of the much bigger grouping of many species that, together with their environment, make up the ecosystems we see in the photographs.

Now think about an area you know well, perhaps near your school or home, and make a list of any species you have seen there. Then think about how they might interact with each other. (At the end of the section, check back to your list and see whether your ideas about how they interact have changed.)

■ What makes up an ecosystem?

In Section 6.1 we tease apart the components of an ecosystem and see what contributes to this structure. We then put it all together into a whole. You also find out about methods you could use to look in more detail at the distribution of plants and animals within the ecosystem being studied.

Population

The term **population** describes a group of individuals of the same species, living in the same place at a certain time.

The bluebells (or primulas) in the photographs each represent a population. In another wood, perhaps 10 km away, there might be another population of bluebells (or primulas). In the background of each photograph, there is a population of certain tree species and on the ground, there are populations of beetles, ants and other small animals.

Community

The term **community** links together the populations of the different species living together in the same place.

Figure 6.1 Pond in Bradfield Woods (Eastern England)

The top-left photograph on page 239 shows a woodland community, with bluebells and other plants on the ground, shrubs and taller trees behind. There are likely to be beetles and other insects, on the ground or flying through the air, and many small invertebrate animals in the soil. In this woodland, there may be some mammals (such as dormice, squirrels or deer) with birds flying between the trees and shrubs. All these are part of the same community, even though you cannot see all the species in the picture.

Figure 6.1 shows a different community in the same woodland as the bluebells. The various species found in the water and on the edges of the pond are different from those in the bluebell picture.

Within each community, there are links or interactions between the species found there. The group of species existing together in a community have certain requirements that they all find in that particular place and the different species in a community are dependent on each other.

Habitat

The term **habitat** is used to describe the place where an organisms lives.

Habitat can be used as the place for just a single species, but often refers to a general area where many species are found. So the woodland in the first photograph on page 239 is the habitat for the various species already referred to (and many more). Similarly, the pond (Figure 6.1) is the habitat for many species (plants, animals and microorganisms) found in the water, but which would not be able to live in the open woodland.

Ecosystem

An **ecosystem** puts all the species, populations, habitats and communities together. The term **ecosystem** is used to describe the whole integrated mixture of biotic (living organisms) components, and abiotic (physical non-living) components that exist in a particular habitat. The biotic components (living organisms) depend upon each other, and on getting the essential abiotic components in order to live and thrive. Abiotic components include things such as water, mineral ions and gases in the air, plus things such as warmth and light and shelter.

Referring again to the same photographs, we can recognise a woodland ecosystem (page 239) and a freshwater aquatic ecosystem (Figure 6.1). Within an ecosystem, there are interactions between the living organisms that are found there. In particular, species are dependent on each other through food chains, which provide a means for the transfer of energy and of various substances required for growth and other activities.

STUDY TIP

Make links to Section 6.3 for information about the non-living components and how they are recycled and to Section 2.1 for detail about the role of magnesium ions.

STUDY TIP

Check details about how energy is transferred through food chains and food webs, and how species are dependent on each other (see Section 6.2).

Biodiversity

The term biodiversity is used to describe the range and variety of living organisms within an ecosystem. It can include all living organisms: plants, animals, fungi and bacteria, and is often used to indicate the richness or number of different species in an area. 'Biodiversity' also takes into account the distribution of each species within the area - whether or not they are spread evenly - as well as the population size of the different species (say of plants or animals) within the area.

Such an area can be defined at different levels: across the whole globe; a desert, a tropical rainforest or a mountain range; or the more limited areas shown in the photographs on pages 239 and 240.

Look again at the photographs on page 239 at the start of this section. At first glance, the two on the left have clear similarities: mainly flowers in the foreground (bluebells or primulas) against a background of trees. They come from different places (very far apart - England in Europe and China in Asia) but we may wish to compare the biodiversity in these two superficially similar communities. Their biodiversity may turn out to be quite different. The methods described below can be adapted to give a measure of biodiversity.

■ Distribution of organisms in their habitats

How quadrats can be used to estimate population size

Let's go back to the photograph of the primulas. Suppose you were asked how many primula plants there were in this field, and whether there were more in this field or in another field on the other side of the valley.

You could try to do this in one of two ways. Either you could count all the primula plants in the area, or you could estimate how much of the area was occupied (or covered) by the primula plants. Both these suggestions would be very tedious to do, take quite a long time and not be very accurate.

The best way to do this is to take a sample area and look at it in detail. To select your sample area you use a piece of equipment called a quadrat (Figure 6.2).

- A quadrat is usually square, but can be any shape.
- A quadrat can be made of any material, but often it is made of wire or string that is pegged into the ground at the corners.
- A quadrat can be different sizes, depending on the size of plant you are looking at. Often quadrats with sides of 0.5 m are used to sample small plants in an area.

A quadrat can also be used to find out about animals in the sample area, provided the animal does not move away quickly.

When you use a quadrat, you must not just put it down on the ground where all the best flowers are growing (or where they are not growing). You need to place the quadrats **randomly** so that you are not showing any bias when collecting your results. One way of getting random positions for a quadrat is to place two long measuring tapes on the ground at right angles, to give a grid. Each 'square' in this grid has a number. Numbers are then chosen randomly and used as the positions for the quadrats.

Figure 6.2 Using a quadrat

STUDY TIP

Remember that using a quadrat is a way of taking a **sample** in the area, rather than counting every plant (or animal).

This practical activity emphasises the use of a sampling technique to make a reasonable estimate of living plants (or animals) in different areas. It is important to choose a size of quadrat that is appropriate to the size of the organism being sampled.

To make your estimate, you look in detail inside the relatively small area of the quadrat. You must decide whether you are going to count the number of plants (for example primulas) or estimate the proportion of ground covered by these plants. Often this is done as 'percentage cover'.

You then repeat the measurements in a number of quadrats. From that you can multiply up the results from all the quadrats you have recorded to give a reasonably accurate estimate of the number of plants in this population in the whole area.

In some habitats, particularly with animal species, equipment other than a quadrat may be used to capture a 'sample' of the animals (see Table 6.1). These include sweep nets (through vegetation), pond nets (in water) and pitfall traps (for small animals on the ground – see Figure 6.3). Often you can devise something that is appropriate for a particular habitat but you need to remember you are taking a sample that is used to represent the distribution pattern in a larger area.

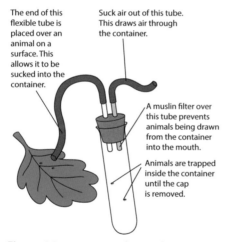

The end of this flexible tube is placed over an animal on a surface. This allows it to be sucked into the container.

Suck air out of this tube. This draws air through the container.

A muslin filter over this tube prevents animals being drawn from the container into the mouth.

Animals are trapped inside the container until the cap is removed.

Figure 6.3 A pooter can be used to collect small animals, for example from vegetation.

Pooter mouthpieces should be sterilised with Milton solution for 30 minutes before use. Each student should have new mouthpiece tubes.

Remember to wash your hands after these ecological activities.

A risk assessment should be carried out if using pond nets to collect small animals from rivers, ponds and lakes.

Table 6.1 Three sampling techniques for invertebrate animals

Sweep net	Pond net	Pitfall Trap
Collects small animals from soft vegetation	Collects small animals from rivers, ponds and lakes	Collects small animals crawling around soil surface
The flat part of the net entrance is dragged across the vegetation trapping animals inside the net.	The mesh net is submerged in the water and then used to scoop up the animals swimming in the water, allowing the water to drain away.	The container is buried up to its rim, with a flat stone used to provide a lid, with space around the edge to allow small animals in, but not large predators.
Cannot be used with stiff woody vegetation like tree branches, and some animals flee before capture!	Lets animals smaller than a certain size pass through mesh, and others swim away before capture!	Requires animals to be killed to avoid them crawling out of the trap, or eating each other! Dilute detergent is usually used.
Surprised animals are collected directly from the inside of the net using a pooter.	Animals tipped into a tray of water for further study	Animals tipped into a tray for further study.

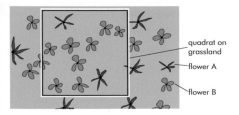

Figure 6.4 How to use a quadrat to estimate the density of two species of flowers on grassland.

■ Using quadrats to estimate the density of two species of flowers in grassland

Figure 6.4 shows two species of flowers (A and B) growing in grassland.

1 Use a quadrat. This can be a square of wire or a loop of string with a stake at each corner. Choose a size of quadrat that is suitable for the size of the plants you are observing.
2 Place the quadrat in the study area, using a random method. This is usually done by using two measuring tapes running at right angles to each other at the border of the area. Random numbers are then used to get a distance along each tape. The intersection of the two distances gives the precise location for the quadrat in the study area.
3 Count the number of flower A and the number of flower B in the quadrat.
4 Then a new quadrat position is randomly chosen. Count the numbers of A and B in this quadrat.
5 Repeat this in several quadrat positions.
6 To calculate the density of each flower, add up the total number of flowers counted in each of the quadrats. Divide this by the total area sampled to obtain the plant density.

■ Practical activity – investigating the biodiversity of two areas of grassland

One measure of biodiversity is to assess how many different animal and plant species are found in a study area. Quadrats, sweep-nets and pooters can be used to assess this in grassland. *You might like to think about what other abiotic differences there could be between the two areas, and how they could be measured.*

Figure 6.5 a)

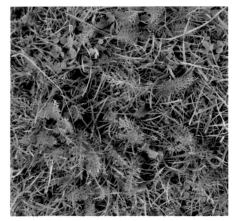

Figure 6.5 b)

Figure 6.5 shows appearance of two areas of grassland around a school playing field. Area (a) is located near the middle of a football pitch, and area (b) is found at the edge of the field, near the fence. Area (a) is mown frequently but area (b) is mown only three or four times a year.

MATHS TIP

Bar charts
Bar charts consist of a series of bars of equal thickness, drawn vertically from the x-axis of the chart. They are used to represent the **frequency** of items in a series of discrete **categories.**

If each category has no logical connection to the others (e.g. different eye colours), the individual bars are often arranged in ascending or descending order of magnitude and separated by small gaps.

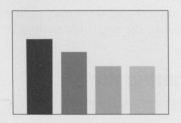

Now go to page 272 to apply this to the presentation of results of some investigations.

PRACTICAL

Pooter mouthpieces should be sterilised with Milton solution for 30 minutes before use. Each student should have new mouthpiece tubes.

Remember to wash your hands after these ecological activities.

Assessing the presence of different plant species using a quadrat

1 Use a random method to take several quadrat samples in both areas. This is usually done by using two tape measures running at right angles to each other at the border of a study area. Random numbers are then used to get a distance along each tape. The intersection of the two distances gives the precise location for a particular quadrat.
2 For each quadrat sample, note the presence or absence of all the different broad-leaf plants (weeds) within the quadrat. *Even if you cannot find the precise name of the plant, you can recognise different ones by their leaf shapes.*
3 Repeat the quadrat sampling in several randomly chosen positions in each study area.
4 Note the total number of different plant species in each area. The area with the larger figure has the greater plant biodiversity.

Assessing the presence of small invertebrate animals using a sweep net and a pooter

1 Collect small invertebrates from both study areas. For each area, drag a sweep net along the grass for a fixed distance (several metres).
2 Grip the entrance to the bag, to ensure the animals are trapped, and then reveal each of them slowly by gradually peeling the bag open. Collect each animal in a pooter, before they fly or crawl away.
3 Once the bag has been emptied, repeat the procedure in the other study area, and use an empty pooter to collect the second batch of animals.
4 Empty each pooter into a shallow tray containing a layer of dilute detergent. This traps and kills the captured animals.
5 Use a small brush to gather the different animals into groups of the same species (that look the same).
6 Note the total number of different species collected from each area. The area with the larger figure has the greater animal biodiversity.

Estimating the population densities of plants and animals in an area

Note the total number of plants of a particular species within each quadrat sample. Then divide this figure by the total area sampled to obtain the plant density in number per unit area.

Note the number of animals of a particular species captured in a pooter and placed on a tray.
Then divide this figure by the total area of grassland swept by the trap:

(total area swept = width of net × distance the net was dragged along the grass)

This gives the density in number per unit area.

■ Factors that affect distribution of organisms in an ecosystem

The organisms that are present in an ecosystem and their distribution within it are influenced by both **biotic** and **abiotic** factors.

The **biotic** factors include all the other living organisms in the ecosystem, how they interact with each other and how they affect each other. These interactions may, for example, be in terms of competition (e.g. for food resources or for water), predator-prey relationships, food chains and food webs or perhaps shading by trees. At any time, such interactions are not fixed, but part of a dynamic and changing system.

The **abiotic** factors are the physical and chemical (non-living) components that contribute to the growth of living organisms in their habitats within the ecosystem. Some abiotic factors are summarised in Table 6.2 with an indication of how they may vary and examples of effects they may have on living organisms in an ecosystem.

Table 6.2 Some examples of abiotic factors – how they may vary and a few examples that illustrate their effects.

Abiotic factor	How it may vary	Example(s) of effect(s)
temperature	hot (40 °C or higher) to below freezing (0 °C)	▶ hot dry desert, plants have adaptations that reduce water loss; animals seek shade or shelter in heat of day
light	▶ bright sunlight to deep shade ▶ different depths in water	▶ different range of plants found in sunny and shady areas of a woodland
pH	acid to neutral to alkaline	▶ lakes with low pH often lack fish populations ▶ plants tolerate different levels of acidity or alkalinity (for example; different plants found in peat bogs and chalky soil)
water	rainfall, available water, moisture in soil, humidity in air	▶ water needed for all ecosystems, but varies from swamps to desert conditions
mineral ions	full range of nutrients available in some, to deficiencies in others	▶ eutrophication (high nutrient) to low levels of nutrient ▶ pollution from very high levels sewage or waste in fish farm
salinity (saltiness)	salty (as in seawater) through estuarine conditions to freshwater river	▶ marine fish cannot live in freshwater (and vice versa)
gases (oxygen, carbon dioxide)	▶ normal oxygen level to low levels (anaerobic) ▶ normal carbon dioxide to enriched levels	▶ low levels of oxygen in polluted rivers and in muddy soil – organisms have adaptations that allow them to survive there
soil	sandy (large particles) to clay (small particles)	▶ sandy soil loose, low in nutrients and does not hold water ▶ clay soil firm but may become waterlogged and lacks air spaces

MATHS TIP

Scatter diagrams

Scatter diagrams provide a visual way of showing whether there is any **correlation** between two different sets (**variables**) of experimental data, plotted on the *x and y axes of a graph.*

A **positive** or **negative correlation**, or **no correlation**, between the variables can be determined, from the **pattern of the points** on the scatter diagram.

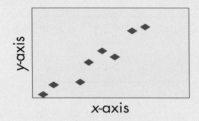

In the example above there is a **mostly positive correlation** between the *x* and *y* values, though there are two anomalous points which do not fit the pattern perfectly.

Now go to page 273 to apply this to the presentation of results of some investigations.

STUDY QUESTIONS

1 Definitions of some terms are given in the table below.
   a) Copy and complete the table by choosing words from the list in the box to match the definitions:

   autotroph; biodiversity; community; ecosystem; habitat; herbivore; population; trophic level

Definition	Term (word)
All the living organisms together with the nonliving components found in a certain area	
The different species (of plants, animals and microorganisms) living together in the same place and interacting with each other	
A group of individuals of the same species (plant or animal or microorganism), living in the same place at a certain time	
The place where an organisms lives	

   b) There are four definitions in the table but there are eight words in the list. Write definitions for the terms you have not used in the table. Extend the table and write your answers in it. You may have to refer to other parts of the book to complete your definitions.

2 Some of your friends were unable to attend the biology class today so missed doing a practical exercise using quadrats.

   Write a set of instructions for your friends so that they would know how to use quadrats to study the distribution of a small plant growing in the grass of a playing field and compare this with its distribution in a grassy area of a park.

   In your instructions, include diagrams so they would know what a quadrat looks like and how to set them out in the field. You should also include a table that would guide them as to what measurements they should take and how to set out their results.

3 Your friends then decide to compare the biodiversity of two areas around the edge of a pond or lake.

   ■ Make a list of the equipment they could use to take samples of both animals and plants in this area.

   ■ For each piece of equipment, say what you might collect or observe with it and how you would use it.

   ■ Make it clear how your comparison would help give a measure of the biodiversity in the area and what might be missed or lead to inaccuracies.

6.2 Feeding relationships

Capture of solar energy

- The rape plants in the fields (in Eastern England) capture energy from the Sun and convert it into the chemical energy contained in plant material – some of it in lipid molecules. When the rape crop is harvested, the seeds are crushed to provide oil, which is used for cooking or as a fuel in motor vehicles.
- The photovoltaic panel outside the ger (in a remote part of Western Mongolia) captures energy from the Sun and converts it to electricity.
- The simple roadside solar collector (in Western China) captures energy from the Sun and directs enough energy to the kettle to boil the water inside it.

But how does this solar energy captured in the rape crop become part of you or me? How is the energy from the Sun distributed through the many plants and animals that live among the crop plants or in the woodland ecosystem beyond? What is the mechanism for energy transfer and how much gets lost on the way?

■ Food and energy

The next time you have a meal, look at your bowl or plate of food and work out the biological origin of the different foods. If you have some vegetables, such as beans, tomatoes or lettuce, it is easy to see that these are parts of plants. You may have some foods high in carbohydrate, such as rice, pasta, potatoes or bread. Again, it is not difficult to trace these back to their plant origins (though the wheat or other grains have been processed to produce the flour that has been made into pasta or bread). If you have meat or fish, you may have to do a little more work to trace what the animal ate back to its plant origins. Herbs and spices may be part of your meal and they also originally come from plants: cinnamon (from the bark of a tree), cloves (dried flower buds from another tree), black pepper (the dried fruit of another plant).

Plants use energy from the Sun in the process of photosynthesis, and

Figure 6.6 Think about the origins of the food on your plate.

STUDY TIP

To make sure you understand the process of photosynthesis check in Section 2.1, and for respiration check in Section 2.3.

synthesise complex molecules from simple molecules. Plants effectively trap and then lock up energy from the Sun, 'storing' this energy as large molecules in their biomass. Later, when an animal eats the plant, some of the energy is passed on in the biomass from the plant to the animal. The energy can then be released (through respiration) so that the animal can move around or carry out different processes in the body. In the same way, you release energy from the food you eat and this energy enables you to carry out various living processes.

■ Trophic levels and food chains

We can now write down a simple food chain, based on a typical human meal.

maize ➡ chicken ➡ human

The arrows show the direction of flow of energy. At the beginning, we can add the Sun to emphasise the source of the energy. So this food chain might be written as:

SUN ➡ maize ➡ chicken ➡ human

We can look at two more food chains, one that might be relevant to someone growing a crop of beans and the other in a more natural ecosystem.

bean plant ➡ aphid ➡ blue tit ➡ sparrowhawk
(insect) (bird) (bird)

green plant ➡ mouse ➡ snake ➡ eagle
(mammal) (reptile) (bird)

STUDY TIP

The term autotrophic is also used to describe the way that green plants feed (see Section 2.1). They make their own food (auto = self; trophic = feeding).

STUDY TIP

The term heterotrophic is also used to describe the way that consumers (animals) feed (see Section 2.2). They take in food that contains large complex molecules (hetero = different; trophic = feeding).

The green plants make (produce) food and are known as **producers**. The animals in the food chain all eat (consume) the food, so are known as **consumers**. At the first level the animals are herbivores (because they eat plants) and are described as **primary consumers** (or **1st consumers**). At the second level they are **carnivores** (because they eat other animals) and are known as **secondary consumers** (**2nd consumers**). At the third level there are more carnivores and they are known as **tertiary consumers** (**3rd consumers**). Sometimes there is even a fourth level, but not often more than that – we discuss this on pages 252 and 253.

For each of these levels we use the term **trophic** level to describe the feeding relationship with other organisms in the food chain. These levels are summarised in Table 6.3, using the food chain above to give examples of the plants and animals involved.

Table 6.3 A simple food chain, used to illustrate trophic levels.

Trophic (feeding) level	Organism	Description
tertiary / 3rd consumer	eagle (bird)	(top) carnivore
secondary / 2nd consumer	snake (reptile)	carnivore
primary / 1st consumer	mouse (mammal)	herbivore
producer	green plant	producer

STUDY TIP

Link 'troph' in trophic levels with 'eutrophication' to remind you the word is to do with feeding or nutrient content.

Energy can also enter a food chain when animals eat dead plants and animals (Figure 6.7). Such animals are known as detritivores and these

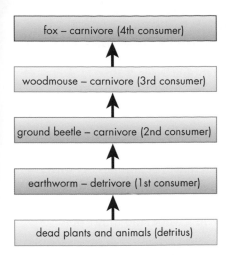

Figure 6.7 Food chain based on detritivores eating dead plant and animal material.

STUDY TIP

For information about natural cycles and recycling of materials, check in Section 6.3.

animals play an important part in utilising energy that was originally trapped by living plants. The term detritivore is derived from the word 'detritus', meaning waste material.

Table 6.4 Summary of the food chain shown in Figure 6.7.

Trophic (feeding) level	Organism	Description
4th consumer	fox (mammal)	(top) carnivore
3rd consumer	woodmouse (mammal)	carnivore
2nd consumer	ground beetle (insect)	carnivore
1st consumer	earthworm (worm)	detritivore
	dead plants and animals	

Organisms in another group are described as **decomposers**. These are mainly fungi and bacteria and have an important role in the decay or breakdown (decomposition) of dead plants and animals. Decomposers can digest the dead material and, through their own respiration, receive energy that had been locked away in the plants (then the animals) and transferred through the food chain. The decomposers break down the dead material to simple compounds and play an important role in releasing mineral ions and carbon dioxide. The mineral ions and carbon dioxide then become available for plants to use again, as part of natural recycling processes (see Section 6.3).

■ Food chains and food webs

Already the picture is becoming more complex and it is less easy to give trophic levels in relation to the feeding patterns of the organisms.

Look at the animals in Table 6.5, which also lists what they may eat. All the animals are likely to eat other organisms in addition to those listed.

Table 6.5 A list of animals linked to some examples of what they eat.

Animal	Group	What they eat
earthworm	worm (segmented)	dead plant material
earwig	insect	dead and living plant material
caterpillar	insect	living plants
rabbit	mammal	living plants
ground beetle	insect	caterpillars, centipedes, earthworms, earwigs
centipede	arthropod	caterpillars, earthworms, earwigs, ground beetles
toad	amphibian	centipedes, earthworms, earwigs, ground beetles
blue tit	bird	caterpillars, earwigs
robin	bird	caterpillars, centipedes, earthworms, earwigs
tawny owl	bird	earthworms, ground beetles, rabbits, woodmice
woodmouse	mammal	nuts and berries (plant material), caterpillars, centipedes, earthworms, earwigs, ground beetles
fox	mammal	earthworms, ground beetles, rabbits, woodmice – also like fruit (such as blackberries)

The information in the table tells us that:
- an animal is likely to feed on several things, rather than relying on a single food source
- animals do not always feed at the same trophic level
- a 'food chain' is a very simplistic representation of a real situation in an ecosystem.

Instead of writing down lots of simple food chains, we can summarise the information given in the table in the form of a **food web** (Figure 6.8). This takes into account the different feeding preferences of animals. As with the simple food chains, the arrows show the direction of energy flow and also indicate what each animal eats. A food web similar to this one is likely to represent some of the feeding relationships in a woodland in Europe.

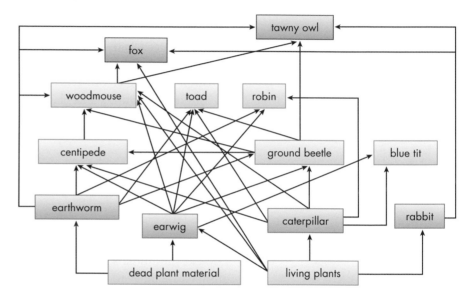

Figure 6.8 Woodland food web, based on the information shown in Table 6.5.

Look closely at the food web. Firstly, you can see a number of food chains within the web.

- Some food chains consist of two organisms – for example, fox eats blackberries.
- Some food chains consist of three, four or even five organisms – for example, rabbit eats plants, fox eats rabbit (= 3 organisms); caterpillars eat plants, ground beetles eat caterpillars, woodmouse eats ground beetles, tawny owl eats woodmice (= 5 organisms).

Secondly, you can see that the organisms do not fit neatly into the trophic levels given above. A woodmouse can be both a primary consumer and a secondary consumer. A toad can be both a secondary consumer and a tertiary consumer. You can certainly find other examples in this food web showing that animals feed at different trophic levels.

Thirdly, the food web draws attention to the way that dead plant material can be brought into the food web and how detritivores utilise energy contained in this dead material.

Overall, the food web emphasises the important interrelationships of plants and animals with respect to feeding patterns.

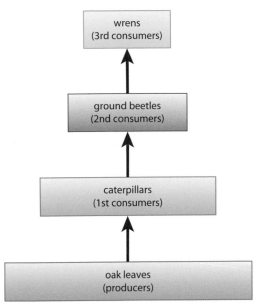

Figure 6.9 Pyramid of numbers. The width of each box gives an indication of the numbers of individuals within that trophic level (but not drawn to scale).

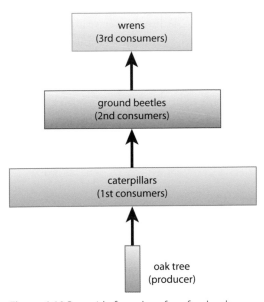

Figure 6.10 Pyramid of numbers for a food web based on a single tree as the producer. The width of each box gives an indication of the numbers of individuals within that trophic level (but not drawn to scale).

■ Numbers, biomass and energy

So far we have described *what* is in a food web but made no attempt to quantify this and say how much of anything is there. We can do this in terms of numbers of organisms, or of living matter (known as **biomass**). We can also make some estimate of the energy flowing through the food web and see what happens to this energy.

Pyramid of numbers

Suppose we could isolate a small area of an ecosystem and look in detail at all the organisms in it. First we could identify all the organisms and decide whether they are producers, primary or secondary consumers, according to what they eat. Then we could count all the organisms in each trophic level for this ecosystem.

We can then represent these numbers in a diagram (Figure 6.9), with producers as the lower layer and working up through the consumer trophic levels. This gives us a **pyramid of numbers**. We give this name to the diagram because the shape is often that of a pyramid with a wide base (many producers) and fewer consumers as you work up the trophic levels.

Sometimes numbers give a different picture. For example, a single tree may support a large number of primary consumers and, in turn, these are food for various secondary or tertiary consumers. The 'number' of producers would be only one, so the 'pyramid' would stand on a very narrow base, as shown in Figure 6.10.

Pyramids of biomass

Consider an area of an ecosystem, as above, but this time make an estimate of the total mass of living material in that area over a certain time. The estimate would be based on the dry mass of material. This is known as the **biomass**.

It may be unrealistic to do this in practice because it would mean that living animals have to be killed, but an estimate does give a measure of the growth of the living organisms within that area. Compared with the pyramid of numbers described above, this gives a more useful representation of the interrelationships of the living organisms with regard to their feeding and what has happened to the energy taken in by the producers in that area over a period of time.

As with the numbers of organisms, we can represent values for biomass in a diagram (see Figure 6.11), with producers as the lower layer and working up through the consumer trophic levels. This gives us a **pyramid of biomass**. A pyramid of biomass has a wide base, like a pyramid.

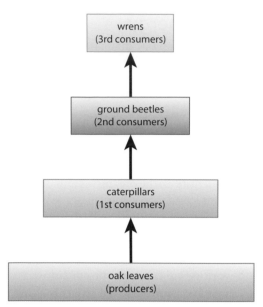

Figure 6.11 Pyramid of biomass. The width of each box gives an indication of the **biomass** of individuals within that trophic level (but not drawn to scale).

Pyramids of energy and energy losses

When an animal (herbivore) eats a plant, or a carnivore eats another animal, each time the animal is taking in food. This food material provides energy for the animal and food substances that the animal needs for growth. These substances include carbohydrates, proteins, fats, lipids, vitamins and mineral ions. In this way, energy and nutrients are passed along the food chain or through the food web to organisms in higher trophic levels.

Let's look more closely at the food materials taken in by a primary or secondary consumer. The primary consumer may not be able to use all the parts of the plant that it eats – woody stems or hard shells of nuts may not be eaten or they may pass through the animal without being digested. Similarly, a secondary consumer may not eat or be able to digest the bones or teeth, nor the fur or feathers of the animal it eats. This means that part of the energy contained within the biomass of the trophic level is wasted and not passed on to the next level. The energy cannot be captured again by other organisms in the food web as we go up the trophic levels – unless the dead material is eaten by a detritivore (or broken down by a decomposer) so that it again enters the food web at the bottom level.

There are other ways that energy can be lost from the food web. A rabbit moves around and a bird flies from tree to tree. Energy is released from food by respiration and this energy enables them to move. Animals excrete and so lose materials from their bodies. Energy is used in living processes, keeping some animals warm and building up materials needed for growth and increase in size.

Scientists have attempted to estimate all the energy within a series of trophic levels, up to the top carnivores. It is difficult to account for everything, but their estimates suggest that only about 10% of the energy is passed from one level to the next. This means that about 90% of the energy is lost between each trophic level. This energy cannot be recovered, so is lost from the system.

In the same way as we did for numbers and biomass, we can represent estimates for energy values in a diagram (see Figure 6.12), with producers as the lower layer and working up through the consumer trophic levels. This gives us a **pyramid of energy**. A pyramid of energy always has a wide base, like a pyramid. The list below illustrates ways in which energy is lost at each trophic level:

- material not eaten by the consumer
- material not digested by the consumer, but is passed out as waste (faeces)
- energy used by the consumer for movement
- energy used by the consumer in body metabolic processes
- energy used in generating heat
- energy lost in urine.

Earlier in the section we looked at food chains of three, four or perhaps five organisms, but indicated that it is unusual to have food chains that are

MATHS TIP

Areas of rectangles

Area (units^2) of any rectangle
= length × width

Now go to page 273 to apply this to
the energy pyramid for Figure 6.12.

much longer than that. You can now understand that the basis of a food chain is that it passes on energy to the next trophic level. You can also see that only about 10% of the energy is available for the next trophic level.

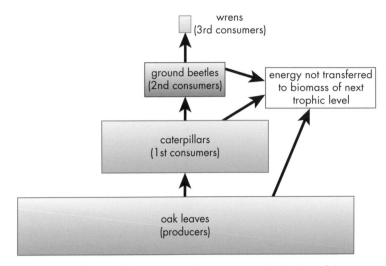

Figure 6.12 Pyramid of energy. The area of each box gives an indication of the **energy** contained in the biomass of individuals within that trophic level (but not drawn to scale).

Suppose there are 1000 units of energy in the producers in the first trophic level. Take one-tenth of that to go to the next trophic level, then one-tenth again for the one above. You can work out that only a very small proportion remains for the 4th and 5th trophic levels so you can see that there comes a point at which there is not enough energy to support another level of organisms within the food web.

Look at the food web on page 250 and use information in this food web in your answers to the questions below.

1 a) Write down **one** food chain consisting of three organisms.
   b) i)   In this food chain, which organism is the producer?
      ii) What is the source of the energy for this producer?
      iii) Which organisms are consumers?
2 a) Write down **one** food chain consisting of five organisms.
   b) In this food chain, which organism is the producer and which are the consumers?
   c) In the food web, what do the arrows represent? Check that you have drawn the arrows correctly in your food chain.
3 Name **one** organism in the food web that is feeding at two different trophic levels.
4 a) Draw the shape of a pyramid of numbers for this food web.
   b) Draw the shape you would expect if a similar food web is drawn up for all the organisms in a food web based on an oak tree as the producer.
5 Suggest how scientists try to collect information to draw up a pyramid of biomass in a limited area. Why is it difficult to do this accurately?
6 What is the source of energy for the whole food web? Explain what happens to the energy between different trophic levels.
7 Explain why it is unusual to have very long food chains (longer than five levels of organisms).

6.3 Cycles within ecosystems

Continuity of supplies for an oak tree

The acorn is nearly 2 cm long and weighs about 4 g. Over a hundred years ago another acorn, similar to this one, fell to the ground and started to grow. It put on a lot of weight and grew into the oak tree, now taller than nearby buildings.

The tree will go on growing, perhaps for another hundred years. Every year the leaves fall off and other parts of the tree break away. But each year the tree continues to put out new growth – new leaves, new shoots, more acorns. When it finally dies, it begins to crumble away through the processes of decay.

Everything that has contributed to this massive structure has been taken in as simple raw materials, either from the soil (water and mineral ions) or from the air (carbon dioxide). The tree cannot run around and find food to build up into the complex structures of the tree – it is, literally, rooted to the spot.

The immediate surroundings could not sustain the supplies needed for the growth of the tree over the many years of its existence unless the various raw materials are replaced. So where do the necessary raw materials come from? How does it balance out?

■ Keeping the balance in ecosystems

The story of the oak tree focuses on how living organisms make demands on the environment. If living communities are to survive, on a local as well as on a global scale, whatever is taken out must be replaced. There must be a balance to allow survival over an extended period of time.

Section 6.3 explores how natural processes recycle and replenish the requirements for growth of the oak tree and other plants within the ecosystem. We also explore how plants and animals are interdependent and emphasise the importance of microorganisms (particularly fungi and bacteria) in the natural cycles within ecosystems.

We look at three natural cycles within ecosystems that help keep the balance: the water cycle, the carbon cycle and the nitrogen cycle.

■ The water cycle

Water is an essential requirement for growth and a major component of cells in all living organisms.

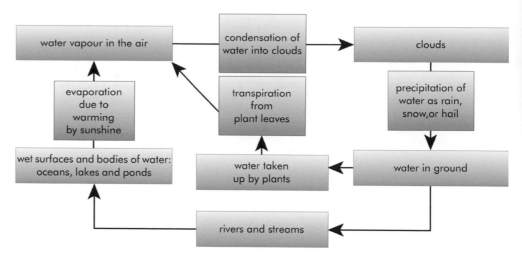

Figure 6.13 The water cycle (water is represented in blue, processes in red)

Figure 6.13 summarises ways that water circulates within an ecosystem, linked to water vapour in the air and large bodies of water in the oceans and on land. The water is shown in blue and processes are represented in red, with more detail below:

■ **Evaporation** – water turns from liquid to vapour when it evaporates. Evaporation of water into the atmosphere takes place from the seas (oceans), from lakes, rivers and other wet surfaces and bodies of water on land.

■ **Transpiration** – this is the loss of water from plants. Water evaporates from the surface of vegetation and a large proportion of this water vapour is lost through the stomata on the leaves.

■ **Condensation** – water vapour in the atmosphere turns back into droplets of water by the process described as condensation. This may be as a result of the air becoming saturated with water vapour, or because of cooling below a certain temperature. Condensation may lead to the formation of clouds, fog or frost.

■ **Precipitation** – water vapour held in clouds falls as rain or snow. The clouds may be blown by wind from the areas where they formed. Precipitation can occur over the land as well as the sea. Some rainfall on land soaks into the ground and is taken up by plants from the ground. Some rainfall soon washes off the land into rivers and then drains back into the sea.

The carbon cycle

The element carbon is found in nearly every molecule and structure within living organisms. Use the diagram in Figure 6.14 to trace possible pathways of carbon through the carbon cycle in natural ecosystems. The compounds containing carbon are shown in green and the processes are represented in red. Fossil fuels are shown in purple to draw attention to the fact that they were formed a long time ago.

STUDY TIP

Check details in Section 6.4 of how human activities produce carbon dioxide and so affect the carbon cycle and how this may lead to enhanced global warming.

Check details in Section 2.3 of how living organisms give off carbon dioxide during respiration.

Check details in Section 2.1 of how plants use carbon dioxide in photosynthesis.

Link the transfer of carbon compounds from plants (producers) to animals (herbivores) and from animals to other animals (carnivores) – see food chains and food webs in Section 6.2.

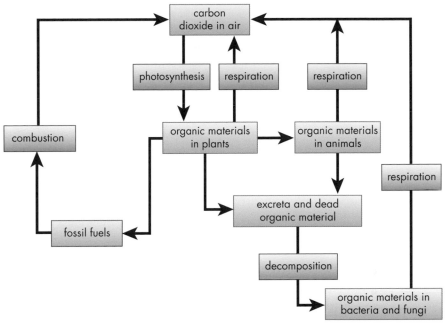

Figure 6.14 The carbon cycle (compounds containing carbon are shown in green and processes in red).

Note in particular, the role played by living organisms (plants, animals and microorganisms) in the recycling of carbon. In Section 6.4 you have an opportunity to see in more detail how specific human activities have an impact on the natural carbon cycle.

More details relating to the processes are given below.

- **Respiration** – all living organisms (plants, animals, microorganisms) respire. The process releases energy and carbon dioxide gas is given off into the air around them. The energy released by respiration is used by the organism in different activities within cells. Respiration breaks down molecules containing carbon – usually sugars, but other compounds (such as lipids) are also broken down during respiration by some organisms. Thus, complex molecules containing carbon are broken down to the small carbon dioxide molecule.

- **Photosynthesis** – this is the very important process, carried out in green plants, that captures energy from the Sun and incorporates it into plant material. The plant takes in carbon dioxide gas from the air and, with the solar energy, uses the carbon dioxide to synthesise sugars. These are then converted into starch and other complex molecules, including lipids and proteins. In this way, the small carbon dioxide molecule is converted into the complex molecules of living organisms. These molecules are then passed to consumers through food chains and food webs and thus circulate to all living organisms.

- **Decomposition** – when living organisms die, the cells begin to break down and decompose. The processes of decomposition are carried out by microorganisms (mainly bacteria and fungi) living in the soil. These processes break down the large complex molecules and release smaller molecules into the surroundings. Carbon is released as carbon dioxide gas from the complex molecules that contain carbon. The carbon dioxide passes into the atmosphere. Other molecules are also released as part of the decomposition process.
- **Combustion** – most of the combustion that takes place is due to human activities, though occasionally spontaneous fires do occur. Materials used for combustion are derived from plants. The plants are either those that have been growing within a relatively recent timescale (as wood from trees in contemporary forests or woodland) or those that existed many millions of years ago and have become fossilised. These include deposits of coal, oil and natural gas and are the 'fossil fuels' used globally by human communities. The carbon had been locked away and was unavailable for release into the atmosphere. The processes of combustion release the carbon as carbon dioxide gas. Note that the rate at which fossil fuels are formed is a very slow process so, within the time-scale of the 21st century, they are not being replaced.

■ The nitrogen cycle

Nitrogen gas accounts for about 78% of the air, yet it is a rather inactive gas with no direct role in the metabolism of living organisms. However, as an element, nitrogen is very important as a component of proteins and nucleic acids, so is present in every living cell.

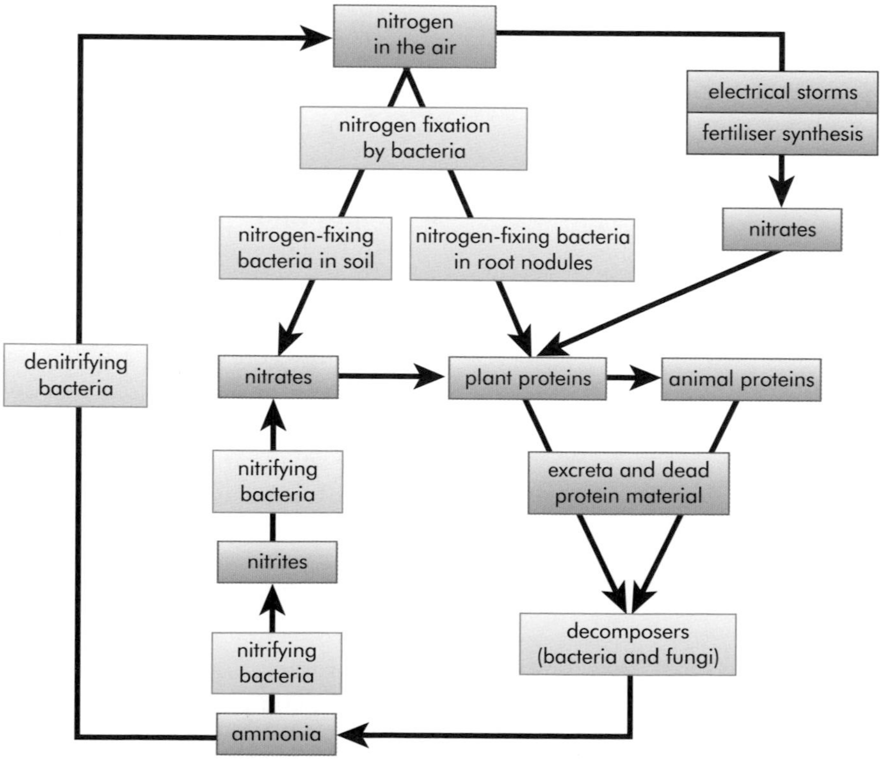

STUDY TIP

You can refer to Section 1.4 to find out about large biological molecules containing nitrogen.

Figure 6.15 The nitrogen cycle (compounds containing nitrogen are shown in green and processes, carried out by microorganisms, in orange).

Use the diagram in Figure 6.15 to trace possible pathways of nitrogen through the nitrogen cycle in natural ecosystems. The compounds containing nitrogen are shown in green and the processes carried out by microorganisms are represented in orange. Note in particular, the role played by living organisms (plants, animals and microorganisms) in the recycling of nitrogen.

Nitrogen in the air is converted into nitrates in the soil. The nitrate ions are taken up by plants and converted into amino acids, then these are built up into proteins and other complex molecules. These molecules are passed to consumers through food chains and food webs and thus circulate to all living organisms.

- **Nitrogen-fixing bacteria** – these bacteria are able to convert nitrogen gas into nitrogen compounds that plants can use to make amino acids and proteins. An important group of nitrogen-fixing bacteria are found in small swellings (called **nodules**) on the roots of leguminous plants (Figure 6.16). This family of plants includes peas, beans, clover and lentils. There are other nitrogen-fixing bacteria that live free in the soil, not within the nodules found on leguminous plants.
- **Decomposers** – when living organisms die, the cells begin to be broken down (decompose). The processes of decomposition are carried out by microorganisms (mainly bacteria and fungi) living in the soil. These processes break down the large complex molecules and release smaller molecules into the surroundings. Nitrogen is released as ammonia or ammonium ions.
- **Nitrifying bacteria** – these bacteria are found in the soil. They convert ammonium ions to nitrites, and the nitrites to nitrates. The nitrates can be taken up by plants and converted into amino acids, then built up into proteins.
- **Denitrifying bacteria** – these bacteria convert nitrates in the soil back into nitrogen gas, which then escapes into the air. These bacteria (also found in the soil), reverse the activities carried out by other bacteria, so these denitrifying bacteria reduce supplies of valuable nitrate in the soil.

Two other processes, also shown on Figure 6.15, are important for helping to build up the level of nitrate in the soil. The first occurs during thunderstorms, when lightning causes a spontaneous reaction between nitrogen gas and oxygen gas. This results in conversion of nitrogen gas into nitrate. The second is the manufacture of fertilisers by chemical processes that provide nitrate for crop plants.

These processes both contribute to the available nitrate in the soil and are included on the diagram of the nitrogen cycle even though they are not biological processes.

STUDY TIP
Refer to feeding relationships in Section 6.2 to understand how nitrogenous compounds are transferred from one living organism to another in food chains and food webs.

Figure 6.16 Roots of a bean plant (leguminous plant) showing nodules that contain nitrogen-fixing bacteria.

STUDY QUESTIONS

These questions check your understanding of the carbon cycle and the nitrogen cycle. (You may need to refer to other parts of this book to answer some of the questions.)

1 **Carbon cycle**
   a) List the stages by which carbon dioxide in the air becomes carbohydrate in the body of an animal. Include reference to the processes and any organisms involved at each stage.
   b) List the stages by which carbon in a carbohydrate molecule in an animal becomes carbon dioxide in the air.
   c) When an animal dies, list the stages by which carbon in the molecules in the animal is released into the air. Include reference to the processes and any organisms involved at each stage.

2 **Nitrogen cycle**
   a) List **three** ways by which nitrogen gas can be converted into nitrate in the soil.
   b) List the stages by which protein in animals can be made available for plants to take in as nitrate. Include reference to the processes and any organisms involved at each stage.
   c) What might happen in a river near some fields if a farmer accidentally adds too much nitrogen fertiliser to the crops in the field?

6.4 Human influences on the environment

A Neanderthal view of the 21st century

The Neanderthals were an early relation of our human ancestors. Imagine you are part of a Neanderthal family, living before they became extinct around 30 000 years ago. You wake up one morning, rub your eyes and amble out of your cave to cast your eye over the landscape below.

You look, blink and rub your eyes again. You just cannot believe what you see. Even though your language skills are limited, you find unfamiliar words are running through your head.

You see a network of roads with cars and trucks buzzing along them, railways, trains, boats on the river beyond. There are lots of houses, some rather tall and precarious, a tangle of wires and cables ... so many people everywhere and it is VERY noisy. Even the nearby hillsides have terraces carved into them with green stuff growing on each level. Where are all the trees and the forests? And where are the bushes you so often visit to gather fruits and nuts? You cannot see any animals roaming freely in the plains below. Oh yes, and it feels strangely warm.

Incredulous, you retreat back into your cave, shaking your head. You hold your head in your hands, curl up on the floor of the cave and fall asleep. Perhaps sleep will help you to wipe out this strange dream that gave you a view of the 21st century world.

■ The 21st century environment

The Neanderthal world and its landscape may still exist in a few isolated parts of the 21st century globe. But for many of us, the imaginary Neanderthal view of the 21st century is closer to the surroundings we are familiar with in our everyday lives.

We can use this Neanderthal view to give us some perspectives on the visible changes to the landscape since the time of the Neanderthals and other early humans. We can probe further into the less visible changes – to the air, the atmosphere and the water – and try to see what has happened. We also consider how far humans are responsible for the changes and whether we can control our environment in the future.

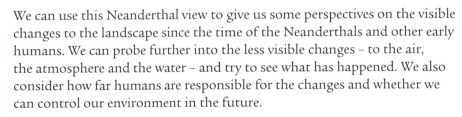

261

Figure 6.17 Factories contribute to air pollution.

■ Air pollution

Various waste gases pass into the air and cause pollution (Figure 6.17). The polluting gases come from industrial processes and from the burning of fossil fuels (such as coal and oil) in factories, power stations, domestic homes and motor vehicles. Two examples of polluting gases are **sulfur dioxide** and **carbon monoxide**.

Sulfur dioxide

Sulfur dioxide is produced from the burning of fossil fuels, especially coal and oil. Sulfur dioxide dissolves in water in the atmosphere to form sulfuric acid. This then falls as 'acid rain' and is known to cause harm to living organisms and to buildings.

Here are some effects that have been observed:

■ Freshwater lakes with low pH (high acidity), as a result of acid rain, have reduced populations of fish – sometimes there are no fish at all. The low pH affects successful reproduction of the fish. Another effect is that the fish produce a lot of sticky mucus on their gills so that gas exchange becomes difficult.

■ Acidified lakes may show a loss of microscopic algae and plant material suspended in the water. The water in these lakes may appear more transparent. The effect of a lack of microscopic algae and plant material spreads through the food webs in the water and alters the balance of species – more of some, reduction of others.

■ Direct damage to plant life has been observed, particularly in coniferous forests in some areas of Europe. Many trees have died, whereas others show poor growth and the needles (conifer leaves) are discoloured. The low pH in the soil probably interferes with the ability of the roots to take in certain necessary mineral ions from the soil.

■ Acid rain can affect buildings using limestone or marble. The acid reacts with calcium carbonate in the stone and starts to dissolve the surface layers. Any carving on the stone may be lost and the stone starts to crumble.

Carbon monoxide

Carbon monoxide in the air comes mainly from the waste gases in exhaust from motor vehicles, as a result of burning petrol. It is also found in cigarette smoke. Sometimes there is a risk of carbon monoxide being produced from domestic heaters when the fuel is incompletely burned. It can reach dangerous levels, particularly if there is inadequate ventilation in the room.

Carbon monoxide is poisonous to humans. It combines with haemoglobin in the blood and thus reduces the amount of haemoglobin available to combine with oxygen. This prevents the formation of oxyhaemoglobin and means that less oxygen is transported from the lungs to the cells of the body. A person suffering from carbon monoxide poisoning is likely to die unless adequate ventilation, with extra oxygen if necessary, is provided quickly.

MATHS TIP

Translating data between tables and graphs

Graphs allow patterns of data to be more easily seen than in **tables** of numbers, but tables are the most efficient way of providing individual values for analysis.

When answering 'describe' questions referring to graphs, describe the overall pattern – for example the variable rises / falls / stays the same, AND give example values, referring to the x and / or y axes of the graph to support your description. Do not just say 'big' or 'small'. Be more precise by reference to actual values.

Now try Question 5 a) ii) on page 280.

Greenhouse gases and the greenhouse effect

The greenhouse effect

First we look at the greenhouse effect, what this means and what contributes to it. The Earth is surrounded by a layer of gases, also called the 'atmosphere'. Solar radiation passes through the atmosphere, towards the Earth. The subsequent events that occur and contribute to the warming of the Earth are summarised in Figure 6.18.

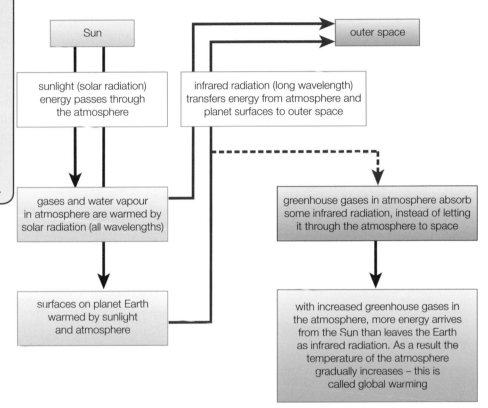

Figure 6.18 Sequence of events that contribute to the greenhouse effect and global warming.

STUDY TIP

Look in Section 7.1 and make a link to how the greenhouse effect is exploited in growing of crops in glasshouses and polytunnels.

This simplified description of the greenhouse effect shows how the layer of gases around the Earth traps radiation. The Earth is probably about 30 °C warmer than it would be without the layer. Without the greenhouse effect, many parts of the Earth would be too cold to support life as we know it.

We can see a similar situation in a glasshouse (greenhouse). The glass is transparent to radiation coming into the glasshouse and it gets warm inside. The panes of glass help keep the warm air inside the greenhouse – a bit like the layer of gases in the atmosphere, though the glass does it in a different way from the gases in the atmosphere.

Greenhouse gases

Certain gases can contribute to the atmosphere layer and increase its warming effect. This is known as the 'enhanced greenhouse effect'. The gases that contribute are known as **greenhouse gases**. The greenhouse gases include: water vapour, carbon dioxide, methane, CFCs (chlorofluorocarbons) and nitrous oxide.

The enhanced greenhouse effect is often referred to as global warming and is linked with 'climate change'. The two terms are often loosely used together, to mean more or less the same thing.

Many people think that global warming has occurred entirely as a result of human activities in a relatively recent timescale (a few hundred years or less), whereas other people recognise that some of the global changes in climate may be due to natural events, occurring over a much longer timescale.

Table 6.6 looks at the greenhouse gases and summarises how far human activities have increased these gases. Any increase in greenhouse gases could have the effect of further enhancing the greenhouse effect and global warming.

STUDY TIP

Make links to cycles and recycling in Section 6.3 and for use of nitrogen fertilisers, look also in Section 7.1.

Table 6.6 Greenhouse gases and how human activities contribute to their generation.

Greenhouse gas	Human activity that generates more of the gas	Comments
water vapour	not relevant	human activities probably do not contribute directly to the level of water vapour in the atmosphere and its influence on the greenhouse effect
carbon dioxide	burning fossil fuels (coal, oil, natural gas)	in factories, power stations, domestic uses, motor vehicles
	deforestation	removal of forests is likely to reduce uptake of carbon dioxide (for photosynthesis) by trees and upset the balance between photosynthesis and respiration, hence affecting the level of CO_2
methane	from cattle and other ruminants	digestion in ruminants produces methane as a waste product – includes domestic animals, such as cattle, sheep and camels
	rice paddy fields	produced by bacterial activity in wet anaerobic conditions, as found in rice paddy fields
	decay of waste materials	anaerobic decay of waste can produce methane, which sometimes leaks out from landfill sites
CFCs	e.g. CFC11 (in aerosol sprays) e.g. CFC12 (cooling agent)	CFCs do not occur naturally and are entirely made by humans for use in aerosol cans, refrigerators, air conditioning and foam packaging but their use is now banned in many manufacturing processes
nitrous oxide	burning fossil fuels (motor vehicle exhaust)	exhaust from motor vehicles contains nitrous oxide, but use of catalytic converters helps convert this to less harmful nitrogen dioxide and to nitrogen
	nitrogen fertilisers	nitrates in the soil (and in fertilisers) are converted by denitrifying bacteria to nitrous oxide (and to nitrogen gas)

The graph in Figure 6.19 shows how **global temperatures** have fluctuated over the past 1000 years. During this time, human communities have responded to the changes and adapted their activities. Global temperatures are now rising and scientists predict that they are likely to continue rising to a level higher than at any time in recent history.

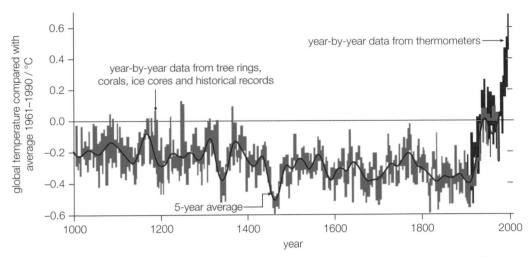

Figure 6.19 Changes in global temperature over 1000 years compared with average temperature from 1961 to 1990. The values shown in blue have been derived from tree rings, corals, ice cores and historical records. The values shown in red are from records taken with thermometers.

Data about greenhouse gases (Figure 6.20) show a noticeable increase in output of carbon dioxide within the last 200 years, since the industrial revolution. Other greenhouse gases (methane, nitrous oxide, CFC12) show similar patterns. Many people claim that global warming is caused by human activities. There is no doubt that human activities *contribute* to global warming, but we cannot *prove* that it is entirely the result of human activities.

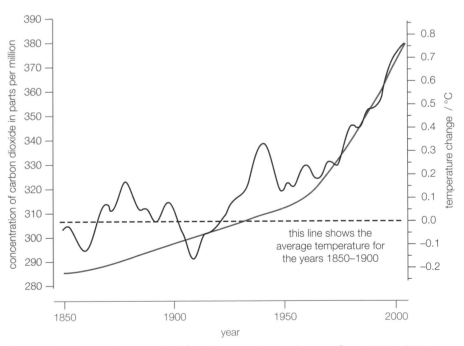

Figure 6.20 Changes in carbon dioxide and temperature in the years from 1850 to 2000. The line in red shows changes in temperature and the line in blue shows changes in carbon dioxide concentration.

STUDY TIP

Make links to cycles and recycling in Section 6.3 and for use of nitrogen fertilisers, look also in Section 7.1.

STUDY TIP

You can find more information about fossil fuels and biomass in Section 6.3.

Many people try to reduce the output of greenhouse gases in an attempt to reduce global warming. Some measures are taken at an individual level, in the home or in the workplace. Other measures form part of legislation at a national or international level. Some examples are listed in Table 6.7.

Table 6.7 Some ways of reducing global warming

Some ways of reducing global warming	How greenhouse gases are affected
recycling — paper, bottles, plastics (reduces manufacturing)	reduced use of fossil fuels, less CO_2 output
design of motor vehicles for greater fuel economy; encourage fewer journeys	reduced use of fossil fuels (petrol and diesel), less CO_2 output
generation of electricity by 'alternative' methods (for example, photovoltaic cells, wind turbines, hydroelectric power)	reduced use of fossil fuels (especially coal), less CO_2 output
use of solar heating (for example, for domestic hot water)	reduced use of fossil fuels, less CO_2 output
design of buildings with a greater emphasis on being 'carbon-neutral' (in terms of building materials, construction costs, integrated energy generation, good insulation) – reducing central heating and air conditioning requirements	reduced use of fossil fuels, less CO_2 output
composting waste plant material (rather than sending it to landfill); harvesting methane (as biogas) often from small-scale units to distribute locally for heating and cooking	reduces methane production to the atmosphere
catalytic converters in motor vehicles	converts nitrous oxide to nitrogen dioxide
legislation to phase out CFCs (in aerosols and refrigerants)	reduced CFCs
alternatives to fossil fuels – in motor vehicles, heating (for example, alcohol from sugar cane, solar panels)	release of CO_2 from present-day sources rather than using fossil fuels from distant times in the past
reference to international targets to reduce greenhouse gas emission	probably intended to affect all, but particularly CO_2

Effects of global warming

The consequences of global warming are widespread and unpredictable. There are considerable concerns about the effect of global warming on life as we know it. Predictions suggest that during the 21st century, temperatures are likely to rise further, perhaps up to 6 °C higher. This may seem a relatively small rise in temperature overall, but the effects may be far-reaching. Here are some examples:

- **Melting of ice in glaciers and polar regions** – this leads to a rise in sea level and increased risk of flooding of land in low-lying areas.
- **Increase in volume of water in the oceans** – due to expansion of water with an increase in temperature (as well as some from melting ice). This gives increased risk of flooding.
- **Less ice in glaciers and less snowfall** – reduces meltwater supplies to areas that have depended on water from melting snow and glaciers, with a knock-on effect on food supply.

STUDY TIP

Section 6.3 gives an outline of the water cycle and consequent links to crop production.

STUDY TIP

Link 'troph' in trophic levels with 'eutrophication' to remind you the word is to do with feeding or nutrient content.

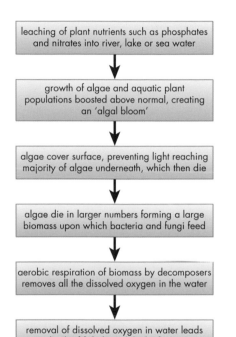

Figure 6.21 Sequence of events following eutrophication.

Figure 6.22 Water hyacinth growing in polluted water make it difficult for boats to pass along the canal.

■ **Changes in weather patterns** – with more likelihood of extremes in temperature, rainfall (droughts as well as heavy rain), winds and severe storms.

■ **Changes in distribution of plants and animals** – effect on natural ecosystems, illustrated by seasonal changes in a particular location (e.g. flowers appear earlier, migrating birds and insects arrive earlier).

There may be changes in the patterns of agricultural crops. In some areas, higher temperatures and higher carbon dioxide levels could result in higher crop yields or successful growth of crops suited to warmer climate zones. In other areas (closer to the equatorial regions), temperatures may become too high for successful growth. It may also mean that in areas where there is only just enough rain for crop growth, drought may severely limit crops. Weeds, crop pests and vectors of disease may be affected in different ways, some of which may lead to increased crop losses.

■ Water pollution and eutrophication

Organisms living in water have the same basic requirements as those living on land. Plants require light and carbon dioxide for photosynthesis. They also require certain mineral ions and they require oxygen for respiration. Animals require a source of food and oxygen for respiration. A range of microorganisms also forms part of an aquatic ecosystem. Sometimes pollution can have a drastic effect on the community of organisms living in the water.

Sewage

Sewage is the domestic waste material from human communities and contains human urine and faeces, together with kitchen and other waste water. Sewage is rich in certain mineral ions (mainly nitrates and some phosphates). It also contains some metal ions and small solid particles suspended in the water.

Suppose some sewage flows into a freshwater lake. Around the entry point of the sewage, there is a sudden rise in levels of nutrients in the water. At the same time, microorganisms in the sewage and in the water start to break down the organic matter in the sewage. This releases more nutrients, particularly nitrates and phosphates. These extra nutrients stimulate a rapid growth and multiplication of algae in the water. One effect of this explosion in population numbers is that the water becomes quite cloudy (or even greenish) because of the algae suspended in the water. This situation is known as **algal bloom**.

A series of events may follow on – summarised in the flow diagram (Figure 6.21).

When fish die due to lack of oxygen, this is known as 'fish kill' and it can occur very suddenly.

This series of events results from a high level of nutrients in the water. The term eutrophication is used to describe this enrichment of the water. The same effect can follow if mineral ions (such as nitrate and phosphate) are washed out (leached) from fields when chemical

STUDY TIP

Look in Section 7.3 for the importance of ensuring there is adequate oxygen in the water for fish in fish farms.

PRACTICAL

This practical activity gives you an opportunity to become familiar with algae and how the population grows. The activity provides a useful illustration of the effects of eutrophication and can be carried out relatively easily with simple equipment.

fertilisers are applied. A similar situation can also arise if the waste from a farmyard (where animals such as cattle, sheep or goats are kept) washes into a lake or river.

■ Practical activity – investigating the effect of different concentrations of fertiliser on the growth of algae

1 Set up 3 test tubes, A, B and C, as shown in the diagram (Figure 6.23). Each tube contains the same volume of algae in the same volume of pondwater.

2 Add different volumes of liquid fertiliser to each tube. A liquid house plant fertiliser is suitable. Add most fertiliser to tube A and least to tube C. Count the drops that you add so that you have an approximate quantitative measure of the concentration of fertiliser in each tube.

3 Do not seal the tubes. Leave the tubes under continuous illumination for a week or more. A windowsill may be suitable, but the tubes should not be left in direct sunlight.

4 Compare the green colour in the three tubes. This gives a measure of the population density of the algae in each test tube. The density can be estimated by using a colour card, or by making a series of diluted samples or in a colorimeter.

The algae reproduce giving different population densities. You can extend the number of tubes in the series, say to 5 or more, and plot a graph showing the effect of fertiliser concentration on algal density. *Students must wash hands thoroughly after handling pond water. There is a risk that harmful bacteria could multiply as well as algae.*

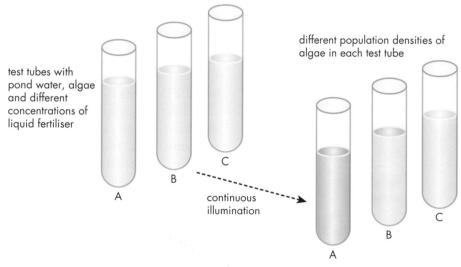

test tubes with pond water, algae and different concentrations of liquid fertiliser

different population densities of algae in each test tube

continuous illumination

Figure 6.23 Investigating the effect of different concentrations of fertiliser on the growth of algae.

Figure 6.24 Unspoilt forest.

Figure 6.25 Forest clearance in Brazil.

◼ Deforestation

For many centuries, people have cut down forest trees for different reasons. When trees are harvested for use as timber and young trees are allowed to take their place, there is no long-term effect on the environment – the balance remains the same.

However, in many locations, huge areas of forest have been cut down to clear the land so that it can be used for other purposes. These include agricultural land for growing crops or for grazing animals. The land is also used for houses or industry and the network of roads and railways that link villages, towns and cities. This permanent removal of trees can have a number of consequences.

Often an early effect of removing trees is that rainwater runs off much more freely from the soil. In a forest, the trees and other vegetation act a bit like a sponge and absorb the water, releasing it slowly to the soil and streams. But when the trees have been cut down, the vegetation in the forest no longer holds back the water. This run-off takes mineral ions from the soil with it – a process known as **leaching**. Soil particles also get washed away. These can accumulate in the valleys below where heavy deposits may block the river channels and cause **flooding**.

Disturbance of the soil can affect the natural nutrient recycling processes. The microorganisms that are a part of the recycling processes tend to get washed away with the soil and there is no longer the reserve of biomass (in the trees and associated vegetation) to break down and help replace mineral ions removed with the harvested trees. The resulting soil is likely to be poor in quality for growing crops in the future.

Removal of trees on a large scale can also affect the **water cycle** by disturbing **evapotranspiration** in the area. In a forest, with many trees together, loss of water by transpiration creates clouds and this can encourage rainfall in the area. This local effect is lost after deforestation and the area may become much drier.

STUDY TIP

More details of natural nutrient recycling processes are given in Section 6.3.

STUDY TIP

More details of the water cycle are given in Section 6.3

STUDY TIP

The term **evapotranspiration** is used to describe the sum of **evaporation** + **transpiration**.

STUDY TIP

Check in Section 2.3 for information on respiration and in Section 2.1 for information about photosynthesis.

Living trees affect the **balance of gases** in the atmosphere. Plants carry out both **respiration** and **photosynthesis**. During photosynthesis, the large number of trees in a forest take up a lot of carbon dioxide (and give out oxygen).

If the trees of the forest are replaced by agricultural land or grassland (say for grazing), the overall quantity of plant material is likely to be reduced, with the result that less carbon dioxide is taken from the atmosphere and less oxygen is given out. This imbalance is made even worse if the former forest is replaced by houses (for towns and cities) or industrial factories.

STUDY QUESTION

1 **Explain why … explain how …**

Each of the following statements or observations are linked to the topics studied in this section.

For each statement, you are asked to 'explain why' or 'explain how' the situation or observation relates to environmental issues or may be of benefit to the environment. For some you may need to look at other parts of this book or use other sources of information to help you.

a) Sometimes the water in a lake with low pH (rather acid) looks very clear.

b) There is an old factory in an area surrounded by forest. The factory is still being used. People noticed that the trees growing in the forest on one side of the factory looked rather poor and many were dying. This effect was much less obvious on the other side of the factory. (Hint: think about the direction of prevailing winds.)

c) A family were using an old gas heater in a room and the windows were tightly closed. In the morning they woke up feeling rather unwell, but soon recovered after they went outside. They were advised to make sure the windows were always left open a little bit.

d) A family always tried to find wood to burn in their stove rather than coal.

e) Another family decided to put photovoltaic panels on the roof of their house to generate a small amount of electricity.

f) Catalytic converters are standard on modern cars but are not necessarily fitted on older cars.

g) People living in villages are encouraged to share car journeys with other people or use public transport when going to work in the nearby town.

h) The water in a lake in an area surrounded by farmland often looks very green and cloudy, particularly in warm sunny weather.

i) A farmer kept cattle and sheep in his yard. He used to wash down his yard with water and let this drain into a nearby river. He then decided to collect the waste. He set up a small digester to produce biogas from this waste and used the biogas for heating and cooking. (Hint: at least **two** environmental benefits in this one.)

j) A group of families cut down the forest on the hillside outside their village so that they could have some extra land to grow more crops. They noticed that the river below started to flood after heavy rain yet the soil in the fields they had created on the hillside was often quite dry and their crops were rather poor. (Hint: several things to explain here.)

Summary

I am confident that:

✓ I know the meaning of the terms population, community, habitat and ecosystem.

✓ I understand the term biodiversity, within an ecosystem and across the whole globe.

✓ I can describe how to use different sampling techniques to find out about distribution of organisms (including plants and animals) in an area.

✓ I understand how to use different practical techniques, including quadrats, to investigate the distribution of organisms in their habitats and to compare the biodiversity of different areas.

✓ I appreciate that biotic and abiotic factors affect population size and distribution of organisms and understand some of the effects of abiotic factors on living organisms in an ecosystem.

✓ I know the meaning of the terms producer and consumer and understand the relationships between organisms at different trophic levels in a food web.

✓ I understand how solar energy is captured by the producers (green plants) and transferred to consumers in a food web, and how energy is lost at each trophic level of a food web.

✓ I understand the importance of decomposers and their role in a food web.

✓ I appreciate the reasons for representing numbers of organisms as pyramids of numbers of organisms, as pyramids of biomass and as pyramids of energy, and understand the limitations of each method.

✓ I can describe the stages of the carbon cycle and understand how living organisms are involved through the processes of respiration, photosynthesis and decomposition.

✓ I can describe the nitrogen cycle, including the roles of bacteria in the soil and the activities of plants and animals that contribute to it.

✓ I know that sulfur dioxide can contribute to acid rain and understand ways that this can be harmful to plants and animals.

✓ I understand the significance of the greenhouse effect for life on Earth.

✓ I appreciate that some human activities increase greenhouse gases and that this may contribute to global warming with various consequences for living organisms.

✓ I know about eutrophication and how this can lead to lack of oxygen in the water and the possible effects on plants and animals living in the water.

✓ I understand how deforestation can have harmful effects on the local ecosystems as well as in the wider environment.

MATHS SKILLS

Percentages
One quantity (n) can be expressed as percentage of another (N), by setting up the first quantity as a fraction of the second and multiplying by 100:

(n/N) x 100 = the percentage of N represented by n.

If n is the magnitude of an increase, or decrease, from the original value N, then the formula above will give the percentage increase / decrease.

Worked example: A sample of fresh soil weighs 50 g. After drying to remove all water present, the sample weighs 25 g. The percentage of fresh soil mass composed of water = (50 – 25/50) x 100 = 50%

A sample of fresh soil weighs 50 g. After drying to remove all water present, the sample weighs 13 g. Calculate the percentage of water present by mass in the soil sample.

74%

MATHS SKILLS

Bar charts
Bar charts consist of a series of bars of equal thickness, drawn vertically from the *x*-axis of the chart. They are used to represent the **frequency** of items in a series of discrete **categories**

If each category has no logical connection to the others (e.g. different eye colours), the individual bars are often arranged in ascending or descending order of magnitude and separated by small gaps.

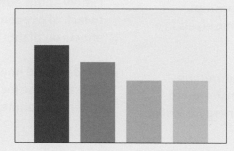

Which of the following investigations would require a bar chart for presenting the results, as opposed to a scatter diagram?

(i)   effects of light intensity on plant growth
(ii)   comparing species diversity in moorland, woodland and grassland
(iii)   comparing diatom populations of three different lakes
(iv)   effects of water velocity on populations of invertebrates

(ii) and (iii)

MATHS SKILLS

Scatter diagrams

A visual way of showing whether there is any **correlation** between two different sets (**variables**) of experimental data, plotted on the ***x* and *y* axes of a graph.**

A **positive** or **negative correlation**, or **no correlation**, between the variables can be determined, from the **pattern of the points** on the scatter diagram.

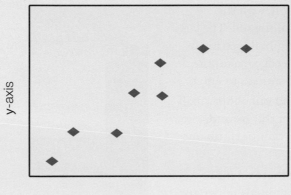

In the example above there is a **mostly positive correlation** between the *x* and *y* values, though there are two anomalous points which do not fit the pattern perfectly.

Draw two versions of the scatter diagram, rearranging the data points to show:

(a)   no correlation

(b)   perfect negative correlation.

b)   orderly descent of points downward from the top left to the bottom right of the chart

a)   completely random scattering of data points

MATHS SKILLS

Areas of rectangles

Area (units^2) of any rectangle = length $\times$ width

Look at the energy pyramid for Figure 6.12.

a)   Use the figure of 1000 units of energy for the producers. Calculate the units of energy present in the other three trophic levels.

b)   Draw a new pyramid of energy for this food chain, using rectangular boxes drawn to an appropriate scale (for example 1 mm^2 = 10 energy units).

a)   1st consumers = 100 units

         2nd consumers = 10 units

         3rd consumers = 1 unit

Example of student response with expert's comments

■ Practical activities

1 The diagram shows a pitfall trap. The container is buried in a hole dug into the soil, with the opening level with the surface of the soil. The large stone over the top stops large animals getting in and the small stones round the edge make sure there are gaps that allow small animals to enter.

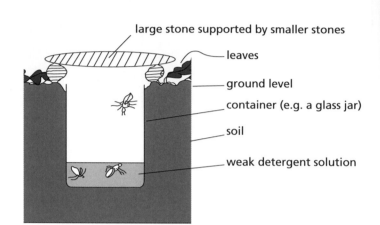

A pitfall trap can be used to find out about small animals in the layer of leaves on the ground. Animals that fall into the trap cannot climb out of the container.

Some students suggested that there might be differences in the invertebrate animals under an oak tree and under a pine tree. They used one pitfall trap under each type of tree and examined the contents of the container after 24 hours. They sorted the animals into groups and counted the number in each group. Their results are given in the table.

a) i) Plot these results in a bar chart so that you can compare them. *(5)*

   ii) Calculate the percentage of the total animals caught that were in the trap under the pine tree. Show your working. *(2)*

b) Suggest why they left the traps for 24 hours before examining the contents. *(1)*

c) One student said that the results show that there is a greater abundance and a greater diversity of animals under the oak tree compared with the pine tree. Describe **two** ways that you could modify or extend the investigation to support this suggestion. *(2)*

(Total = 10 marks)

Group of animals	Total number of animals caught in 24 hours	
	Under oak tree	Under pine tree
Beetles	12	10
Springtails	31	5
Spiders	7	2
Harvestmen	6	0
Woodlice	2	3
Slugs	2	0

Student response Total 11/13	Expert comments and tips for success
a) i) 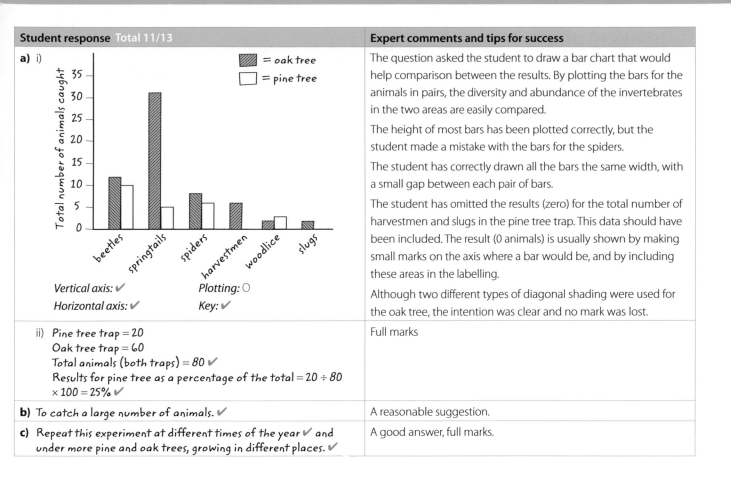 Vertical axis: ✔ Plotting: ○ Horizontal axis: ✔ Key: ✔	The question asked the student to draw a bar chart that would help comparison between the results. By plotting the bars for the animals in pairs, the diversity and abundance of the invertebrates in the two areas are easily compared. The height of most bars has been plotted correctly, but the student made a mistake with the bars for the spiders. The student has correctly drawn all the bars the same width, with a small gap between each pair of bars. The student has omitted the results (zero) for the total number of harvestmen and slugs in the pine tree trap. This data should have been included. The result (0 animals) is usually shown by making small marks on the axis where a bar would be, and by including these areas in the labelling. Although two different types of diagonal shading were used for the oak tree, the intention was clear and no mark was lost.
ii) Pine tree trap = 20 Oak tree trap = 60 Total animals (both traps) = 80 ✔ Results for pine tree as a percentage of the total = 20 ÷ 80 × 100 = 25% ✔	Full marks
b) To catch a large number of animals. ✔	A reasonable suggestion.
c) Repeat this experiment at different times of the year ✔ and under more pine and oak trees, growing in different places. ✔	A good answer, full marks.

Applying principles

1 The food web shows some feeding relationships for marine organisms living in polar seas. Phytoplankton is made up of microscopic organisms that can carry out photosynthesis, including many algae.

a) Use organisms shown in the food web to complete the table. Organisms may be used once, more than once or not at all. *(3)*

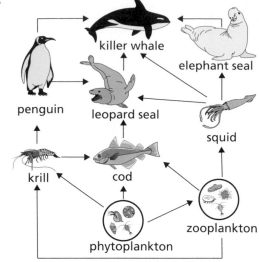

Description	Name of organism
one group of organisms that are producers	
two organisms that are both primary and secondary consumers	
two organisms that are tertiary consumers	

b) i) Explain what is meant by **biomass**. *(1)*
   ii) Which group of organisms has the greatest biomass in this food web during the whole year? *(1)*

c) A region of the polar seas, with organisms in this food web, suffers from pesticide pollution. The pesticide is not biodegradable (does not break down quickly). An analysis of samples of tissues from organisms in this food web showed the highest concentration in the killer whale. Suggest an explanation for this. *(2)*

d) Fishing of cod increased in the region until it became 'unsustainable'. This means that the fish (cod) were removed at a faster rate than they could be replaced by natural breeding.
   If cod became extinct as a result of over-fishing, predict and explain **two** ways that this might affect the food web. *(2)*

e) Polar regions have a good supply of mineral ions, carried in sea currents, such as the Gulf Stream.
   i) Which group of organisms benefits directly from this supply of mineral ions? *(1)*
   ii) Suggest **two** mineral ions that are important for this group. *(2)*

Student response Total 12/12	Expert comments and tips for success
a) phytoplankton ✔ cod and krill ✔ penguin and leopard seal ✔	To find organisms that were both primary and secondary consumers, the student looked for organisms that were the first consumers in one food chain and the second consumers in another. e.g. phytoplankton ⟶ krill phytoplankton ⟶ zooplankton ⟶ krill
b) i) Biomass is the total mass of living organisms in, e.g. a wood. ✔	A clear answer.
ii) The producers. ✔	Correct answer.
c) The pesticide is taken in from the water <u>by the producers and passes to the primary and secondary consumers, going up the food chain to the killer whale.</u> ✔ The <u>pesticide doesn't break down</u> ✔ (it is not biodegradable) so it stays poisonous and <u>builds up to dangerous levels</u> (✔).	The student applied understanding of food chains to the information about the pesticide (not biodegradable) in the question. She could have gained a third mark for saying the pesticide builds up in the killer whale, but the maximum for the question had been reached.
d) There would be <u>more zooplankton because they would not be eaten by cod.</u> ✔ There would be <u>more squid, as more food (zooplankton) for them.</u> ✔	Full marks for a good description of two possible outcomes if cod became extinct.
e) i) The <u>phytoplankton.</u> ✔ They are the producers in the food web and like all green plants they need mineral salts e.g. magnesium.	Do not waste time by writing more than is necessary. Always check the mark allocation to judge how much information is expected. Strictly speaking algae are not plants, but the student has already answered the question and this part of the answer is not needed.
ii) Magnesium ✔ and phosphates ✔	In this question, any correct answer is given credit, even if it is based on knowledge beyond that required by the specification (e.g. phosphate).

■ Extended writing

1 Explain how deforestation can cause flooding and affect the fertility of the soil in the deforested area. *(6)*

Student response
When trees are cut down the soil is easily eroded. Nutrients are also washed out of the soil so it is less fertile. If there aren't any trees the leaves don't fall to the ground and so the bacteria in the soil don't break them down and release the nutrients. Sometimes there is flooding because the trees don't take up the water and it stays in the ground.

Mark scheme
The mark scheme shows how marks are awarded.
(1) vegetation / soil on ground exposed / no longer protected by trees / eq
(2) less water taken up by trees from soil / eq
(3) water can run off more easily (if on slope) / eq
(4) (vegetation) does not bind / hold soil / eq
(5) soil eroded / washed away / eq
(6) (washed away soil) accumulates in rivers / blocks waterways in valleys (leading to flooding) / eq
(7) soil can become saturated (because trees do not take up the water), leading to flooding / eq
(8) mineral ion content (of soil) washed away / removed / leaching / eq
(9) microorganisms (in soil) washed away / eq
(10) (therefore) disturbance to soil recycling / eq
(11) less vegetation on surface / eq that would break down / decompose / eq
(12) mineral ions / eq not released from decaying vegetation / eq
(13) soil less fertile / eq
Total: 6

Student response Total 6/6	Expert comments and tips for success
When trees are cut down the soil is easily eroded. ✔	The student gains mark (5) with correct use of the term 'eroded'.
Nutrients are also washed out ✔ of the soil so it is less fertile. ✔	The first sentence gains marks (8) and (13). 'Nutrients' is an acceptable alternative (as 'eq') to 'mineral ions'.
If there aren't any trees the leaves don't fall to the ground and so the bacteria in the soil don't break them down and release the nutrients. ✔	Several points contribute to mark (11) but no mark for mention of bacteria (9), as this is awarded only in the context of being washed away. The student correctly links less vegetation on the soil with being broken down (or decomposed).
Sometimes there is flooding because the trees don't take up the water ✔ and it stays in the ground. ✔	'Trees don't take up the water' earns mark (2). No mark for mention of 'flooding' as this is in the question but it is correctly linked to water in the ground, equivalent to mark (7).

See General Advice for Extended Writing questions on page 79.

Exam-style questions

1 The diagram shows a sweep net.

the sweep net is attached to a long handle

A sweep net can be used to capture insects and other small animals in soft vegetation such as long grass. The net is dragged through the grass with the flat part of the entrance ring pressed close to the ground. As it is swept forward, small animals are caught in the cotton bag. The trapped animals can then be collected from the bag, identified and counted.

Some students made a study of invertebrate animals in an area of grassland. They wanted to draw up a food web that represented feeding relationships of the community in the area.

The net was swept through the vegetation for a total of 4 m. The table shows information about the animals captured.

Organism	Description	Numbers caught in one sweep
aphids	feed on sap of plants	240
plant bugs	feed on sap of plants	12
snails	feed on algae on surfaces and on plant leaves	4
spiders	eat insects	2
harvestmen	eat insects	1
flies	feed on decaying matter	10

a) i) Explain why the net catch may not accurately represent the actual community of animals present in the grassland. [2]

ii) What additional steps could the students take to ensure greater reliability for the results? [2]

b) Draw a simple grassland food web, using the species caught. [4]

c) i) Name **one** herbivore and **one** carnivore in the catch. [2]

ii) Suggest why the catch shows fewer carnivores than herbivores. [2]

d) Suggest possible effects on the grassland web community if a pesticide led to the death of the aphids. [2]

[Total = 14]

2 The diagram shows a food web in a pond.

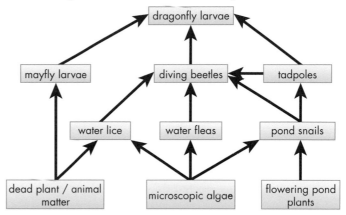

Use information in this food web to answer the questions that follow.

a) Name **two** producers in this pond ecosystem. [2]

b) i) How many organisms are there in the longest food chain in this web? [1]

ii) List the organisms for **one** food chain with this number of organisms and state the trophic level of each organism in the chain. [4]

c) If all the tadpoles in the pond died from the effects of a new fungal disease, predict what might happen to the populations of the other pond species. [3]

d) A pair of predatory (carnivorous) fish is introduced to the pond. (The predatory fish are secondary consumers.) Describe and explain the possible consequences to the balance of the whole ecosystem. [3]

[Total = 13]

3 a) i) The following pathways represent different parts within the nitrogen cycle.

Which pathway is not dependent on the activities of microorganisms? [1]

A nitrates in the soil ➜ plant proteins ➜ animal proteins

B nitrogen in the air ➜ plant proteins ➜ animal proteins

C ammonia in the soil ➜ nitrites in the soil ➜ nitrates in the soil

D nitrogen in the air ➜ fertiliser synthesis ➜ nitrates in the soil.

b) Explain why leguminous plants (such as beans, clover and lentils) are often grown by farmers and gardeners to help maintain the fertility of the soil. [3]

c) Human activities may affect the nitrogen and carbon cycle. Explain how the use of fossil fuels may link into the carbon cycle. [2]

[Total = 6]

4 Describe and explain the shortest route by which nitrogen atoms present in the proteins of an animal's body might become part of the proteins in a wheat plant. **[Total = 6]**

5 Some scientists were looking at the effects of pollution. They did a study of about 1500 lakes in Norway. They measured the pH of the water and made estimates of the numbers of fish in each of the lakes.

They found that some lakes had very few or no fish in them, while others had quite good numbers of fish. Some lakes had very good populations of fish. They divided their results into categories, according to the estimated number of fish in the lake.

Their results for two groups of fish population numbers are given below.

	Percentage of lakes (with the pH given)	
pH	No fish	Good populations of fish
less than 4.5	73	2
4.5 to 4.7	54	5
4.7 to 5.0	38	22
5.0 to 5.5	24	32
5.5 to 6.0	8	58
more than 6.0	2	86

a) The bar chart below shows the results for the lakes with good populations of fish.

i) Make a copy of the graph on a graph grid. On the same grid, plot the results for percentage of lakes with no fish. [4]

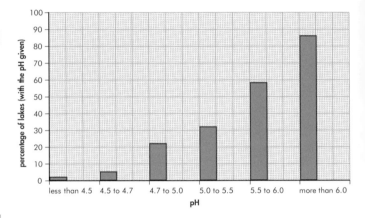

ii) Describe the effect of pH on the numbers of fish in the lakes for no fish in the lake and a good population of fish in the lake. [2]

b) The scientists noticed that fish in the lakes with the lowest pH (very acid) appeared to have a lot of mucus around their gills. Suggest how this might affect the numbers of fish in the lake. [2]

c) The scientists linked the pH of the lake to the presence of sulfur dioxide in the air.

i) Explain how sulfur dioxide could affect the pH of the water in the lake. [2]

ii) Suggest **one** possible source of the sulfur dioxide in the air. [1]

d) Suggest why the scientists included 1500 lakes in their study. [2]

[Total = 13]

EXTEND AND CHALLENGE

1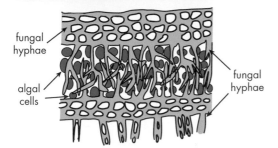

These brightly coloured crusts are lichens, growing on the bitterly cold rocks found on the Antarctic peninsular. They are an example of one of the toughest organisms found on Earth. They survive the intense cold, dryness and long winter nights in this region.

Lichens are living organisms, composed of a mixture of fungal and algal cells. The fungus forms a tough mat on the rocks, which forms a space inside for the resident algal cells. The algae make food by photosynthesis and they share some of this with the fungus. Together the two organisms in the lichen are able to survive and grow in this inhospitable place, where nothing else can survive. The growth rate of a lichen is very slow compared to that of green plants – a colony may grow only a few millimetres in a year.

a) Two ecological terms used for a mutually supportive relationship between two species are 'mutualistic' or 'symbiotic'.
  i) Explain the roles of each of the two partners in the lichen.
  ii) Find out about **three** other examples of mutualism (symbiosis). For each example, suggest what the benefits are to each organism in the relationship. (You can find out about at least one example in this book.)

b) List environmental factors that would affect growth in Antarctic regions. Explain why the growth rate of lichens here is so slow, compared with green plants growing in other regions of the world.

c) Herds of caribou (reindeer) are found in northern tundra, for example in regions of Canada, Greenland and Norway. Lichens growing in the tundra form a substantial part of the diet for caribou herds. The caribou dig through the snow and scrape the lichen off the rocks underneath. The people who herd the caribou are nomadic, moving with the seasons rather than living in one place. Why do they have to move in this way?

2 In the 21st century, a lot of publicity is given to the effects of deforestation, particularly the loss of tropical forests. But global change is not new. Some of the effects of deforestation were recognised by Plato, who wrote this passage 2500 years ago. He was writing about Attica, in Greece.

'There are mountains in Attica which can now keep nothing but bees, but which were clothed not very long ago, with fine trees producing timber suitable for roofing the largest buildings … There were also many lofty trees, while the country produced boundless pastures for cattle. The annual supply of rainfall was not lost as it is at present, through being allowed to flow over the denuded surface to the sea, but was received by the country … where she stored it … and so was able to discharge the drainage of the heights into the hollows in the form of springs and rivers with an abundant volume and a wide territorial distribution.'

Look at the information in the passage and discuss how far Plato's words are relevant today. Use the following questions to guide you. You may need to refer to other sections in the book and use other sources of information.

a) List some uses of timber that is harvested from forests.

b) List reasons why forests are cut down and how the land is used after deforestation.

c) What were Plato's concerns about the rainfall and what happens to it? Describe the events that occur with respect to rainfall when forests are cut down in the 21st century and the problems that may result. How far are these similar to Plato's view of the situation?

d) Describe how deforestation can lead to poor quality soil.

e) In this passage, Plato makes no mention of changes in the balance of gases in the atmosphere. Explain, to a non-scientist, how deforestation may contribute to global warming and climate change.

f) Suggest how deforestation might affect biodiversity. Find out information about tropical rainforests, but also consider forests in temperate regions.

g) Draw up a list of recommendations that could help reverse the effects of deforestation. In your discussion, include measures that can be carried out by individuals as well as by governments and on a wider international scale.

7

Use of biological resources

TO THINK ABOUT . . .

Make a list of ways that human populations use biological resources (plants, animals and microorganisms) for food. Include humans from the past as well as from the modern world.

Food – past, present and future

For most of the very early history of humans (from about 200 000 years ago) the total human population of the world is estimated to have been relatively small. At one stage it dwindled to a band of a few hundred. But after the development of agriculture (about 10 000 years ago), the population began to swell until by 2012 it was estimated that there were 7 billion people in the world. One UN estimate predicts that, by 2050, the population will rise to 9 billion people.

Probably just about all the useful agricultural lands and seas of the world are already in use, feeding the estimated 7.4 billion people alive today. If the population continues to rise, to ensure that sufficient food will be available for the extra population expected in the future, humans need to improve the efficiency (productivity) of the soil and the seas.

Scientists have a crucial role in finding better ways of using plants and animals to produce more food. In Section 7 of the book, you learn about some of the older traditional practices and newer ways which humans are adopting in their attempt to ensure that there will be food, and enough of it, for all people in the future.

How do you think future generations will manage to produce enough food for increasing populations of people and what do you think the priorities should be?

7.1 Using crop plants to produce food

Pushing the boundaries for growing crops

The photographs show polytunnels in two different but extreme conditions. The first is in Almeria in Spain, an area where conditions can be hot and dry, almost desert-like. As far as you can see across the plain, the landscape has been taken over by a mass of polythene tunnels for growing crops (mainly for salads). The second is on a snowy day in Eastern England on a small organic farm. Winter salad crops are growing inside.

A polythene covering may not seem very substantial, but these polytunnels each provide a protected environment that allows crops to grow inside them, even in these extreme conditions. So how can this thin transparent material make enough difference to allow crops to be grown inside? What else can growers do to increase the yield of their crops?

■ Why people grow crops

Crops are grown mainly to provide food – food for people and food for animals. Other kinds of crops are also grown – some to provide energy (including trees and some substantial grasses), or flowers or materials (such as cotton or hemp). Growing crops is a commercial activity, intended to provide products beyond the requirements of a family or small community. The crop grower expects to make a profit.

Growers want to make more money and, as human populations increase, there is a need for more food to feed people. So over the years (and over the centuries since agriculture began), people have looked at ways of increasing the yield from the crops they grow to achieve bigger, better harvests.

If plants are to grow successfully, they require **water** and **mineral ions** from the soil, **carbon dioxide** and **light** (and water) for photosynthesis, and a temperature that is warm enough for enzyme activity. Growers are able to manipulate these factors in different ways to increase the yield of their crops. Another way to improve yields at the time of harvest is to reduce losses of the crop. These losses can result from competition by pests and weeds, and from disease.

STUDY TIP

Check in section 2.1 and make a list of the basic things that plants need to grow.

◼ Glasshouses and polytunnels

Glasshouses and polytunnels provide protection for crops and an enclosed environment in which the grower can, to some extent, control the climate inside. Use of plastic polytunnels, in particular, has increased considerably in recent years. They are used on a small scale – in gardens, family vegetable plots (such as allotments) and small farms – and now they are also used in relatively large areas by commercial growers of fruit (e.g. strawberries and tomatoes), vegetables (e.g. peppers, cucumbers, aubergines and lettuces) and flowers.

The protection allows the grower to achieve increased yields. Another benefit is that the protection enables early and out-of-season crops to be grown at a time when market prices are higher. The photographs on page 283 emphasise how a suitable environment can be created and maintained inside a polytunnel, in areas that would otherwise be unsuitable for growing crops.

Protected environment and controlled climate

The features that contribute to improved growth are listed below. They are similar for glasshouses and polytunnels, but attention is drawn to some differences, where relevant.

- **Light** – the transparent cover (glass or polythene) allows **sunlight** to pass through, giving some benefit from the greenhouse effect. The light radiation warms the soil and plants, and the trapped heat energy raises the temperature inside. Artificial lighting can be used to extend the day length and, in some cases, boost the intensity of light in the polytunnel. This means that there is extra production in the plant from photosynthesis.
- **Temperature** – inside the polytunnel the **temperature** warms up quickly when the Sun comes out. Extra heat can be used at certain times if needed. The effect of the raised temperature is to increase the rate of enzyme reactions, including those in photosynthesis, hence the rate of growth of the plants. However, it is important that there is enough ventilation or shading to guard against temperatures rising too high.
- **Supply of water and control of humidity** – crops grown inside must be irrigated. This must be controlled – too little or too much (water) can reduce plant growth. Care must be taken to ensure humidity of the air is not too high as this could encourage spread of diseases, caused by fungi and bacteria.
- **Wind** – protection from wind means the temperature does not fall as much and loss of water by transpiration from the growing crops is reduced. Growing plants are also protected against damage that may result from strong winds.
- **Frost** – protection from frost may be important in some areas, early and late in the growing season. Ice crystals produced in frozen cells make them burst. Polytunnels can help reduce frost damage to sensitive crops (for example, cucumbers).

Figure 7.1 Salad crops growing inside a polytunnel in Eastern England with snow outside.

STUDY TIP

Look in Section 2.1 to remind yourself how light intensity affects the rate of photosynthesis.

STUDY TIP

Look in Section 1.4 to remind yourself about how temperature affects enzyme activities. How is this important for growth of crops in glasshouses or polytunnels?

STUDY TIP

Name the process by which plants lose water. What factors affect the rate of water loss? Check in Section 3.4 to make a link with the conditions that would affect plants growing in glasshouses or polytunnels.

STUDY TIP
Make a link between the conditions in a polytunnel at different times of day and how carbon dioxide concentration affects the rate of photosynthesis (Section 2.1). You can also look in Section 3.1 to make sure you understand about gas exchange in the leaf of a plant at different times of day.

- **Carbon dioxide** – the level of carbon dioxide inside the glasshouse or polytunnel needs to be monitored carefully. Carbon dioxide is used by plants in photosynthesis and a shortage of carbon dioxide would limit the rate of photosynthesis. On a warm sunny day, actively growing plants use more carbon dioxide than they give out, so the level of carbon dioxide in the polytunnel can drop. This would reduce the rate of photosynthesis of the plants. Adequate ventilation to help replace the carbon dioxide from the air outside is important to prevent this reduction. Carbon dioxide can also be supplied artificially to restore the normal levels. In addition, many growers supply extra carbon dioxide, above the level in the atmosphere outside, to encourage increased growth of the crop plants. This is more effective in a glasshouse than in a polytunnel.

It is important that conditions inside are carefully monitored, particularly the temperature, humidity and level of moisture in the soil. Often this is done automatically so that adjustments can be made as required (Figure 7.2).

Figure 7.2 A control panel used to monitor conditions inside a commercial greenhouse.

PRACTICAL

Using half a polythene bottle in this way gives a simple model that shows the benefits of a polytunnel and demonstrates the greenhouse effect. This is a very simple activity, but one that helps support your understanding of both these topics.

■ Practical activity – using a polythene bottle as a model polytunnel

1 Cut a plastic drink bottle into two equal halves.
2 Use one half of the bottle as a polytunnel and lay it on some soil.
3 Place one thermometer underneath the bottle and one on the soil outside the bottle.
4 Read the temperature from both thermometers at regular intervals.

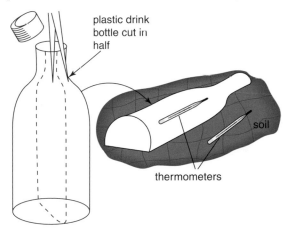

Figure 7.3 Using a polythene bottle as a model polytunnel and demonstration of the greenhouse effect.

■ Use of fertiliser to increase crop yield

Growing plants require a range of mineral ions. These are taken into the plant by the roots from the soil. Mineral ions need to be replaced if crop plants are grown in the same field year after year. Replacement of mineral ions is usually through supplies of **fertiliser**, which can be in the form of chemical fertiliser or 'organic' fertiliser.

STUDY TIP

Check in Section 6.4 and draw a flow chart to show what would happen if excess fertiliser is washed out from a field into a nearby river.

STUDY TIP

Refer to Section 2.1 to find out why nitrate and magnesium ions are important for the growth of crop plants. You should also be able to work out why phosphate ions are important.

PRACTICAL

You can use this simple equipment to investigate the effect of chemical fertilisers on the growth of grass seedlings. This activity helps you understand the use of fertilisers on a larger scale by farmers in the field with their crops.

MATHS TIP

Sampling and scientific data

A single scientific measurement may be representative of the real situation, or it might be an anomaly. To avoid a wrong conclusion based on few anomalous readings, it is best to take replicate measurements, and calculate an average value for each batch of data.

Now go to page 322 to apply this to some data collected in a growth experiment.

Use of fertilisers to replace mineral ions taken from the soil by crops is probably as old as agriculture. **Farmyard manure** and **compost** are examples of organic fertilisers used by farmers and growers. These materials replace mineral ions used by the crop plants in order to maintain fertility of the soil. **Chemical fertilisers** usually contain nitrate, phosphate and potassium ions (known as NPK). Other minerals are also included in chemical fertilisers used when growing crops. These chemical fertilisers are often applied to the soil as dry granules but the chemicals can also be sprayed on in liquid form.

■ Practical activity – investigating the effect of chemical fertiliser on the growth of grass seedlings

Many chemical fertilisers contain nitrate, phosphate and potassium ions and are known as 'NPK' fertilisers. By using fertiliser supplied as small pellets, this simple investigation gives a good indication of the effect of adding different numbers of pellets (different 'quantities' of fertiliser). The NPK pellets release their nutrients slowly, as the seedlings grow.

1   Use a seed tray with chambers, or similar small containers with holes in the base. Add materials to each chamber as shown in Figure 7.4.
2   Vary the number of NPK pellets in multiples of 5.
3   Stand the chambers in a tray with a shallow layer of distilled water.
4   Leave the tray in the light for 1 to 2 weeks, watering as necessary with distilled water.
5   Measure the average height of the tallest leaves of the grass seedlings.

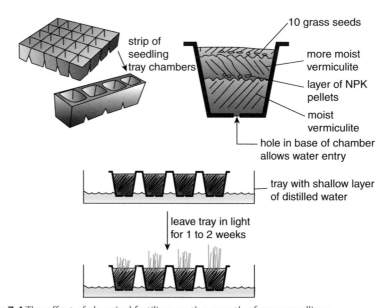

Figure 7.4 The effect of chemical fertiliser on the growth of grass seedlings.

■ Pest control – pesticides and biological control

An important approach to increasing the overall yield from crops is to reduce losses from attack by pests or competition from other plants.

Here are some examples of pests on plants commonly grown in glasshouses or polytunnels:

- large white (cabbage white) caterpillars eat cabbage leaves
- slugs and snails eat many plants (especially young seedlings)
- aphids cause direct damage by piercing the phloem and sucking out sap (e.g. on cucumbers and tomatoes).

Damage caused by these, and other pests, means poorer growth by the plants, resulting in reduced yield. Figures 7.5 to 7.7 show examples of damage causing loss of yield or crops becoming unsuitable for market.

Figure 7.5 Chinese mustard leaves with holes from beetle damage.

Figure 7.6 Seedlings of Chinese mustard with damage from leaf beetles. The beetles are only about 2 mm in length yet cause severe damage to the crop.

Figure 7.7 Damage to a courgette flower with snail visible inside. The flower is unlikely to be pollinated and develop into a courgette 'fruit'.

Chemical pesticides can be used to help reduce pests and, increasingly, different methods of **biological control** are used.

When **chemical pesticides** are used, the chemical substances have been chosen because they are toxic and kill the pest organism. Chemical pesticides are usually sprayed onto the crop, sometimes by hand from a small tank, but in other cases by machinery pulled by a tractor or from a small aircraft flying closely above the crop in a field.

Often the chemicals in the pesticide kill a wide range of organisms (such as insects or small mammals), as well as killing the pest. Sometimes the pesticide lasts just a short time but others are taken into the plant and have their effect for a longer time.

Biological control exploits the relationship between the pest and another living organism. The pest organism is the 'prey' and the control organism is the 'predator'. By introducing suitable numbers of the predator (control organism), it is possible to reduce the numbers of the pest. Sometimes the control organism is a parasite, which weakens or causes disease in the pest.

Here are some examples of biological control agents used to control insect pests on crop plants grown in glasshouses and polytunnels. Crops affected by these pests include cabbages (for the caterpillars) and peppers, tomatoes and cucumbers (for aphids, whitefly and red spider mite).

Table 7.1 Some examples of biological control agents used with different pests.

Biological control agent	Pest	Comments
Bacillus thuringiensis (a bacterium)	caterpillars (larvae) of some butterflies	the bacterium produces toxins that kill the caterpillars
Encarsia formosa (an insect)	whitefly	*Encarsia* is a very small wasp. The adult wasp lays its eggs in and feeds on the whitefly which kills it
ladybird, for example, *Adalia punctata* (an insect)	aphids	both the larvae and the adults are voracious feeders on aphids – ladybird larvae seek the aphids as soon as they have hatched

■ Advantages and disadvantages of chemical and biological control methods

To understand the advantages and disadvantages of each of these approaches, it helps if you know a bit more about each method.

Chemical control (with pesticides)

- Chemical pesticides work quickly and destroy the pest.
- Chemical pesticides may kill or harm a wide range of organisms and not just the pest (including farm workers and other people who come into direct contact with the pesticide).
- Sometimes the chemical substances in the pesticide break down quickly so that spraying must be repeated to continue control of the pest.
- Some chemical pesticides do not break down quickly and so persist (as residues in the soil or in the organisms). This means the chemical substances may be passed along the food chain to other organisms. Sometimes they accumulate so that levels of the pesticide become toxic to larger animals in the food web (including humans).
- The toxic effects of the chemical pesticide may be lost if the pest organisms evolve resistance to the toxic chemical.
- The chemical pesticides may be washed out of (leached from) the soil into water courses (such as streams and rivers), harming organisms in the water.

STUDY TIP

Check in Section 5.6 and describe how natural selection might lead to pests becoming resistant to the toxic chemicals in pesticides.

Biological control (with predators)

- Biological control organisms are often quite specific in that they target only one (or just a few) species (the pest).
- It is important to introduce the predator (the biological control organism) at the right time so that enough of the pest organisms are available for the predator to feed on.
- It takes time for the predators to become established and reduce the pest population.
- If the predators are too successful in eating the prey, there will be nothing left for the predator to feed on – the predator has removed its own food supply and might switch to another (non-target) prey.
- A small population of prey (the pest) needs to be left so that the biological control continues to be effective. This probably means that the pest is never completely eliminated.

Biological control often works well in a confined space (such as glasshouses and polytunnels) but becomes less effective in fields of crops. Sometimes strips of particular plants can be grown close to crops in fields and these plants encourage natural predators by providing food plants. Organic growers have strict rules about using chemicals (or not using them) so biological control is often the only option to controlling pests on their crops.

For both the biological and the chemical approaches, commercial growers need to consider the cost of application (of the chemical or the biological control agent), the frequency of application, how effective it is in controlling the pest and the wider knock-on effects of its use.

STUDY QUESTIONS

1 a) Make a comparison of the advantages and disadvantages for a grower of controlling pests using chemical pesticides or using biological control. Consider a range of examples of crops and their pests. Include crops grown in fields as well as those grown in polytunnels or glasshouses.

   You can discuss this with a group of students and aim to present your response in a table. Remember to explain why you may wish to control the pests.

   b) When you have done this, decide what advice you would offer to a grower in each of the following situations and justify your advice. You need to consider both approaches to pest control and think about the timescale.

   (i) A fruit farmer notices an outbreak of caterpillars damaging leaves of some fruit trees just as the fruit is forming. What should the fruit farmer do?

   (ii) A grower of flowers in polytunnels hears that infestations of aphids are being predicted later in the season. What should the flower grower do?

2 Make a list of the benefits of using polytunnels to grow crops. For each item in your list, give biological reasons to help explain why there is a benefit.

7.2 Using microorganisms to produce food

Microorganisms and some useful fermentations

Too often stories about microorganisms tell of disease and decay . . . but why not look at some of the good stories about the microbial fermentations that produce foods and drinks?

Fermentations with microorganisms have been part of the human way of life from the very beginnings of civilisation. The first yoghurt may have come from Central Asia where nomadic people carried milk from their animals on their journeys. It was warm and the milk soon turned sour, developed different but quite pleasant flavours and became thicker. They had made yoghurt and this could be kept for much longer than fresh milk.

Today many people are familiar with yoghurt from cow's milk, but across the world milk from buffalo, camel, goat, sheep and other mammals is fermented into different foods and drinks. The first photograph shows 'kumiss' – fermented mare's milk – being poured into a bowl inside a nomadic tent in Mongolia; the second shows dadih, and both these traditional yoghurts are a big jump from the modern industrial production of yoghurt (below).

Here are some other traditional foods: 'chal', 'dadih' (shown in the second photograph), 'sho' and 'shubat'. Where do they come from? What is the basis of yoghurt-based drinks such as 'dhalla', 'lassi' and 'yoghurt smoothies'? Can you add any more to the list?

■ Introducing fermentations

We often use the term **fermentation** for processes carried out by microorganisms. This term is sometimes used as equivalent to anaerobic respiration, but it is also used in a wider sense, and may include reactions that involve aerobic respiration. You should understand that a number of different processes carried out by microorganisms (including bacteria and fungi) are described as fermentations, and these may be aerobic or anaerobic. The photograph on the left shows modern industrial fermentation to produce yoghurt.

> **STUDY TIP**
> Check details about bacteria and fungi and their features, and where they fit in the variety of organisms (see Section 1.2).

STUDY TIP

Link this with details about anaerobic respiration in Section 2.3.

STUDY TIP

Look in Sections 1.4 and 2.2 to check the name of the enzyme that converts starch to maltose and maltose to glucose.

PRACTICAL

An experiment to investigate carbon dioxide production by yeast in different conditions is given in Section 2.3 (Respiration).
This experiment helps you understand how alcohol is produced in anaerobic respiration and the importance of yeast in the production of bread.

■ Production of bread using yeast

In Section 2.3, you study respiration and learn about **anaerobic** respiration. Yeast is a single-celled fungus. It can carry out both aerobic and anaerobic respiration. When yeast carries out anaerobic respiration, it produces an alcohol (ethanol). Here is the equation for anaerobic respiration:

$$\text{glucose} \xrightarrow{\text{yeast}} \text{carbon dioxide} + \text{alcohol (ethanol)} + \text{energy}$$
$$C_6H_{12}O_6 \xrightarrow{\text{yeast}} 2CO_2 + 2C_2H_5OH + \text{energy}$$

Making of bread is an example of a fermentation using yeast. The art of making bread was known more than 2500 years ago and was certainly part of the daily life of early Egyptian, Greek and Roman communities. As long as 6000 years ago, a simple flat bread was made by crushing cereal grains to make a coarse flour. This was mixed with water to form a paste, which was then dried in the Sun to make the bread. But when 'something' from the air got into the mixture, the paste expanded and the product was easier to chew. We now recognise the 'something' as being wild yeast and the lighter paste was the beginning of dough, the basis of what has become a wide variety of breads.

Modern breadmaking uses mainly wheat flour, though other flours, such as rye and barley, are also used. The flour is mixed with water and yeast, plus some sugar and salt and a little fat or oil. The mixing process is described as **kneading** and thoroughly works the ingredients together. Traditionally, kneading was done by hand but now, on a commercial scale (and increasingly in the home), the kneading is done using machines.

Sometimes the yeast is first added to water with a little sugar and kept warm (at about 25 °C). The yeast starts to respire the sugars in the water and bubbles of gas form a froth on the surface. This is carbon dioxide released as a waste product of yeast respiration.

The frothing liquid is added to the flour mixture and, as it is kneaded with the flour, the mixture becomes more elastic and a dough is formed. Mixing ensures all the yeast cells are evenly spread throughout the dough. Its elasticity allows for spaces to develop inside the dough, resulting from the carbon dioxide given off by anaerobic respiration of the trapped yeast cells (there is little or no oxygen in the dough), as seen in Figure 7.8. Further sugars are released from enzymatic breakdown of starch in the bread flour, allowing the yeast to continue to respire anaerobically. Some modern baking yeasts can be added as a dry powder to the flour and other ingredients, to which warm water is then added, to make the dough.

Rising of the dough (known as **proving**) is due to pockets of carbon dioxide released by the respiring yeast cells. Proving generally takes place over a few hours, kept at a warm temperature (say about 25 °C). During this time, the dough increases to a volume four or five times that at the start. The risen dough is then cut into pieces (often to fit in tins) before being baked. **Baking** is done in a hot oven, usually at about 230 °C, for about 15 minutes, though times vary with different types of flour

During the baking, any ethanol – formed as a waste product of anaerobic respiration – remaining in the dough is evaporated in the heat. The yeast is killed by the high temperature so there is no further respiration. Once cooled, the bread is ready to be eaten.

Figure 7.8 The surface of a toasted slice of white bread, showing the spaces formed by trapped carbon dioxide inside the dough before baking.

■ Practical activity – investigating the effects of temperature on the proving of dough

Increases in the volume of dough during the proving stage of making bread are due to pockets of carbon dioxide, released as a waste product of anaerobic respiration in the yeast.

1   Add water to a mixture of flour, sugar and yeast in a mixing bowl and work with your hands to make it into a single elastic lump of dough.

2   Roll the lump of dough into a long sausage shape of even diameter.

3   Measure the length of the sausage shape and divide it into portions of equal length (one for each incubation temperature).

4   Label a series of transparent containers (such as plastic drinking cups) with the intended incubation temperature.

5   Force one portion of dough into each container and mark the maximum height of the dough on the wall of the container.

6   Place each container in a different temperature (for example, fridge, lab bench, incubator, warm oven) and leave for at least 1 hour.

7   Mark the new height of the dough in each container and calculate the increase in height for each. This gives a measure of the yeast activity at each temperature.

This method could be adapted to investigating the importance of individual ingredients in the dough mixture (for example, the mass of yeast), keeping the temperature of incubation the same.

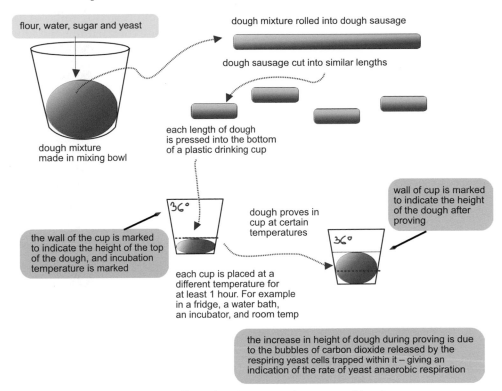

Figure 7.9 Investigating the effects of temperature on the proving of dough.

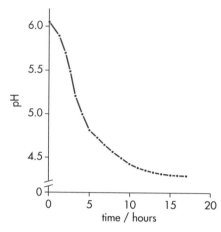

Figure 7.10 A graph showing the pH changes when UHT milk is incubated with yoghurt starter at 37 °C over a period of hours.

MATHS TIP

Determining the slope of a graph
The slope (gradient) of a straight line on a graph shows how the magnitude of the *y*-variable changes in relation to the *x*-variable.

Now go to page 322 to apply this to the graph linked to the production of yoghurt.

PRACTICAL

This investigation using resazurin dye is a useful way of comparing numbers of microorganisms in samples of milk. It is a helpful activity to support your understanding of growth of populations of microorganisms.

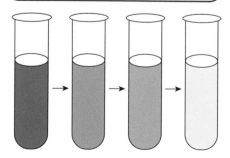

Figure 7.11 Investigating the effect of temperature on the growth of microorganisms in milk, using resazurin dye.

■ Production of yoghurt

Making yoghurt from milk is another example of a fermentation that uses microorganisms to make a food product. This is a traditional process that was used as a way of preserving milk. In modern times we recognise that it is a way of extending the shelf life of a food. It is pleasant to eat and some people like it because it is more digestible than the original milk.

For production of yoghurt, a bacterium called *Lactobacillus bulgaricus* is involved. *Lactobacillus* bacteria can use **lactose** (a sugar found in milk), and convert this to **lactic acid**. During the fermentation process, a bacterial starter culture, including *Lactobacillus*, is added to the milk. The milk is then kept at a temperature between 37 °C and 44 °C for about 3 to 6 hours.

As the lactic acid is produced, the pH falls (becomes more acid). The acidity makes the protein in the milk coagulate (go solid). This gives the yoghurt thick or almost solid texture. The acidity also helps prevent growth of other microorganisms that may be harmful, so acts as a preservative. This means the dairy products can be kept for a longer time.

When making yoghurt in the home (or laboratory), the milk is first heated to a temperature just below boiling point. This sterilises the milk, killing other bacteria that might be harmful or undesirable in the milk during the fermentation process. The milk is then cooled to between 40 °C and 46 °C before adding some yoghurt from a previous batch of 'live' yoghurt. This contains *Lactobacillus* bacteria and acts as the starter culture.

Fruit or other flavours can be added to the yoghurt, usually towards the end of the process. Different milks can be used, including skimmed milk and dried milk or milk from different mammals.

Do not eat any yoghurt made in a science laboratory.

■ Practical activity — investigating the effect of temperature on the growth of microorganisms (using resazurin dye)

When a population of microorganisms grows, the number of microorganisms increases. The number of bacteria in milk can be measured using a blue dye called resazurin.

When the bacteria respire they release hydrogen (H⁺) ions. The H⁺ ions are taken up by molecules of the dye, causing it to change colour. The dye starts a blue colour, then changes through mauve to pink and finally to white. This colour change sequence indicates increasing numbers of the bacteria. The colour changes are shown in Figure 7.11.

1 Place 10 cm³ of milk in a test tube and add 1 cm³ of resazurin dye. With the dye, the milk is a pale blue colour.
2 Place the test tube in a water bath at 50°C and watch as the colour changes from blue, through mauve to pink and finally colourless. You can time how long it takes for the milk to reach a certain colour (say white or pink).
3 Set up a series of similar tubes but place them in water baths at different temperatures (e.g. 15 °C, 20 °C, 25 °C and 45 °C). *Note – avoid temperatures between 35 °C and 40 °C because of the risk of incubating human pathogens.*

4 Record the time for the milk to reach the same colour (for example, pink or white) at each temperature. This gives a measure of the number of microorganisms in the milk sample. A faster time indicates a higher number of microorganisms.

This procedure can be adapted to compare the number of microorganisms in different milk samples, but kept at the same temperature.

■ Industrial fermenters

A fermenter is any container used for fermentations. In the laboratory, a flask or even a boiling tube can be used as a small-scale fermenter, whereas for industrial fermentations, vessels of several thousand litres may be used. Figure 7.12 shows an industrial fermenter.

The microorganisms grown in the fermenter require nutrients and certain conditions are necessary for successful growth. During the fermentation, the population of microorganisms increases inside the container. At some stage, the products of the fermentation are harvested – these may be substances produced as a result of the metabolism of the microorganism or the microbes themselves may be harvested. Sometimes products are removed during the fermentation. In other cases, all the products are harvested at the end.

Figure 7.13 shows the essential features of an industrial fermenter. Think about what microorganisms need for growth, and you will understand the

Figure 7.12 The stainless steel containers that are used in the industrial fermentation of beer can be effectively sterilised between batches.

MATHS TIP

Substitute numbers into equations using appropriate units

Useful scientific theories often generate equations which can be used to understand biological processes. When new measurements are put into an equation, the results make biological sense, supporting the original theory.

Now go to page 322 to apply this to a calculation about population growth of microorganisms in an industrial fermenter.

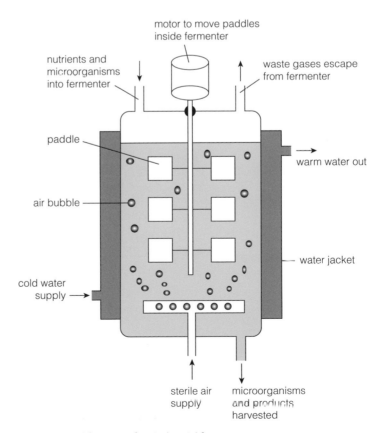

Figure 7.13 Essential features of an industrial fermenter.

need for the features shown on the diagram. In addition to the comments shown on the diagram, note the following:

- The **sterile air** supply ensures conditions are aseptic and prevents contamination. This means that the only microorganisms growing inside the fermenter are the ones being used in the fermentation.
- **Oxygen** is provided in the air if the fermentation is aerobic, but **NOT** if it is **anaerobic**.
- A water jacket surrounds the fermenter to help control the temperature. Respiration (and other metabolic reactions) gives off heat so some sort of cooling is needed to prevent the reaction mixture getting too hot and to keep it at the optimum temperature for enzyme activity.
- Usually the pH is monitored and adjusted if necessary – this is not shown in the diagram.
- The paddles help to agitate (or stir) the contents to ensure uniform conditions and distribution of nutrients (and oxygen if required) throughout the container.

STUDY QUESTION

1 Making bread and producing yoghurt are both fermentations that produce food.

Compare these two process, using these questions to help you. Summarise your answers in a table.

For each process:

a) Name the microorganism(s) involved.

b) List the starting material(s) and name the biological molecule(s) of importance for the fermentation in these materials.

c) Describe the reactions that take place during each fermentation. (It is helpful to write an equation, in words.)

Name any enzymes involved in changing the large molecules in the original material into a molecule the microorganism uses in the fermentation.

d) Give an approximate temperature for the fermentation and the time it takes to be completed.

e) What happens to stop the fermentation? (Hint: this may, for example, be something chemical or a change in temperature or pH.)

f) Describe how the end product is different from the starting materials.

g) List any benefits that result from producing these two foods.

7.3 Producing food in fish farms

Wild fish or farmed fish?

Over the 60 years from 1950 to 2010, global fisheries data show a 120 times increase in production of fish in freshwater fish farms. The increase is even greater for marine fish. Here, nearly 650 times more fish was produced in fish farms in 2010 compared with 1950.

We can also make a comparison between catches of wild fish and production of fish in fish farms. Over the same 60 years, production of farmed fish in fresh water has increased so that it is now three and a half times more than catches of wild fish. However, for marine fish production of farmed fish still lags behind and the total of marine fish caught in the wild is still much higher.

The photographs show a traditional fishing boat for small-scale fishing and a nearby fish farm, close to the coast.

What has led to these considerable changes? Is it a reflection of increased world population or is it because more people now eat fish and eat more of it? Is wild fishing sustainable and can natural wild stocks last to feed the global population as it increases? Is fish farming as an industry able to fulfil the needs of global populations in the future?

▨ Fish farms

Farmed fish can make an important contribution to the provision of protein in a human diet and also as feed for other animals. Fish farming can be done on quite a small scale, with low inputs in terms of cost and management, or on a relatively large scale.

Examples of farmed fish include salmon, tilapia and trout. Other aquatic animals, particularly prawns and shrimps, are also farmed on a large scale, under the wider term 'aquaculture'.

Fish farms can be in fresh water or in sea water (salt water), depending on the fish being reared. Usually the farmed fish are kept in ponds, large tanks or in mesh cages anchored in the water.

Figure 7.14 Pond at a freshwater trout farm (Southern England), showing circulation of water, which helps with aeration of the water, and a net over the pond to protect against predatory birds.

Figure 7.15 Tanks for growing fish in a fish farm in China.

Figure 7.16 Aerator in use at a fish farm.

■ Maintenance of water quality

On a fish farm, a lot of fish are kept in a relatively small space. It is important that the quality of water is maintained to ensure the fish obtain materials they need for growth and to remain healthy.

Oxygen

Oxygen in the air dissolves in water. Fish use oxygen in respiration and they use the oxygen in the water. If the fish are very crowded, there may not be enough oxygen for all the fish in the container (pond, tank or cage). There are different ways of helping to keep the water oxygenated on a fish farm.

In **ponds** or **tanks** the water may be:
- agitated mechanically so that air mixes with the water and more oxygen dissolves in it
- recirculated, making use of a pump
- oxygenated artificially by bubbling air or oxygen through it.

For **cages in more open water** (including the sea, estuaries, rivers and some lakes), the natural movement of water helps to bring fresh oxygen to the cages. This movement of water may be the result of tides or from currents in flowing water.

Plants in the water, including rooted plants and free-swimming microscopic algae, give off oxygen during photosynthesis and this increases the oxygen in the water.

Waste materials

Waste materials can accumulate in the water. These come from the nitrogenous excretory products of the fish (mainly ammonia) and also from excess food materials left in the water if it is not consumed by the fish. Ammonia (and ammonium ions) is toxic to fish and high levels can be dangerous.

Another reason why it is important to prevent the levels of nitrogenous waste and other pollutants becoming too high is that **eutrophication** may occur. This could lead to serious losses of fish through lack of oxygen. Fish may appear normal and active then, in a short time, suddenly die if oxygen levels fall too low. This mass mortality of fish is known as 'fish kill'.

STUDY TIP
Make a link with the excretion of nitrogenous materials in humans, described in Section 4.2.

STUDY TIP
Check details about water pollution and eutrophication in Section 6.4.

Removal of waste materials can be done by circulation of the water, either artificially (with a pump) or by natural flow of water (as a result of tides or currents). This is similar to the methods described for maintaining oxygen levels (see above). Circulation of water in this way helps to prevent waste materials from accumulating in the water surrounding the growing fish.

■ Competition and disease

Interspecific predation

Sometimes nets are used to cover an open pond to prevent large predatory birds from catching the fish. This is an example of interspecific predation (one species eating another species).

Intraspecific predation

If the fish are very crowded in the water, they may try to eat each other. This is an example of intraspecific predation (one individual eating another of the same species). The best way to prevent this is to make sure the number of fish in the water (the density of the fish) is kept fairly low and to provide adequate food for the fish.

Disease

Diseases and other infections do occur and it is important to maintain a healthy stock to ensure there is a good yield of fish. Sometimes diseases are controlled by adding chemicals to the water or a fish farmer may use drugs (such as antibiotics) in the feed for the fish. Low density of fish in the water helps reduce the chances of disease spreading amongst the fish.

■ Feeding – quality and frequency

Diet

In the fish diet, there is a requirement for high protein, but also a need for some fat and carbohydrate (for energy), and for a range of vitamins and mineral ions. Generally, higher proportions of protein are required during the younger stages when the fish are growing rapidly.

Feed

Sources of protein for the feed may be derived from smaller fish and sometimes waste from fish-processing industries (for example, filleting of fish) is used. The usual way of delivering fish feed is in the form of pellets, made from remains of fish or other food materials.

Certain fish are more likely to be sold if the flesh is pink in colour (for example, trout and salmon), so pigments may be added to the feed. Sometimes enough colour may be obtained from certain microscopic animals found naturally in the water.

Feeding

Feed can be delivered automatically into the ponds or cages, though often the fish farmer does some feeding by hand. This also gives the fish farmer a daily opportunity to check the fish, ensuring they are healthy and that there are no other problems.

STUDY TIP

There is concern about the widespread use of antibiotics in agriculture practices. Look in Section 5.6 (Resistance of bacteria to antibiotics) to remind yourself about the development of, and danger to public health, of resistance to antibiotics.

Figure 7.17 Feeding fish in a marine fish farm in Scotland, UK.

Figure 7.18 A computerised feeding system sprays feed out of a storage bin and shuts off automatically when infrared sensors below detect particles falling to the bottom of the pen.

The quantity of feed given and the frequency of feeding may need to be adjusted according to the size of the fish and their activity. Young fish are given food more frequently than the more mature fish.

Enough food needs to be available for the fish to enable growth to occur at a suitable rate. However, if excess feed is delivered, it is wasteful, leading to possible pollution problems and unnecessary cost.

■ Selective breeding and improving the quality of fish populations

An important way of improving the yield of fish from a fish farm is to select superior individuals and to use these for breeding to provide the next generation of fish. Another way is to obtain good stock from those who supply the eggs or young fish for growing on in the fish farm.

There are also ways of controlling the sexes of fish stock and so increasing the yield when the fish are harvested. Older male fish tend to become aggressive and, after reaching sexual maturity, the quality of the flesh usually declines. It is possible to have fish stocks that are all females or that are sterile. This means that populations with fish that are all female or sterile can give higher yields than from normal mixed sex populations. Artificial methods of changing the make-up of the fish populations in this way are increasingly being used to improve the yield of fish harvested from fish farms.

STUDY QUESTIONS

1 a) What can a fish farmer do to maximise the growth of the fish and increase the yield when harvested? In your answer, refer to each of the following:
   ■ the diet of the fish
   ■ feeding methods
   ■ water quality
   ■ competition
   ■ control of disease
   ■ having a good-quality stock of fish.
   Make sure you emphasise the practical aspects of what the fish farmer does in each case.
   b) Explain how fish farming may have harmful effects on the environment. What can a fish farmer do to minimise these harmful effects?

2 Describe how eutrophication could lead to 'fish kill' (sudden death of the fish) in a fish farm.

7.4 Selective breeding

Domesticated cattle and how they have changed

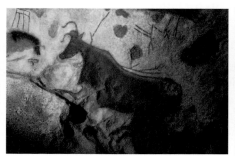  

Cattle have been associated with human communities from early Neolithic times. There is evidence for domesticated cattle in Asia around 8000 years ago and of production of milk from cattle in Egypt around 6000 years ago. Early cave paintings (shown from a location in France) give us an idea of what cattle looked like around 20 000 years ago.

Modern breeds of cattle have clearly changed from their ancestors, but they also differ from each other. The Prespa cow (middle photograph) and the Highland cattle illustrate a tiny selection from the range of modern breeds of cattle.

This brief overview prompts us to ask the questions: 'Why did the populations change?' . . . 'How did the changes take place?' and 'How do people use domesticated cattle today?'

Figure 7.19 A Charolais bull with massive muscular body.

■ Variation in domestic breeds of cattle

A small herd of Prespa cattle lives by the shores of Lake Prespa, on the border between Greece and Albania. The Prespa breed of cattle are few in number with probably no more than 500 animals remaining in the herd and the cattle are small in size – about 1 m in height (or waist high for a normal adult person). They have small horns and a Prespa cow produces between 1 and 4 litres of milk a day at limited times during the year.

By contrast, Charolais cattle (Figure 7.19) are large animals, approaching a man's shoulder height and with massive muscular bodies. A Charolais bull (male) has a body weight of around 1000 kg. The Charolais breed is popular in Europe for beef production.

Other breeds of cattle are favoured for milk production. For example, British Holstein cows can produce 36 litres of milk a day or a yield of over 10 000 litres per year. The Jersey breed (Figure 7.20) produces less (around 4000 litres) but it is valued for its high-quality milk – rich in nutrients (lipid and protein) and regarded as having superior flavour.

Figure 7.20 The large udder of a Jersey cow gives high yields of milk. Compare the Jersey cow with the Prespa cow (above) and see how different they are.

Figure 7.21 A humped form of cattle, used for ploughing – shown here in Myanmar.

STUDY TIP
Refer to Section 5.5 to remind yourself about the way characters are inherited and make sure you understand the terms phenotype and genotype.

Yet another form of cattle is shown in Figure 7.21. The prominent hump allows these cattle to be used for ploughing. We could describe more types of cattle around the world, large and small, humped and without humps, some with horns, some without. These descriptions emphasise the variation found in domesticated cattle around the world. Section 7.4 explores how, through the centuries, humans have influenced these changes and have developed breeds of cattle that supply different needs for farmers today.

■ Selective breeding – how and why?

In Section 5.5 you study how one character is inherited, or passed from one generation to the next. You see how a diploid parent produces haploid gametes but, at fertilisation, you cannot predict which allele (in the male gamete) will join with which allele (in the female gamete). It is a matter of chance. So you cannot be sure what the overall 'appearance' of the offspring will be when you look at all the characteristics together. Also, many genes are often involved in determining different characteristics seen in the phenotype, so you can see that it would take several generations to establish a change in the population.

Let's return to the story of cattle. Suppose you are a farmer and you want to produce cows that give more milk. You would probably choose cows from your herd that give a lot of milk and use these as your 'breeding stock' to produce calves for the next generation. You would also mate the cows with bulls that had previously given daughters with high milk yield. Another farmer may be more interested in beef production so would select cows and bulls that show desirable features, such as muscular body and low fat, for meat production. A third farmer may wish to use cattle for ploughing so the main feature of interest is the hump for dragging the plough. In each case, the farmer selects the 'best' individuals for a certain character and uses these for mating. Over several generations there is likely to be an improved herd of cows, for milk yield, beef production or whatever desired characteristic the farmer is aiming for.

This illustrates the basis of selective breeding. From very early times in agriculture, with plants and with animals, people selected suitable 'parents' for the next generation, then from these offspring again the 'best' were selected as parents for the next generation and so on.

Selective breeding – some more examples

A few species of grasses found in the region between the Tigris and Euphrates rivers are probably related to modern day wheat cereal crops. Over 10 000 years ago, our Neolithic ancestors began to adopt a settled lifestyle and established the principles of modern agricultural methods. This same region, in Western Asia, known as the fertile crescent, was probably one of the areas where earliest agriculture began to develop. So these grasses may be similar to the ancestors of our modern wheat crops. Through the centuries, progressive changes, generation by generation, have produced wheat crops with very different characteristics when compared with the early grass species.

The practice of selection has continued up to modern times. Different characters may be considered important for different reasons. For example, in a wheat crop, some farmers want wheat with shorter stems so that the crop is less likely to be damaged by bad weather (including wind and heavy rain) and so the crop is not difficult to harvest. But if a farmer wanted to use the straw from the wheat, it might be better to have wheat with longer stems. Similarly, with some cattle it is a priority to have more milk, but others may be kept for a high yield of meat.

We can look at other examples. In modern times, the quest for better crops, higher yields and more productive animals continues. Selective breeding stimulates a highly competitive world, for example with fancy breeds of chickens (Figure 7.22) or ever more exotic orchids.

Table 7.2 gives a few more examples of plants and animals that have changed from their original or ancestral type, as a result of selective breeding over tens, hundreds or even thousands of years. The table also gives an indication of the modern range of varieties and some of the 'desired' characteristics that have been selected.

a

b

c

Figure 7.22 Different types of chicken that have been selectively bred for their appearance. **(a)** White Silkie, **(b)** Rockin Rooster and **(c)** White Yokohama.

Table 7.2 Some features displayed in a range of modern varieties (chickens, fruits, vegetables).

Plant or animal	Features in modern range of varieties
chickens	• modern varieties have been selected for meat (heavy birds) or high number of eggs • fancy breeds have been selected for show qualities (such as colours, cockerel tail, comb)
apples	• modern varieties of apples have been selected for flavour, colour, time of ripening and harvesting, keeping qualities, cooking qualities, suitability for making cider or juice
tomatoes	• modern varieties come in many shapes, sizes and colours – they have been selected for shape, colour, flavour, keeping qualities, use in cooking or as salad
potatoes	• modern varieties come in many shapes, sizes and colours – they have been selected for shape, colour, cooking qualities (floury, waxy), storage qualities, early or late maturing

■ Timescale for change in the populations

In early agriculture, the timescale for population change was likely to have been hundreds or even thousands of years, eventually establishing the small changes that produced the crops or animal breeds that we have today. The changes were probably random initially, though the selection helped to direct the trend of characters towards the desired outcome.

Modern techniques in breeding include use of artificial insemination (AI) for domestic animals and controlled pollination for plants. In each case, the parents are chosen and the technique for crossing has been refined, rather than relying on chance mating or random cross pollination. These and other techniques have surely speeded up the process of change and focused on more specific characteristics.

Genetic manipulation techniques take the process a stage further in that more precise changes can be developed within the population.

By selective breeding over a period of time, the crop or animals may be 'improved' so that more of the individuals in the population show the desirable characteristics. This process is also described as artificial selection. The changes (in the population) are by human choice rather than by chance.

STUDY QUESTIONS

Use this as an opportunity to revise or bring together relevant topics from different parts of the book, linking them into the ideas developed in Section 7.4 on selective breeding. You can follow some of these links through the study tips given throughout this section.

1 a) How does random fertilisation produce genetic variation in offspring so that the offspring are different from each parent? (You can show this for one character and it may help you to draw a genetic diagram.)

   b) Explain how selecting parents with certain characteristics could help increase the chances of the offspring in the population having the desired characteristics.

2 a) Look at the varieties available for one fruit or vegetable in a local market or supermarket to see the range of different characteristics they offer. Which features would you select as someone about to purchase this fruit or vegetable? Give reasons for your answers.

   b) Now make a list of some characteristics a grower might want in a range of varieties to be grown in a vegetable or fruit crop. (Think about features that a grower might want, such as successful growth, free from disease or blemishes, yield, size of product, time of harvest, storage qualities and other qualities required for marketing.)

3 a) Through the centuries, how have people been able to change the characteristics of domesticated animals in their herds? Why have they wanted to change the populations?

   b) Choose **one** example of a domesticated animal. Find out where the herds live and how the features of the animals help people keep them successfully in this location.

4 Suppose you find an unusual variety of orchid. Describe how you could produce many similar plants with identical characteristics.

7.5 Genetic modification (Genetic engineering)

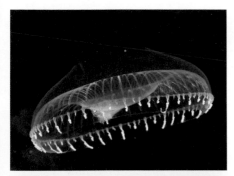

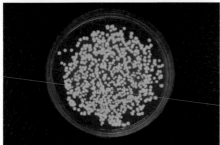

Mice glow a green colour

In 2008, the Nobel Prize for chemistry was awarded to Japanese scientist Osamu Shimomura along with two other colleagues. They had developed a way of making living cells glow green in UV light.

To do this, they transferred a gene found in a Pacific jellyfish (*Aequorea victoria,* top photograph) into the cells of different species, such as *Escherichia coli* (*E. coli*, middle photograph) and mice (bottom photograph). The green glow of cells where the gene is present is due to the production of GFP protein (Green Fluorescent Protein). The GFP gene has been used extensively by scientists as a way of labelling cells that have had the DNA of another species successfully transferred into them. The green glow of GFP marks the precise cells that have become 'transgenic' – incorporating and using the foreign DNA in their cells. So how can this technique be used elsewhere . . . for the benefit of people in their crops or their medicines?

■ Manipulating genes

The **genome** of an organism is the genetic information carried on the chromosomes inside each cell. This information is carried in the DNA molecule, in the form of sequences of bases that make up each **gene**. Genetic modification (GM) involves making changes to the DNA molecule.

Examples of different kinds of genetic manipulation include:
- **moving** a gene from one species into a different species – this then becomes a transgenic species and this species has been given a new property (characteristic)
- **changing the base sequence** of a gene – so altering the structure of the protein for which it codes
- **deleting** a gene, by removing it from the chromosome.

The movement of genes between different species does not happen in nature, so scientists have developed some clever tricks that allow them to manipulate and change DNA. When the DNA is moved from one species to another, **transgenic** organisms are created.

■ Using restriction enzymes and ligase enzymes to make recombinant DNA

Scientists have developed a set of **enzyme** 'tools' that can be used to 'cut and paste' DNA molecules. One of these tools is called a **restriction enzyme**. Hundreds of different kinds of these enzymes have been collected from different species of bacteria. Each type of restriction enzyme recognises a particular sequence of bases in a DNA molecule (known as the **restriction site**), and completely cuts the DNA double helix at this site.

> **STUDY TIP**
> Look in Section 5.4 for more information about use of the term **genome.**

> **STUDY TIP**
> Look in Section 5.4 to remind yourself that a gene is a section of a DNA molecule. Check the names of the bases and how they pair up in the two strands of a DNA molecule.

A restriction enzyme usually cuts the DNA double helix in a staggered fashion, leaving overhanging single-stranded lengths of DNA (sticky ends). Table 7.3 shows some examples of DNA base sequences recognised by different restriction enzymes and the results after cutting. All the restriction enzymes are given names but these are not included in the table.

Table 7.3 Some examples of how and where certain restriction enzymes cut a sequence of bases in DNA.

Restriction enzyme	Sequence recognised (showing base pairs on a DNA molecule)	Effects of cutting the sequence	
restriction enzyme type 1	GAATTC CTTAAG	G CTTAA	AATTC G
restriction enzyme type 2	GAATTC CTTAAG	GAA CTT	TTC AAG
restriction enzyme type 3	GAATTC CTTAAG	GAAT CT	TC TAAG
restriction enzyme type 4	GAATTC CTTAAG	GA CTTA	ATTC AG

When different pieces of DNA, with matching 'sticky ends', are mixed together in a solution, the sticky ends of one piece pair up with the sticky ends of another piece. This creates a single **recombinant DNA** molecule. This recombinant DNA is a hybrid, composed of the different original pieces of DNA linked together by a temporary join.

The temporary join between the two DNA pieces in a recombinant DNA molecule can be made permanent with the use of another enzyme. This is **DNA ligase**, also found in bacteria. DNA ligase makes a strong bond in the region where the recombinant DNA molecules have been temporarily joined together. The series of steps to produce recombinant DNA is summarised in Figure 7.23.

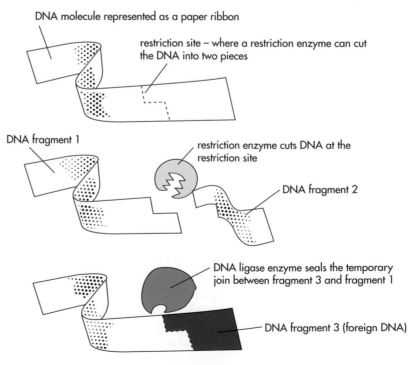

DNA molecule represented as a paper ribbon

restriction site – where a restriction enzyme can cut the DNA into two pieces

DNA fragment 1

restriction enzyme cuts DNA at the restriction site

DNA fragment 2

DNA ligase enzyme seals the temporary join between fragment 3 and fragment 1

DNA fragment 3 (foreign DNA)

Figure 7.23 Steps involved in cutting and pasting DNA to create recombinant DNA.

STUDY TIP
Check in Section 1.2 for details of the structure of a virus and the nucleic acid (DNA or RNA) it contains.

Plasmids and viruses as vectors

A **vector** can be used to carry genes into cells to create transgenic organisms. A piece of DNA, such as a **plasmid** or DNA from the inside of a **virus**, can move into cells under certain conditions and act as a vector.

Using recombinant plasmids as vectors

Plasmids are small circles of DNA present in bacteria. They usually carry just a few genes and are much smaller than the main DNA chromosome of bacteria (with hundreds of genes). Plasmid genes include those that help bacteria resist attack by antibiotics such as penicillin (antibiotic resistance). Plasmids are naturally exchanged and swapped between bacteria. This provides scientists with a way of introducing foreign genes (i.e. genes from other species) into bacteria.

Scientists extract plasmids from bacteria and then cut them open with restriction enzymes. Some foreign DNA that has been cut with the same restriction enzymes is mixed with the plasmids so that they join up to create **recombinant plasmids**. The temporary joins in the recombinant plasmids are made permanent with DNA ligase enzyme.

The recombinant plasmids are then mixed with bacteria lacking plasmids. Some of the bacteria take up the recombinant plasmids, becoming transgenic. These GM bacteria now have some new property. One example is the transgenic bacteria that have been used to synthesise useful medical drugs such as insulin (Figure 7.24).

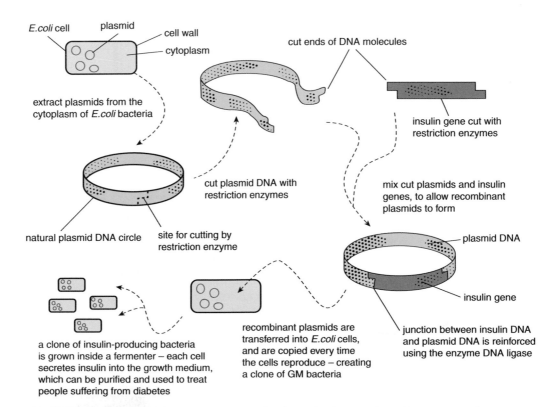

Figure 7.24 Using plasmids as vectors in gene technology. The diagram shows how the gene for human insulin can be inserted into *E. coli* bacteria. Note that the same restriction enzyme is used to cut the gene (for insulin, as illustrated) and the plasmid into which the gene is inserted. When the insulin gene has been inserted, the transgenic *E.coli* bacteria are able to produce human insulin.

Using bacteriophage vectors

Another way of creating transgenic bacteria is by using bacteria-specific viruses called **bacteriophages**. Bacteriophages ('bacteria-eaters') are a group of viruses that target bacteria. In the course of the infection, bacteriophage DNA is injected directly through the bacterial cell wall into the cytoplasm of the bacteria. Here the bacteriophage DNA reprograms the activities of the cell, causing it to synthesise hundreds of new bacteriophage particles. These eventually burst out of the cell and infect others.

Scientists can delete a number of the harmful genes on the bacteriophage DNA and replace them with one or more foreign genes. This creates recombinant bacteriophage DNA. When recombinant bacteriophages infect bacteria, the recombinant DNA is injected into the bacterial cytoplasm. The foreign gene reprograms the cell to do something useful, such as making a valuable medicine.

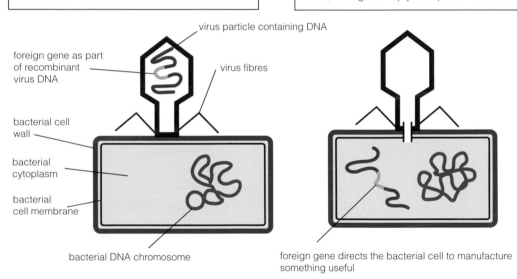

Figure 7.25 Using bacteriophage viruses as vectors to transfer foreign genes into bacteria.

STUDY TIP

In Section 7.2 you can find details of industrial fermenters and how they are used to produce large quantities of a microorganism or to harvest products from microorganisms grown in them.

■ Applications of GM bacteria

Transgenic bacteria are stable and can be cultured on a large scale in **fermenters**. Useful products from such bacteria can be separated from the bacterial cells in the fermenter and purified for use. Products from GM bacteria grown in fermenters include medicines, enzymes and hormones (such as insulin, described above). Use of fermenters to grow bacteria on a commercial scale and harvest their products is described in Section 7.2.

One of the first commercially successful products of this technology was created by Herbert Boyer in 1978 in the USA. He produced transgenic *E.coli* bacteria, which were given human genes coding for the synthesis the hormone **insulin**.

Insulin is a hormone that helps to control the **blood sugar** levels in humans. Some humans do not produce enough insulin and have the condition known as **diabetes**. Treatment of diabetes involves daily injections of insulin. Before 1978, the supply of insulin for injecting diabetic patients was mainly obtained from the pancreas of pigs. Transgenic *E.coli* provide a much cheaper, more reliable and safer source of the insulin for treating patients. A further advantage is that the insulin gene in the bacteria codes for the human form of the hormone. This is more effective in controlling blood sugar than the form of insulin from the pig, which is slightly different.

Table 7.4 Some examples of products made with GM bacteria.

Product	Comments
insulin	Hormone required by diabetes patients to help control blood sugar levels. Originally extracted from the pancreas of pigs.
human growth hormone	Hormone required to treat small stature (dwarfism) in humans. Originally extracted from pituitary glands of human corpses, which presented the danger of passing diseases from corpse to patient.
chymosin	Enzyme that causes the clotting of curds during cheese making. Originally extracted from the stomachs of young calves. Vegetarian cheeses are made using chymosin produced in transgenic bacteria and yeasts and no calves are killed.
blood clotting factors	Proteins required by patients suffering from haemophilia, when blood does not clot properly. Originally extracted from donated blood. The large quantities of donated blood used meant that infections such as HIV were passed from donors to haemophilia patients.
washing powder enzymes	Digestive enzymes added to 'biological' detergents allow washing machines to clean difficult stains at low operating temperatures. Originally, the enzymes were extracted from animal faeces. Now the enzymes have been further modified to make them work at higher washing machine temperatures without denaturing.
vaccines	Proteins from the walls of microbial pathogens which are injected into the bloodstream of a healthy person, making them immune to the pathogen. Originally the whole (dead or inactivated) pathogen would be injected – but this always carried the risk of live microbes causing the disease. The vaccines produced by transgenic bacteria do not carry this risk, because whole pathogen cells are not used.
alpha amylase	Useful enzyme that can digest starches to sugars which can be used to grow yeast, producing alcohol as a fuel.
protease	Digestive enzyme used in the food industry to break down meat products during the manufacture of baby foods.

■ Investigating the use of enzymes in washing powders

Prepare some standard stained pieces of fabric (for example, pieces of linen, 5 cm × 5 cm, with a dried on spot of tomato ketchup or curry sauce).

1 Set up two beakers with 250 cm³ water in each. Label one beaker D + E and the other D.
2 Place each beaker on a tripod with gauze, above a Bunsen burner (unlit).
3 Light the Bunsen under each beaker and gently raise the temperature of each to 40 °C, stirring and checking with a thermometer. Stop heating when the target temperature is reached. Alternatively, a water bath could be used as it is then easier to control the temperature and not exceed 40 °C.
4 Add 1 teaspoonful of *non-enzyme* detergent to the beaker labelled D and 1 teaspoonful of *enzyme* detergent to the beaker labelled D + E. Use a glass rod to mix the detergent into the water until it has dissolved.
5 Add one piece of similarly stained fabric to each breaker. Use the glass rod to swirl each around in the detergent mixture for about 30 seconds, every 5 minutes.

When some coloured substances (for example, grass and blood pigments) become attached to the fibres of a fabric, they cannot be washed off easily by water and plain detergent. These substances are stains. Often the pigments are bound to the fabric by proteins and lipids – these form an organic 'glue' that prevents the detergent molecules from carrying the stain away. Many modern detergents have enzymes (extracted from genetically modified bacteria or fungi) added to them. These enzymes break down this organic 'glue', allowing the stain to be washed away.

6 After 30 minutes, use forceps to remove each piece of fabric from the beakers. Place them on a white tile.
7 Place a piece of unwashed, but moistened, stained fabric beside the two washed pieces of fabric. Record the extent to which the washing process seems to have worked (and removed some or all of the stain). Note whether there is any difference between the two treatments and whether the presence of enzymes in beaker D + E has removed the stain from the fabric significantly more the from the fabric in beaker D.

■ Producing GM plants

Some of the world's plant crops, such as soybeans and cotton, have now been genetically modified. The reasons for foreign genes being introduced to crop plants include:

- increasing plant resistance to infectious diseases caused by fungi, bacteria and viruses
- improving plant resistance to pests (especially herbivorous insects)
- giving plants immunity to chemical herbicides (used to kill weeds).

Scientists are also trying to produce GM crops that can grow in stressful conditions. These include areas with little rainfall, or where the soil is infertile or polluted. Other GM plants are being developed to be used as a means of cleaning up soil contaminated with pollutants (bioremediation).

Bacterial plasmids as vectors for gene manipulation in plants

The plant cell wall is a major physical barrier to getting foreign DNA into a plant cell. One solution is to use the plasmids of a natural bacterial plant pathogen called *Agrobacterium tumefaciens*. These bacteria send their plasmids into the nucleus of plant cells. Once inside the nucleus, the plasmid inserts itself into the plant chromosomes.

Scientists can use these plasmids as vectors for putting useful foreign genes into the plant chromosomes. This creates a transgenic plant. Here are the steps they take (see also Figure 7.26):

- Scientists extract the plasmids from bacteria and substitute disease-causing genes with a useful foreign gene. This creates a recombinant plasmid.
- The recombinant plasmids are then mixed with plant cells.
- Some plant cells take up the recombinant plasmids and the plasmids are taken into the plant chromosomes to become part of them.
- These genetically modified plant cells are now transgenic. They can be cloned by tissue culture to produce whole plants, with roots, stems and leaves.
- Techniques of micropropagation are then used to produce large numbers of GM plants. They all carry the foreign gene, which gives them some useful property.

Section 7.6 gives you details about tissue culture in plants and how micropropagation can be used to produce large numbers of plants that are clones of the original plant.

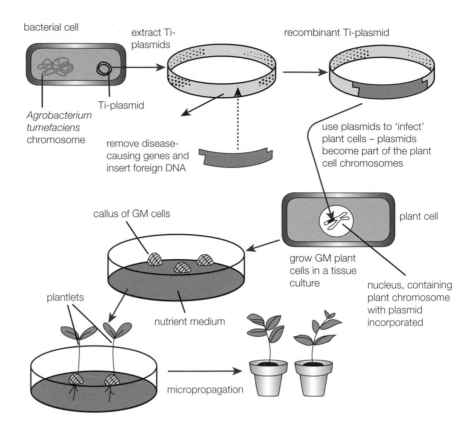

Figure 7.26 Using plasmids (and *Agrobacterium tumefaciens*) to infect plant cells.

Developing insect-resistant GM plants

The development of GM plants that are resistant to insect pests began in 1992. Certain bacteria produce a toxin, known as *Bt*. The *Bt* gene was originally obtained from bacteria and inserted into plant cells, resulting in transgenic plants with the *Bt* gene inside their cells. This gene codes for the production of *Bt* toxin (a protein), which destroys the intestines of herbivorous insects that eat the plant tissues. This means the GM plants produce their own insecticide inside their cells, so these GM crops do not require as many applications of chemical insecticides as ordinary (non-*Bt*) crops. Fortunately, the *Bt* toxin is harmless to humans.

Figure 7.27 A genetically modified (GM) cotton fruit.

Figure 7.28 Some genetically modified (GM) potatoes. This variety has been modified to have improved resistance to late potato blight.

So far, it has been estimated that planting genetically modified Bt crops has led to a 25% reduction in the use of chemical pesticides against herbivorous insects. Some GM crops intended for human consumption, such as potatoes, have been approved in the USA. Further examples of GM crops are given in Table 7.5.

Table 7.5 Some examples of development of GM crops and their properties.

Crop	Genetically modified property
soybean	resistant to chemical herbicides
maize	resistant to insects, improved nutrient content
cotton	resistant to insects (Bt technology)
tomatoes	longer shelf life (Flavr-Savr) as fruit softens slowly
Hawaiian papaya	resistant to papaya ring spot virus
apples	cut apple flesh does not go brown on contact with air (arctic apples)

■ Consumer concerns about GM crops

Consumers, particularly in European countries, have shown concerns about possible dangers of GM technology, largely as result of anti-GM media campaigns. The case of the **Flavr-Savr tomatoes** (see Table 7.5) is notable. These tomatoes had been genetically modified to allow them to become ripe without softening. This meant the tomatoes would keep well without becoming damaged and develop a good flavour. No *foreign* DNA was introduced to the tomato plants – a gene for an enzyme was simply reversed on the plant chromosome. But anti-GM campaigns stopped the sale of these GM tomatoes (and products from them) to the public. Note that Flavr-Savr tomatoes were not transgenic as no foreign DNA was put into them.

However, the extensive and increasing use of GM crops globally shows that many nations believe the theoretical and unproven risks of GM plant technology are far outweighed by the known benefits. In the future, scientists are likely to develop more crops that are genetically modified to provide a range of products and improved food supply.

STUDY QUESTIONS

Make a summary of different aspects of gene technology under the headings in the list below. Section 7.5 gives some of the information in flow diagrams, so present your answer in a different way rather than just copying the diagrams.

1 List the steps involved in cutting and pasting DNA to make recombinant DNA.
2 Name **three** vectors used in gene technology. For each vector, list the steps that show how the vector is used to insert foreign DNA into another organism.
3 List **three** examples that show how gene technology has been used and, for each, describe how the GM organism is of benefit to people.

7.6 Cloning

Why are some twins identical?

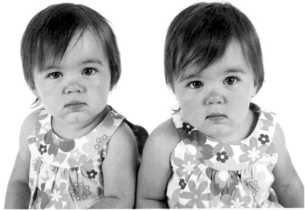

You may have met some identical twins. Such twins occur when a single fertilised human egg develops into two embryos, each of which becomes a baby. The twin babies are genetically identical, as they both have exactly the same alleles from their shared original fertilised egg. This is an example of a natural clone. A clone is a group of individuals and all of them have the same alleles.

As the twins grow up, they often look similar, speak in a similar way, and even behave in similar ways. As they grow older, they will become increasingly different. These differences result from different environmental forces acting upon each twin. Although clones share exactly the same set of alleles (their genotype), their appearance (phenotype) is the result of the interaction of their genotype and the effects of their individual environments.

Make yourself a list of the environmental factors that you think might cause two identical twins to become different in appearance.

■ Cloning in the natural world

Cloning is the process by which genetically identical organisms are produced from a single parent organism. This is achieved naturally by various **asexual reproduction** mechanisms, which are widespread throughout the natural world.

The tiny little plantlets growing on the edge of the *Bryophyllum* plant leaf (Figure 7.29) eventually drop off onto the ground. If they are lucky enough to find a suitable space and moisture to grow, they develop into a natural clone of the parent plant.

Figure 7.29 The plantlets along the edge of the *Bryophyllum* leaf fall off and develop into a clone of the parent plant.

In plants and fungi the cloned offspring often remain physically attached to the original parent. One of the world's largest organisms, in terms of total size, has been produced in this fashion. A forest of aspen trees (*Populus tremuloides*) in the USA (Figure 7.30) is a clone covering an area of over 40 hectares and weighing 6000 metric tonnes (about twice the mass of a giant redwood tree). The original aspen tree must be well suited to this region and, although these aspens do all produce flowers and seeds, it is the cloned trees that grow best here.

Figure 7.30 A clone of aspen trees in the USA.

Figure 7.31 The aphids in this cluster are all part of a clone.

■ Artificial cloning in plants and animals

Parts of plants, such as leaf or stem cuttings, can be grown artificially to produce clones of the original parent and this is described in Section 5.2.

Groups of cells of both plants and animals can be isolated and grown, in artificial conditions, to produce a whole new organism. We often call this tissue culture or may describe it as *in vitro*. Tissue culture needs to be done under laboratory conditions. Generally the cells are grown on an artificial medium (such as agar) until they are large enough to exist as independent organisms.

Plants or animals produced by tissue culture are clones of the parent, and are genetically identical to each other, and to their single parent. The specialised techniques for plant tissue culture and for animal tissue culture are described below.

Micropropagation (plant tissue culture)

Plant scientists exploit the ability of very small fragments of plants to regrow into whole plants, using the techniques of **tissue culture**. This method of propagating plants is known as **micropropagation**. In this way, large numbers of genetically identical plants can be produced. It is often used as a way of producing a clone from a parent plant that is valuable for a particular reason.

The fragments of plants are called **explants**. The stages for tissue culture of these explants are as follows:

1   A number of **explants** are removed from the selected parent plant.
2   The surface of each explant is **sterilised** with a disinfectant solution then rinsed with sterile water. This reduces the chance of contamination (from microorganisms) during the early stages of growth.

3 The sterilised explant is placed inside a sterile **tissue culture vessel**, containing a **sterile growth medium**.
4 The growth medium contains nutrients that encourage the growth and division of the plant tissue cells to form a mass of cells, known as a **callus**.
5 The callus is transferred to a fresh growth medium containing a mixture of **plant growth regulators** where it develops roots, stems and leaves and becomes a 'plantlet'.

These stages are summarised in Figure 7.32.

A typical tissue culture growth medium contains:
- agar to create a solid jelly
- a source of energy (sugar) for the callus, though light may also be provided to allow photosynthesis to take place in green tissues
- a mixture of mineral ions (for example, nitrates, phosphates)
- water
- sometimes certain vitamins are needed.

It is important that the explants are free from contaminating microbes, which would otherwise quickly use up the nutrients in the glassware, and infect and destroy the explants. Sometimes antibiotics or disinfectants are incorporated in the growth medium to prevent contamination by fungi or bacteria. The problem of maintaining sterile conditions is one reason why tissue culture is carried out by trained personnel and requires laboratory facilities. It is not easy to do without specialised equipment and suitable precautions.

At first the young plantlets require a protected environment and are kept in carefully controlled conditions of light, temperature and humidity before being grown on into larger independent plants, able to survive outside the laboratory conditions.

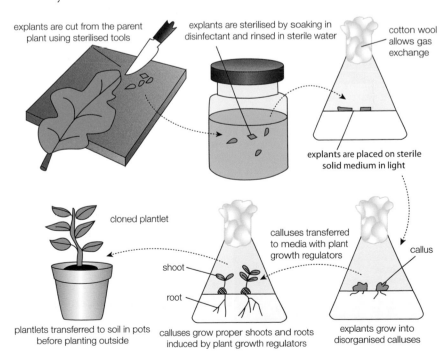

Figure 7.32 This flow chart illustrates the sequence of steps taken in micropropagation, from cutting the explant from the parent plant, through to production of the cloned plant growing in a pot. This is a simplified version and can work fairly well, but in a commercial situation, rigorous care is taken to ensure sterile conditions at the start and adjustment of external conditions when the plantlet is ready for growing on.

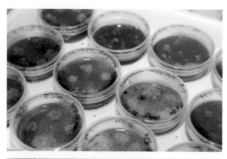

Figure 7.33 Early stages of micropropagation are carried out in a laboratory under sterile conditions. The explants are cultured on a suitable medium and grown under lights.

Figure 7.34 At a later stage, the young plantlets are transferred to modules in a glasshouse with high humidity and allowed to adjust to the conditions.

Figure 7.35 Orchids produced by micropropagation, in a nursery in Thailand, being grown until they are ready for sale.

When the plantlet is large enough, it is removed from the glassware and planted in a small container of suitable compost or soil. This container is eventually moved out of the sterile laboratory into a glasshouse. When the tiny plantlets have grown larger, they are transferred to soil in bigger pots. Eventually they will be large and robust enough to be planted out in a garden or field. Figures 7.33 to 7.35 show stages of adjusting to external conditions – firstly in a glasshouse in which humidity is kept at a high level, and later ready to be planted in the outside world.

You can see that, in a commercial situation, micropropagation offers the potential for producing enormous numbers of plants from a selected desired parent plant. The original explant may consist of only a few cells, so it is possible that a single valuable plant could have hundreds or even thousands of explants taken from it. And even more explants could be generated, by removing explants from the explants taken previously. This is a major advantage of tissue culture over taking cuttings. Another advantage is that the resulting plants can be free of disease (providing the original explants were taken from a disease-free part of the parent plant).

Many plants, such as orchids, used to be collected from the wild for sale to the public. This trade led to the destruction of habitats and risked the extinction of wild populations of popular plants. The use of tissue culture and micropropagation techniques has meant that wild plants should now be left alone. Plants species that were at risk from extinction, such as the Madagascar Periwinkle (*Catharansus roseus*), have been micropropagated and sold widely to the public, making extinction much less likely.

Here is a summary of some of the advantages and disadvantages of micropropagation, compared with more conventional methods of propagation.

Advantages of micropropagation:
- It can produce thousands of plantlets (from the desired parent plant), while conventional techniques might produce only a fraction of this number in the same time.
- It produces rooted plantlets ready to grow on to mature plants, saving time for the grower when seeds and cuttings are much slower to establish and grow.
- The plants are genetically identical so retain the desired character (or set of characters) of the selected parent plant.
- It can be used to produce disease-free plants.
- It can be used to produce plants at any time of year.
- A greater number of plants can be produced per square metre and they can be stored for longer and in a smaller area .
- Production can be geared so that plants are ready for the outside market when required.
- It is the only way of regenerating whole plants from genetically modified cells and has important applications in plant breeding.

STUDY TIP

Look at Section 7.5 and make a list of some examples of GM crop plants. Then make a link with Section 7.4 to suggest ways micropropagation could be applied in plant breeding.

PRACTICAL

This is a simple demonstration of cloning and gives useful support to help your understanding of the stages required for micropropagation and how it is carried out under more controlled laboratory conditions.

Remember to wear eye protection.

Disadvantages of micropropagation:

- It requires sterile laboratory facilities and trained personnel.
- It produces genetically identical offspring, so if one plant is defective or susceptible to a disease, all the plants will also be susceptible.
- Planting fields with genetically identical crops might result in considerable damage from pests if they invade the field where the crop is growing.

■ Practical activity — a simple demonstration of cloning with mustard or cress seedlings

Young seedlings of mustard (or cress) are easily grown from seed in a container on damp filter paper or on another suitable medium. A shoot is then cut as shown in Figure 7.36 and taken through the stages to demonstrate how a new plant can be grown artificially on agar, in a specimen tube.

1 Use a clean specimen tube with plain agar gel (4% agar in distilled water).
2 With clean forceps, transfer a shoot cut off of a mustard or cress seedling to the specimen tube.
3 Use the tip of the forceps to puncture the gel and ease the stump of the cress shoot into the gel, so that the shoot is standing up in the tube.
4 Leave the cress explant under continuous illumination for several days.
5 After a few days, a new root system will have developed, and new leaves will have grown on the cress shoot.

The lack of nutrients in the agar gel discourages the growth of microbes, but the shoot makes its own food by photosynthesis.

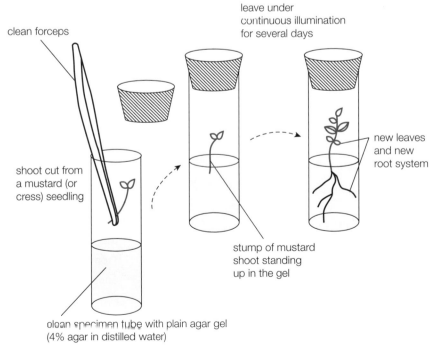

Figure 7.36 Cloning with mustard seedlings.

A stem cell is one that can divide by mitosis to produce more cells, each of which can develop and specialise, becoming any one of the 200 different types of cells found in the human body.

STUDY TIP

Look in Section 1.3 for more information on stem cells.

Artificial cloning of animals

Two techniques have been developed by scientists for the artificial cloning of mammals: embryo splitting and cloning from an adult body cell.

For many years, scientists have been able to take a growing embryo from the womb of a valuable animal and split it into two or more pieces. They then allow each piece to grow into a complete embryo and implant the embryos into different **surrogate mothers**. Eventually this results in a group of cloned (genetically identical) offspring, even though the offspring are different from their surrogate mothers. This technique uses the fact that the **embryonic stem cells** (taken from the embryo) naturally have the ability to divide and grow into complete embryos.

More recently, a new form of cloning has become possible in which the genetic material is removed from an adult body cell (a somatic cell). This is then inserted into another cell (from which the nucleus has been removed) and allowed to grow into an embryo.

The stages in this procedure are as follows (summarised in Figure 7.37).
1 A nucleus (with its diploid set of chromosomes) is removed from an **adult body cell**.
2 This nucleus is inserted into an **enucleated egg cell** (a cell from which the original haploid egg nucleus has been removed).
3 The egg cell is then allowed to grow into an embryo, on an artificial medium.
4 The embryo is implanted into a **surrogate mother**.
5 Eventually an animal is born. This 'offspring' has no genetic material from its surrogate mother but got its genetic material from the animal whose adult (diploid) cell provided the nucleus.

The offspring is a clone of the animal whose cell provided the donor nucleus.

The adult (somatic) cell cloning of a mammal was first achieved in 1996 by Ian Wilmut and colleagues at the Roslin Institute in Scotland, UK. The cloned mammal was a sheep called Dolly.

Table 7.6 gives some examples of mammals that have been cloned and when the first clone was produced.

Table 7.6 Examples of first times various mammals have been cloned.

Mammal	First cloned example
sheep	Dolly – born in 1996 in UK from a skin cell
mouse	Cumulina – born 1996 in Hawaii
cat	CopyCat – born in 2001 in USA
goat	Megan – born 2001 in USA
racehorse	Prometheus – born in 2003 in Italy
dog	Snuppy – born in 2005 in South Korea from an ear cell

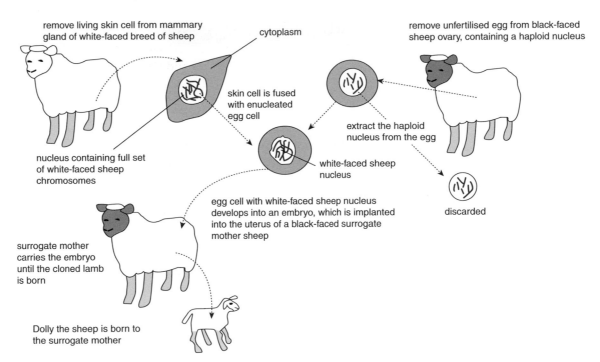

remove living skin cell from mammary gland of white-faced breed of sheep

cytoplasm

remove unfertilised egg from black-faced sheep ovary, containing a haploid nucleus

skin cell is fused with enucleated egg cell

extract the haploid nucleus from the egg

nucleus containing full set of white-faced sheep chromosomes

white-faced sheep nucleus

discarded

egg cell with white-faced sheep nucleus develops into an embryo, which is implanted into the uterus of a black-faced surrogate mother sheep

surrogate mother carries the embryo until the cloned lamb is born

Dolly the sheep is born to the surrogate mother

Figure 7.37 Stages involved in producing Dolly the sheep. Dolly the sheep was the first cloned mammal, and was genetically identical to the sheep from whose mammary gland her chromosomes had come.

STUDY TIP

Look at Section 5.4 and explain the difference between haploid and diploid cells in the human life cycle.

STUDY TIP

Look at Section 3.5 and explain the process leading to blood clotting.

STUDY TIP

Look at Section 5.5 and calculate the chances of two carriers of cystic fibrosis having a child affected with the disease.

The production of such clones requires immense scientific expertise and technical resources, so it is it is not undertaken for trivial reasons. The majority of attempts do not succeed in producing a cloned animal. Dolly was achieved only after 276 attempts.

The main reason why the scientists who produced Dolly at Roslin wanted to be able to clone sheep, and other large mammals, was because of their research aimed at producing medicines in the milk of such mammals. If a transgenic animal producing a useful medicine in their milk is created, cloning could give a whole herd or flock of mammals all producing the same medicine. (This practice is now known as 'pharming'.) Examples of medicines that have been produced in this way include:

- blood clotting factor IX – for treating haemophilia (a blood clotting disorder)
- alpha-1-antitrypsin – for treating patients with cystic fibrosis
- antibodies that can target cancer cells in humans, or fight specific diseases.

In theory, this technique could allow an adult human to be cloned – producing children who carry a complete copy of their genome. However, this practice is currently banned by International Treaty.

STUDY QUESTION

1 Sometimes a novel variety of a plant is discovered, resulting from a random mutation. An example might be a rose with blue flowers. You may have to look in other parts of this book to help you with the answers to these questions.

a) What is a 'mutation' and what might increase the chance of a mutation occurring?

b) Why might cross-pollinating blue rose flowers with flowers from a normal red-flowered rose plant not produce any blue-flowered plants? Use a genetic diagram to help explain your answer.

c) If a gardener had a single plant with this mutation, what methods of artificial propagation could be used to produce a group of plants to grow around a garden?

d) If a plant breeder got hold of just a single rose plant with blue flowers, how could artificial propagation methods be used to produce a large number of rose plants with blue flowers for sale to the public? In your description, indicate how the plant breeder could arrange for the rose plants to be ready for a particular date in the year.

Summary

I am confident that:

✓ I understand how glasshouses and polytunnels can be used to provide a protected environment and how this can be used to increase crop yield.

✓ I understand the effects of increased carbon dioxide and increased temperature on the yield of crops grown in glasshouses.

✓ I know how fertilisers are used and how they increase crop yields.

✓ I can describe examples of ways in which insect pests cause a fall in crop yield.

✓ I can discuss differences between chemical and biological pest control and the advantages and disadvantages of each.

✓ I know that yeast carries out aerobic and anaerobic respiration and that yeast is used in the production of bread.

✓ I know how to investigate the role of anaerobic respiration by yeast in different conditions.

✓ I know that fermentation reactions of *Lactobacillus* bacteria in milk produce lactic acid and that this lowers the pH and causes the coagulation of proteins to produce the thick texture and flavours characteristic of yoghurt.

✓ I understand the use of industrial fermenters and know the reasons for the different features that are included and their importance for the successful growth of microorganisms.

✓ I can describe the way fish are farmed in freshwater or marine situations and understand the factors that need to be considered (such as maintenance of water quality, feeding and use of selective breeding) to ensure successful production of fish in the fish farm.

✓ I can describe examples of how plants and animals with desired characteristics have been artificially selected and used to provide the future generations of improved crops or herds of animals.

✓ I can describe the stages involved in cutting DNA and joining pieces together in genetic modification techniques and appreciate how DNA can be manipulated to incorporate DNA from another organism, including the role of restriction enzymes and ligases in these processes.

✓ I am familiar with the use of plasmids and viruses as vectors and can describe how they take up pieces of DNA, which are then incorporated as recombinant DNA into other cells.

✓ I can discuss examples of how useful products (such as human insulin) have been manufactured from genetically modified bacteria grown in a fermenter.

✓ I can describe the sequence of stages involved in micropropagation of a plant with desired characteristics and how this technique has become an important commercial industry.

✓ I can describe the stages for production of a cloned animal, such as Dolly the sheep, and understand the genetics of the sequence of events that takes place.

✓ I understand that the term transgenic refers to organisms produced as the result of the transfer of genetic material from one species to another.

MATHS SKILLS

Sampling and scientific data

A single scientific measurement may be representative of the real situation, or it might be an anomaly. To avoid a wrong conclusion based on few anomalous readings, it is best to take replicate measurements, and calculate an average value for each batch of data.

Here is set of measurements of grass seedling leaf heights using different numbers of fertiliser pellets in the plots:

Number of fertiliser pellets per growth plot	Maximum height of seedling leaves / cm					
30	7.2	11.3	9.0	13.6	10.1	9.9
15	12.0	10.1	6.9	9.8	10.0	12.0

a) Work out the average value for both fertiliser treatments, with appropriate number of decimal places.
b) On the basis of this data, which number of fertiliser pellets would you advise a farmer to use?
c) Explain the basis for your answer to (b).
d) If you were planning further investigations which variables would you change?

d) try other numbers of pellets per plot, and more replications of each treatment
c) there is no significant gain in growth with double the number of pellets
b) use 15 pellets / plot
a) 30 pellets = 10.2 pellets / plot ; 15 pellets = 10.1 pellets / plot

MATHS SKILLS

Determining the slope of a graph
The slope (gradient) of a straight line on a graph shows how the magnitude of the y-variable changes in relation to the x-variable.

Look at the steepest part of the slope of the graph in Figure 7.10 and estimate the gradient as the change in pH units per hour.

approximately 0.4 units / hour

MATHS SKILLS

Substitute numbers into equations using appropriate units

Useful scientific theories often generate equations which can be used to understand biological processes. When new measurements are put into an equation, the results make biological sense, supporting the original theory.

In optimal growing conditions, such as those found inside an industrial fermenter, microorganisms can double their population with every new round of cell division. This type of population growth is called exponential growth and can be expressed by the following equation:

$x_t = x_0 (1 + r)t$

where x_0 is the population of x at time 0, x_t is the population at time t, and r is the number of rounds of cell division per hour.

Use the equation to work out the population of microbes present in 1 microlitre of fermenter culture after 1 hour (t = 1), if the initial population at time 0 was 50, and r = 3.

Example of student response with expert's comments

Using and interpreting data

1 Radishes are plants in the cabbage family. As the plant grows, the roots swell. These roots are often eaten as a vegetable.

An investigation was carried out in a school laboratory into the effect of fertiliser containing different concentrations of nitrogen on the growth of radish plants.

The radishes were grown in small plastic pots, as shown in the diagram. Each nutrient solution was identical except for different levels of nitrogen (supplied as potassium nitrate). The pots were kept in the light throughout the investigation.

One radish seed was pushed into the sand in each pot. After 18 days the radishes were harvested. Each plant was removed from the pot and washed carefully in water to remove any sand. The plants were then separated into the tops (leaves) and roots. Each part was dried at $110\,°C$ for 24 hours and weighed to obtain the dry mass. The results for all the plants (from 18 pots) at each nitrogen concentration were combined and a mean value was calculated.

The results for the whole plants and for the roots only are shown in the graph.

a) i) In this investigation, what was the highest dry mass of whole radish plants obtained and at what concentration of nitrogen? *(2)*

  ii) Describe the effect of nitrogen concentration on the growth of roots compared with the growth of the whole plants. Make reference to similarities and differences shown in the patterns of growth. *(3)*

b) The table shows the percentage of the dry mass of the whole plant that was found in the leaves at each nitrogen concentration. Note that the percentage dry mass for a nitrogen concentration of 112 arbitrary units is missing from the table.

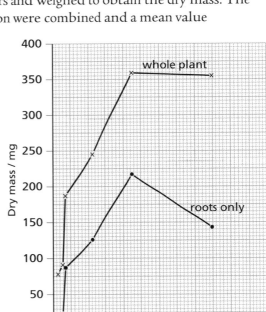

whole radish plant

hole

black plastic pot
(3 cm x 5 cm)
- contains sand washed in distilled water
- stands in tray with nutrient solution

Nitrogen concentration / arbitrary units	7	14	28	56	112	224
Percentage of dry mass in leaves at harvest (%)	74.0	75.0	53.0	48.5		59.7

  i) Use information in the graph to calculate the percentage dry mass in the leaves grown at a nitrogen concentration of 112 arbitrary units. Show your working. *(3)*

  ii) What does the investigation show about the proportion of material in the leaves of plants of different sizes (compare smaller plants with larger plants)? *(2)*

c) A grower needs to encourage good root growth to obtain a high yield of radishes to sell in a local market. From the results of this investigation, what advice would you give to the grower? *(2)*

(Total = 12 marks)

Student response Total 9/12	Expert comments and tips for success
a) i) 355 mg O at 112 nitrogen units ✔	The student has misread the scale but does give the units (for dry mass). The value for nitrogen is correct and 'units' is acceptable here. You should check the scale carefully — every graph is different, and here two squares (on the vertical axis) represented 10 mg (dry mass).
ii) Both roots and whole plant growth are stimulated ✔ by nitrogen, up to a point, but after that point the roots seem to shrink ✔ whilst the whole plant seems to stay still. ✔	The student makes three correct points so gains full marks. (Often more marks are available than the total for the question part.) 'Shrink' is just acceptable for 'less'. 'Up to a point' is a bit vague and you are advised to quote figures where possible to help support your answer.
b) i) Roots = 217 ✔ so leaves = 355 - 217 = 138 mg ✔ % difference = 138 ÷ 355 × 100 = 38.9% ✔	The student has approached the calculation sensibly, by reading the values for whole plants and roots from the graph, then subtracting the roots from the whole plant to find the dry mass of the leaves. The student next worked out the percentage of the plant contained in the leaves. However, the student had already made an error in reading from the graph in (a)(i), but examiners do not penalise you twice for the same error. So marks were awarded even though an incorrect value (for whole plant dry mass) had been used. This emphasises the importance of showing your working in calculations. If the student had not shown working, the answer would not have gained any marks.
ii) The bigger plants have a greater proportion of their mass in the form of roots than smaller plants. ✔	A correct interpretation and is acceptable for 1 mark (though the mark scheme presents it the other way round).
c) Use 112 units of nitrogen because this gives you the biggest percentage of dry mass in the roots, ✔ as opposed to the leaves.	The student makes an attempt to give a reason in terms of wanting to have the biggest percentage of dry mass in the roots. The value of 112 units of nitrogen is likely to be quite different when growing the crop in the field. Even though this answer is not exactly as given in the mark scheme, it is accepted as 'eq'.

■ Extended writing

Example

1 Describe how a fish farmer produces large numbers of healthy fish for sale from a freshwater fish farm. *(6)*

Student response

The farmer needs to sell fish all year round, so he buys new fish in batches. Each batch is kept in a different container, and is fed protein-rich food in order to speed their growth. The oxygen consumed by the respiration of the fish must be replaced, and this is achieved by pumping bubbles of air and using paddles to stir the water. The wastes produced by the fish must be removed, and if any diseased fish are spotted, they must be removed before all the fish become infected.

Mark scheme

The mark scheme shows how marks are awarded.

(1) fish contained in pond / tank / cage in water / eq

(2) water kept oxygenated by paddles / aerators / sprinklers / recirculated / eq

(3) net / eq covers container / eq to prevent predation / interspecific competition / eq

(4) prevent overcrowding / watch fish density / separate into other tanks / eq

(5) separate fish of different sizes / large ones become aggressive / ref intraspecific competition / eq

(6) regular feeding in small amounts / appropriate food / for example, high protein especially in younger stages / eq

(7) removal waste nitrogenous material / excreta / eq / prevent pollution / eq

(8) check for disease and remove unhealthy fish / use antibiotics in food / eq

(9) select good breeding stock / eq

(10) containers / eq with fish of different ages to allow continuity of supply / eq

(11+) *accept other points relating to water quality (e.g. ref temperature, pH) + quality of fish stock + detail of fish feed (for example, canthoxanthin for flesh colour)*

Total: 6

Student response Total 5/6	Expert comments and tips for success
The farmer needs to sell fish all year round, so he buys new fish in batches. Each batch is kept in a different container, ✔	This is a suitable description for 'continuity of supply' and gains mark (10). There is no description of the container so the student does not gain mark (1).
and is fed protein-rich food ✔ in order to speed their growth.	The student gains mark (6) with this reference to feeding.
The oxygen consumed by the respiration of the fish must be replaced, and this is achieved by pumping bubbles of air ✔ and using paddles to stir the water.	The student gains mark (2) — 'pumping bubbles of air' is accepted for aerators and 'using paddles to stir' is another way of getting the same mark.
The waste produced by the fish must be removed, ✔	Even though the student did not say 'nitrogenous' material, saying 'produced by the fish' gives enough to gain mark (7).
and if any diseased fish are spotted, they must be removed ✔ before all the fish become infected.	The student gains mark (8).

See General Advice for Extended Writing questions on page 79.

Exam-style questions

1 A small market garden in Eastern England supplies vegetables direct to customers in their houses, rather than to shops or a market. They grow their crops inside polytunnels and also outside in a field.

Here are their records (in 2011) for a variety of pea, known as 'Mangetout'. In this variety, the whole pod and the peas inside are eaten. They pick the pea pods at 2- or 3-day intervals, over a period of several weeks, when they are ready. They weigh them in batches, as harvested.

Inside the polytunnel, the peas are sown in single rows, 25 m in length. Outside the polytunnel, the peas are sown in double rows and each row had a length of 100 m. They were sown at the same time, inside and outside, during March and April.

Harvest data, in kg, as extracted from the grower's notebook, are as follows:

> One row, inside polytunnel:
> 0.35; 3.0; 2.0; 1.7; 3.5; 6.7; 4.2; 2.1; 1.4; 1.2
> Double row, outside in field:
> 2.4; 0.8; 10.1; 3.4; 7.8; 1.35; 1.1; 4.2; 1.4; 2.5; 1.3;
> 1.0; 1.1; 1.3; 1.1; 0.25; 0.25

Harvest dates inside the polytunnel: from 27 May to 19 June.

Harvest dates in the field outside the polytunnel: from 2 June to 27 July.

Data courtesy of Longs Farm, Hartest.

a) Draw a suitable table and organise the harvest data in your table so that you can compare the yield of pea pods grown inside and outside. [4]

b) The total yield for pea pods grown outside is 0.21 kg per metre of row.

   i) Calculate the increase in yield, per metre of row, for the pea pods grown inside the polytunnel. Show your working. [2]

   ii) Suggest **two** reasons for the higher yield inside the polytunnel. [2]

c) The harvesting dates show that the pea pods were picked at different times, though there was some overlap. Suggest why this might be an advantage for the grower in her market garden. [2]

[**Total** = 10]

2 A student wanted to investigate the effect of starter culture volume on the thickness of yoghurt. First she did a trial experiment, as follows:

   1 She heated some fresh milk to a temperature of 95 °C then allowed it to cool.

   2 She added some yoghurt to the fresh milk, as shown in diagram A. She used the yoghurt as her starter culture.

   3 She incubated the milk with the starter culture at 37 °C for 3 hours.

   4 She took the yoghurt that had formed and poured it into a filter funnel, as shown in diagram B.

   5 She timed how long it took for all the yoghurt to pass out of the filter funnel. She used this time as a measure of the viscosity (thickness) of the yoghurt.

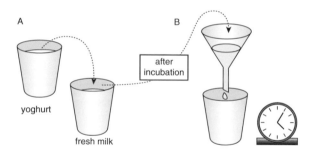

a) i) Why did the student heat the fresh milk to 95 °C? [1]

   ii) Name **one** microorganism that should be in the culture if she is to be successful in making yoghurt. [1]

   iii) Explain why the milk becomes thicker (more viscous) as it turns into yoghurt. [2]

b) Describe how she could use this procedure to investigate the effect of starter culture volume on the thickness of yoghurt. [4]

c) Suggest limitations in her experimental procedure or any difficulties she might have when carrying out the investigation. [2]

[**Total** = 10]

3 Some students made a visit to a freshwater fish farm. When they came back they decided to do an experiment to investigate how numbers of fish in the water affected the concentration of oxygen. They were able to use the apparatus shown in the diagram and were given several small fish.

When doing the experiment, the concentration of oxygen in the water declines steadily as the fish take up oxygen through their gills. Before the oxygen level falls to zero, the water is stirred to restore it to 100%.

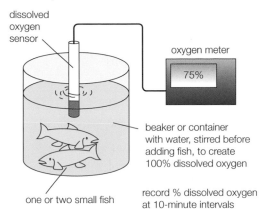

a) i) Design an investigation the students could do to find out the effect of fish density on oxygen concentration in the water. In your plan, include any precautions that they should take. [6]

ii) Predict the results you would expect them to obtain. [1]

iii) Suggest **one** precaution they should take when doing the investigation. [1]

b) i) Suggest how the outcome of this experiment might have implications for a fish farm. [2]

ii) Suggest **two** measures that may be used in a fish farm to ensure that oxygen is kept at a suitable level in the water. [2]

[Total = 12]

4 Explain what is meant by the term **selective breeding** (artificial selection) and explain how this differs from the process of evolution by natural selection. [6]

[Total = 6]

5 a) The table gives descriptions of terms or processes that are used in gene technology.

Copy the table. Choose words from the list below it to match each description and complete the table. Words may be used once, more than once or not at all.

bacteriophage; clone; ligase; plasmid; recombinant DNA; restriction enzyme; sticky end; tissue culture; transgenic

Description	Word(s)
A genetically modified organism that contains foreign DNA from two different species.	
These enzymes join sticky ends of DNA.	
A type of virus that can be used as a vector for transferring genes into bacteria.	
These enzymes recognise a base sequence in DNA and make a cut through the DNA at this site.	
A small circle of DNA found in bacteria.	

[5]

b) Describe the steps involved in producing a piece of recombinant DNA that will allow a bacterial cell to manufacture a human hormone, such as human growth hormone for treating dwarfism. [6]

[Total = 11]

6 The diagram shows three mice (A, B and C) involved in an adult cell cloning experiment.

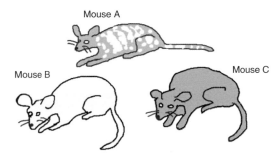

Mouse A was a female mutant mouse with an unusual coat colour, which had been bought from a pet shop. No other similar looking mice could be found to breed with it and the owner decided to clone it.

As part of the cloning process, a skin cell was taken from female mouse A and female mouse B was used as the donor of an unfertilised egg. Female mouse C was used as a surrogate mother.

a) The normal diploid number (2n) of chromosomes in mice is 40 (20 pairs of chromosomes).

i) How many chromosomes are there in the skin cell removed from mouse A for the cloning process? [1]

ii) How many chromosomes would there be in the unfertilised egg removed from mouse B before it is used in the cloning process? [1]

iii) How many chromosomes will there be from mouse B in the egg at the end of the nuclear transfer? [1]

b) The cloned mouse will have grown from an egg from mouse B and is carried in the womb of mouse C. Which mouse will it look like when it is born? Explain your answer. [3]

[Total = 6]

EXTEND AND CHALLENGE

1 Worldwide, pests cause considerable losses in crops. Scientists work with farmers to find ways to control pests, many of which are insects, and reduce these crop losses.

Bacillus thuringiensis is a bacterium, known as Bt. It contains a toxin that is harmful to many insects, particularly the caterpillars of butterflies and moths. The Bt toxin harms insects by interfering with their digestion. Pesticides containing Bt have been developed and used for many years. They are applied as a liquid spray to certain crop plants. Farmers also use chemical pesticides.

It is now possible to use gene technology to introduce the bacterial gene that produces the Bt toxin into a crop plant. One example is the development of genetically modified (GM) cotton. The aim is that plants containing the Bt gene produce their own Bt toxin, so that the plants are protected against insects and farmers do not need to apply other pesticides to control insects.

Use this passage and other sources of information to find out about different approaches used in the attempts to protect crops from insect pests and some of the problems that may arise.

a) Use of chemical pesticides
  i) List some disadvantages of using chemical pesticides, including their effects on food chains and food webs.
  ii) Explain why populations of insects can become resistant to chemical pesticides.

b) Development of GM cotton
  i) Explain why the term transgenic is used to describe GM cotton.
  ii) Outline the stages that would be used to produce a GM plant, such as GM cotton. Include reference to the enzymes involved in cutting and joining the DNA.
  iii) Find out more information about GM cotton, such as:
  - who was involved in the development
  - where it is grown now
  - any reports of its success or otherwise, compared with other methods of pest control.

c) Possible benefits of GM crops
  i) Scientists predict that development of GM crops should reduce the need for using chemical pesticides, so that natural predators are likely to return to the area.

- Explain why there should be less use of chemical pesticides.
- Explain the benefit of natural predators and why they might return to the area.
- Try to find out if there is any evidence from particular situations that suggest natural predators are returning where GM crops are being grown.

  ii) Find out about **two** other GM crops that are being grown. Choose examples that are not aiming to reduce crop losses from insect pests.
  - What genes were incorporated into the examples you chose and what are the benefits of growing these GM crops?
  - Where are these GM crops being grown?

d) Some people disagree with the use of GM plants, particularly for food crops.
  i) List reasons why some people disagree with the use of GM plants.
  ii) List reasons for supporting the development of GM crops by scientists.

2 At the start of Section 7.2 (page 290), reference is made to different ways in which microorganisms affect our lives. From the human point of view, there are good stories and bad stories.

Find out more about the role of microorganisms in the following. Prepare your information either as a story or a presentation and in a way that can be understood by someone who is not a science specialist. Make sure that the science is correct, that you name the microorganisms involved and describe the part they play in the process.

a) More fermentations:
  i) production of silage (fermented grass) or sauerkraut (fermented cabbage)
  ii) production of soya sauce (fermented soya beans) or cheese (from milk).
  In your story, include details of the processes. Suggest why these fermentations were important to people over the centuries and how they have remained of value (or become part of the diet) in modern societies.

b) Find out about the story behind the production of mycoprotein from the fungus *Fusarium* and how it is used, often as a substitute for meat. Suggest how this could be important in the diet of different people. Think about whether you have had opportunities to eat mycoprotein and find out what other people think about it.

(continued)

c) Industrial cultivation of GM (genetically modified) bacteria such as:

   i) production of chymosin (for vegetarian cheese)

   ii) production of human insulin for diabetics.

In your story, include an outline of how the GM bacteria could be produced and details of production of large quantities of the GM bacteria on an industrial scale. What benefits to people have resulted from these two processes?

d) Follow up a 'bad story' about microorganisms, such as an example of disease or decay. How can people overcome any problems arising from the situation you describe?

3 At the start of Section 7.3 (page 297), you are given some general information about the increase in farmed fish compared with catches of wild fish, over a period of 60 years (from 1950 to 2010).

Think about some of the reasons for these changes, for both freshwater and marine fish, and consider how fish can make a contribution to food requirements for increasing human populations in the years ahead. Use figures in the table to help you answer the questions that follow. You may need to use other sources of information to help you with your answers.

Note that figures in the table refer to global catches of wild fish or global estimates of farmed fish in the years 1950 and 2009. All figures are in tonnes (1 tonne = 1000 kg).

Year	Freshwater fish		Marine fish	
	Wild caught	Farmed	Wild caught	Farmed
1950	1 754	251	14 087	3
2009	8 907	30 635	65 263	1 949

a) i) Which method of catching fish gave the highest yield for the following categories and in which year(s)?
   - for freshwater fish
   - for marine fish (in sea water)

   ii) Calculate the increase in global catch for each category of fish between 1950 and 2009. Put your answers in a table and give the figures in tonnes (rather than converting to percentages or fractions).

b) Suggest reasons for the increases:

   i) in total catches of fishing (freshwater and marine)

   ii) in fish farms.

c) In fish farms, the grower provides the stock (parent fish or eggs) for the next generation of fish and cares

for the young stages until they grow into adults.

   i) In the wild, where does the next generation come from?

   ii) What happens if there is 'over-fishing' (unsustainable fishing) in a particular area?

   iii) Find out about measures that can be taken to reduce the effects of over-fishing. Can you find some examples of where this is effective or when the regulations have been disregarded?

   iv) What are the long-term implications of unsustainable fishing in an area?

   v) Suggest the effects that fishing might have on food webs (or food chains) in an area, particularly if unsustainable.

d) As a consumer, would you prefer to eat fish that have been caught in the wild or raised on a fish farm? List reasons for your preference. What about other people – do they all have the same preferences as you? If not, give reasons for their preferences.

e) Find out about the wider range of species raised by 'aquaculture'. List some species raised in this way and find out whether they are raised on a small scale (for consumption by a local community) or on a larger scale, for export worldwide.

f) Some people suggest that agriculture is becoming stretched to its limit so farming from the sea is 'what it's all about for the future'.

List some of the issues that lie behind such a comment and consider how far this suggestion may help with providing food for future generations of people.

4 Design a plant or an animal

This is your opportunity to design a plant or an animal, of your choice, but with some limitations.

a) Choose a plant or an animal that you wish to 'improve'. The plant can be a crop plant, such as maize or soya beans. The animal can be a domesticated farm animal or one that is an 'exotic' or fancy breed.

b) For your chosen plant or animal, list at least **four** characteristics that you wish to improve. For each characteristic, give reasons why the desired improvement would be of benefit to you (or your community). You may get some ideas from the illustrations included in Section 3.

(continued)

c) Collect images that show the appearance of a range of some modern varieties of your chosen plant or animal. If you cannot find suitable images, write descriptions that highlight the differences between the different varieties.

d) Select **one** of the characteristics you listed in (b) and plan a breeding programme to produce your designed plant or animal. In your plan, include how you would:
- select the parents
- carry out a genetic cross between the parents
- select the offspring
- ensure the 'improvement' spreads through the plant or animal populations.

e) Make sure you indicate the numbers of the plant or animal that would be used in your breeding programme and give some idea of the timescale. It may take several generations to establish the improvement in the population, so you need to find out how long it is from one generation to the next.

5 Find out about some examples of how (and why) plant tissue culture is used in scientific research and in commercial situations.

Here are some suggestions to get you started:
- maintaining a rare variety of potato
- conserving a rare wild species of orchid
- establishing plantations of oil palm with high yield from the crop
- providing roses for the market for special occasions during the year.

There are plenty more examples, so in your searches you can either explore one or two specific examples in some depth or you can make a list of a lot of examples to show the wide benefits of tissue culture being used in one or more of the following areas:
- agriculture
- horticulture
- plant breeding
- plant protection and conservation.

6 Micropropagation has sometimes been described as 'factory gardening'.

Plan how you would set up a commercial enterprise to be your 'factory' for growing large quantities of a particular plant for the market.

In your plan, describe what you would need for each stage in the process, including the following:
- buildings and equipment
- people and their training
- some idea of the timing of the entire process of micropropagation of your plants.

Remember to explain why you want to produce this particular plant.

Experimental design (CORMS)

Example of student response with expert's comments

1 When bananas ripen, their skin turns from green (for unripe bananas) through a series of colours to yellow (for ripe bananas). For fruits, a gas called ethene helps with the ripening process. Ripe fruits give off ethene.

Design an investigation to find out if placing a ripe tomato with green bananas speeds up their ripening.   (6)

Student response
CORMS: Outline: Put green bananas into bags — some with a ripe tomato. Record time for the bananas to go yellow. O: I would use bananas from the same bunch, so they were the same variety, the same size and the same colour green (so the same ripeness). The tomatoes must also be as similar as possible. C: Put the bananas into identical polybags, transparent, to show the colour of the bananas clearly. Five bags would contain one banana and five would have one banana and one tomato. The bags would be sealed with a double knot to make sure no gas escaped. R: I would observe 5 sets of bags to make sure my results were reliable. S: I would keep the bags in a dark cupboard, where the temperature was the same for all the bags. M: I would check the bags twice a day, recording the date and time, and the colour of each banana. I would use a photograph of a ripe banana to check the final yellow colour. I would record the time, in hours, for each banana to become yellow and compare the results.

Mark scheme	
The mark scheme shows how marks are awarded. The meaning of the letters C, O, R, M, S is explained below. (1) C – bananas in container with and without tomato (2) O – same variety of banana / same stage of ripeness / same colour at start / eq (3) R – repeats with several bananas / several batches bananas / eq	(4) M1 – measure colour of banana (e. g. match on colour scale) / eq (5) M2 – time <u>stated</u> to reach this colour (6) S1 – same container around same number of bananas + tomato / eq (7) S2 – same temperature / same light / same humidity / same other variable / eq <div align="right">Total = 6</div>

Student response Total 6/6	Expert comments and tips for success
C O R M S	Always start your answer by writing down the word 'CORMS'. This will remind you to check that your answer covers all relevant aspects. Another good idea is to write these letters next to the descriptions so you can check that your answer is complete.
Outline: Put green bananas into bags — some with a ripe tomato. Record time for the bananas to go yellow.	Planning the method in outline before you write the answer will help you when you write in more detail about each part.
O: I would use bananas from the same bunch, so they were the same variety ✔ (O), the same size and the same colour green (so the same ripeness). The tomatoes must also be as similar as possible.	A good answer, listing several aspects that need to be kept constant in order for the comparison to be fair. Do not worry about losing marks if you get the CORMS letters wrong – the examiner only marks the written text.

(continued)

Student response Total 6/6	Expert comments and tips for success
C: <u>Put the bananas into identical polybags,</u> ✔ (S1) transparent, to show the colour of the bananas clearly. <u>Five</u> bags would contain one banana and <u>five</u> ✔ (R) would have one banana and one tomato ✔ (C). The bags would be sealed with a double knot to make sure no gas escaped.	A good, concise answer. Mark S1 awarded for the use of identical polythene bags with the same numbers of bananas in them. Mark R awarded for mentioning the need for repeats (5 of each bag). Mark C awarded for explaining that the control is a bag with just a banana in it, while the test bags contain a banana and a tomato. Note that although the student put all of this under 'C', the examiner awarded marks as appropriate.
R: I would observe 5 sets of bags to make sure my results were reliable.	The R mark was allocated in the previous paragraph.
S: I would keep the bags in a <u>dark cupboard,</u> ✔ (S2) where the temperature was the same for all the bags.	Mark S1 awarded for keeping all bags in the dark (i.e. light kept the same) but could also have been awarded for keeping the temperature the same.
M: I would check the bags twice a day, <u>recording the date and time, and the colour of each banana. I would use a photograph of a ripe banana to check the final yellow colour</u> ✔ (M1). I would record <u>the time, in hours,</u> ✔ (M2) for each banana to become yellow and compare the results.	Student gave sufficient detail for assessing the final yellow colour, so mark M1 awarded. Sufficient detail given for measuring the time, M2, (daily checks and suitable unit of time – hours) but maximum of 6 marks already reached.

Experimental design questions – general advice

These are the 'experimental design' questions, often known as the 'CORMS' questions. This name helps provide a framework for your answer. The letters are given in the mark scheme and it helps you to use them in your answers. In this way you can check that you have included reference to all the necessary factors that should be considered in designing the investigation or experiment asked for in the question.

We now look at each letter in turn.

C = what is being **C**hanged (or **C**ompared) in the experiment. This is the independent variable and also covers the idea of a **C**ontrol. In this question on bananas ripening, the change is the presence (or not) of ethene from the ripe tomato.
O = the **O**rganism being used and some statement about it to make sure the investigation is valid. Often this is covered by reference to using the same species or variety so that the effect of the change can be judged fairly. So, in this question, all bananas should be of the same variety or from the same batch and starting from the same stage of ripeness.
R = **R**eplication or repeats, so that several results are obtained rather than relying on a single measurement or observation. This is good practice in any experimental work.
M1 + **M2** = the **M**easurements taken. This is the dependent variable, because it 'depends' on what the change is when setting up the experiment. Often this may refer to a change in mass, or height or something you can measure in numbers. For the question on bananas, you decide on an 'end-point' on the colour scale (which represents the stage of ripeness). A second M mark is usually given for reference to a time scale – in this case, how long it takes the banana to reach the colour chosen for the end-point. It is important to suggest an actual time – this shows you are thinking about whether the change occurs in seconds, or hours or perhaps weeks. You may not know the correct time, but make a sensible attempt. To give some other examples, cabbages treated with fertiliser are likely to take several weeks before reaching a stage to be weighed, whereas a yoghurt investigation is likely to be completed in a few hours.
S1 + **S2** = a variable that must be kept the **S**ame or controlled in this experiment. Such factors may include the quantity used (same volume, same mass) or other variables such as same temperature, same humidity or whatever is appropriate for the investigation. Usually there are plenty of factors you could choose, but make sure it is relevant to the investigation.

Lastly – the topic in the question may be a novel one – so perhaps not something you have already studied in your specification, but you should be able to apply and adapt these principles to any of the questions set. Spend a couple of minutes thinking how you would plan the investigation, then try to follow through the CORMS letters to make sure you cover all essential aspects of the design.

Index